# Introduction to the Theory of Collisions of Electrons with Atoms and Molecules

# PHYSICS OF ATOMS AND MOLECULES

## Series Editors

**P. G. Burke**, *The Queen's University of Belfast, Northern Ireland*
**H. Kleinpoppen**, *Atomic Physics Laboratory, University of Stirling, Scotland*

### Editorial Advisory Board

**V. V. Balashov** (*Moscow, Russia*)
**U. Becker** (*Berlin, Germany*)
**K. Burnett** (*Oxford, U.K.*)
**C. W. Clark** (*Washington, U.S.A.*)
**J. C. Cohen-Tanoudji** (*Paris, France*)

**C. J. Joachain** (*Brussels, Belgium*)
**W. E. Lamb, Jr.** (*Tucson, U.S.A.*)
**M. C. Standage** (*Brisbane, Australia*)
**K. Takayanagi** (*Tokyo, Japan*)

---

*Recent volumes in this series:*

COINCIDENCE STUDIES OF ELECTRON AND PHOTON IMPACT IONIZATION
Edited by Colm T. Whelan and H. R. J. Walters

DENSITY MATRIX THEORY AND APPLICATIONS, SECOND EDITION
Karl Blum

ELECTRON MOMENTUM SPECTROSCOPY
Erich Weigold and Ian McCarthy

IMPACT SPECTROPOLARIMETRIC SENSING
S. A. Kazantsev, A. G. Petrashen, and N. M. Firstova

INTRODUCTION TO THE THEORY OF COLLISIONS OF ELECTRONS WITH
ATOMS AND MOLECULES
S. P. Khare

NEW DIRECTIONS IN ATOMIC PHYSICS
Edited by Colm T. Whelan, R. M. Dreizler, J. H. Macek, and H. R. J. Walters

PHOTON AND ELECTRON COLLISION WITH ATOMS AND MOLECULES
Edited by Philip G. Burke and Charles J. Joachain

PRACTICAL SPECTROSCOPY OF HIGH-FREQUENCY DISCHARGES
Sergei A. Kazantsev, Vyacheslav I. Khutorshchikov, Günter H. Guthöhrlein, and
Laurentius Windholz

RELATIVISTIC HEAVY-PARTICLE COLLISION THEORY
Derrick S. F. Crothers

*A Chronological Listing of Volumes in this series appears at the back of this volume.*

---

# Introduction to the Theory of Collisions of Electrons with Atoms and Molecules

## S. P. Khare

*Chowdhary Charan Singh University*
*Meerut, India*

Springer Science+Business Media, LLC

Library of Congress Cataloging-in-Publication Data

Khare, S. P. (Satya P.)
    Introduction to the theory of collisions of electrons with atoms and molecules/S.P. Khare.
        p.  cm. — (Physics of atoms and molecules)
    Includes bibliographical references and index.
    ISBN 978-0-306-47241-1       ISBN 978-1-4615-0611-9 (eBook)
    DOI 10.1007/978-1-4615-0611-9
        1. Collisions (Nuclear physics).  2. Electrons—Scattering.  3. Atoms.  4. Molecules.  I.
Title.  II. Series.

QC794.6.C6 K53 2002
539.7′57—dc21

                                                                                2002022129

ISBN 978-0-306-47241-1

©2001 Springer Science+Business Media New York
Originally published by Kluwer Academic / Plenum Publishers, New York in 2001

http://www.wkap.nl/

10  9  8  7  6  5  4  3  2  1

A C.I.P. record for this book is available from the Library of Congress

*To Pushpa*

# *Foreword*

Collisions of electrons with atoms and molecules provide a unique diagnostic probe of the fundamental interactions of many-electron systems and are the basic physical processes that determine the behavior of ionized gases, ranging from those created for plasma processing technologies to the plasma existing in the early universe after the first few seconds. Early experiments on electron collisions played a central role in the development of quantum mechanics. The demonstration of diffraction of electron beams by gases confirmed the quantum mechanical duality of waves and particles and measurements of the energy losses in electron collisions in gases established the discrete nature of the energy level structure of atoms and molecules.

To understand and to predict quantitatively the behavior of ionized gases produced by electrical discharges in lighting systems and by lightning or created in fusion plasmas or found in astrophysical environments requires development of the theory of electron collisions and the construction of mathematical methods that enable reliable calculations of the critical collision parameters identified by the theory. Experiments provide essential benchmarks to test the reliability of the theoretical concepts and calculations but cannot hope to produce the vast array of collision data that enter into plasma modeling.

Photon interactions are of equal importance both for the fundamental information uncovered by studies of the effects of radiation on atoms and molecules and because many kinds of plasma are created by the absorption of photons and reveal their properties through the emission of photons.

Professor Satya Khare has made many notable contributions to the theory of electron and photon collisions with atoms and molecules and in this book he presents a systematic unified introduction to the still evolving theory that is needed to interpret the wide range of physical phenomena that occur when electrons and photons collide with atoms and molecules.

Alexander Dalgarno F. R. S.
*Phillips Professor of Astronomy*
*Harvard University*

# *Preface*

The present book deals with nonrelativistic quantum mechanical theories for the collision of microprojectiles with potential fields, atoms, and molecules. The spinless particles, the electrons (with occasional reference to its antiparticle i.e., positron), and the photons are taken as projectiles. This introductory book is the outgrowth of lectures I delivered at Meerut University, the University of Western Ontario, London, Ontario and Wayne State University in Detroit. It contains a lot of new information and refers to many papers published in the 21$^{st}$ century. The prerequisites for understanding this material are introductory courses in atomic physics and quantum mechanics.

An attempt has been made to develop the subject matter in a very systematic manner. The basics of collision physics are introduced in Chapter 1. As the physical state of a free particle changes in a collision, in Chapter 2 we discuss the motion of a free particle highlighting its characteristics such as energy, momentum, and wave function along with its partial wave expansion in terms of its angular momentum. This is followed in Chapter 3 by a discussion of the collision of a spinless particle with a potential field. To facilitate our understanding of the effect of open inelastic channels on elastic scattering in the case of electron–atom collisions (Chapters 7 and 8) we have considered a complex absorption potential field. Thus the concept of the absorption cross section is introduced and it is shown that the optical theorem is a consequence of the conservation of incident flux. The various approximate methods for evaluating scattering amplitude using integral and differential approaches are described.

In Chapter 4 the spinless particle is replaced by an electron. With the use of nonrelativistic theory, the spin–orbit interaction potential is obtained and with its help spin-flip scattering is discussed. Readers are introduced to the concept of polarized electrons, and the impossibility of their production by a Stern–Gerlach type of experiment is demonstrated. With the help of the density matrix the polarization of unpolarized electrons due to scattering and the scattering parameters are discussed. The measurements of the Shermann function and the degree of polarization of an electron beam by a Mott detector are described. In Chapter 5 collision between two particles has been considered. It is shown that the two-body collision is equivalent to the collision of a single body with a potential field. Thus all the methods developed earlier become applicable. For identical particles, the symmetry condition imposed on the wave function of the system gives the exchange scattering along with the direct scattering. Collisions between two bosons and two fermions are discussed.

In Chapter 6, collision of photons with multielectron atoms is dealt with. Expressions for the excitation and ionization cross sections and the Einstein's $A$ and $B$ coefficients are obtained. The concept of polarized photons in terms of their spins and the spin states of photons are discussed. The Stokes parameters, which are required to completely determine the polarization of a mixed photon beam is introduced. The density matrix and $I(\alpha)$, the intensity of the transmitted beam when a mixed beam of intensity $I$ moving along the $z$-axis is passed through a Nicol prism, whose axis of complete transmission makes an angle $\alpha$ with the $x$-axis, are given in terms of the Stokes parameters and $I$. The theory of the production of polarized electrons by photoionization of unpolarized atoms by circularly polarized light (Fano effect) is described.

Chapters 7 and 8 deal with the collision of electrons with atoms. In Chapter 7 a number of approximate methods derived from the integral approach are described. It is shown that the various approximate methods are obtained by taking different approximate forms of the exact free-particle Green's function. The relationship between photon and electron impact collisions is highlighted. The recent successful method I developed for inner-shell ionization is presented. The next chapter deals with the approximate methods obtained from the differential approach. The origins of the static field, local exchange, polarization, and absorption potentials are explained. The usefulness of reducing a many-body problem to a one-body problem with the help of an optical potential and the construction of the optical potential are discussed. The spin–orbit potential is also included to obtain polarization $S$, $T$, and $U$ parameters for atoms. The electron impact excitation of atoms using the electron–photon delayed coincidence technique is described. The usefulness of the Stokes parameters (described in Chapter 6) in that technique to study collision dynamics is discussed. In both chapters the theoretical results obtained with the help of the different approximate methods are compared with the experimental results for a good number of atoms.

In Chapter 9 the collision of electrons with multicenter molecules is considered. It is a formidable problem. Its reduction to tractable forms, which yield reasonable results, is described. Application of, e.g., the first Born, the second Born, and the modified Glauber approximations are discussed. Recently developed models to evaluate ionization (including dissociative ionization) cross sections of molecules that are due to electron impact are given. The independent-atom model and its modifications such as the modified additivity rule, and single-center charge density method, which utilize the differential approach, are also discussed. The theoretical results are compared with the available experimental data to demonstrate the usefulness of the various approximate methods.

References are cited in the text and a set of problems is provided after each chapter.

I acknowledge the benefit I derived from fruitful discussions with my colleagues and students at the universities mentioned above. I am indebted to my teacher and mentor, Prof. A. Dalgarno, who has kindly written the foreword for this book. Useful discussions and correspondence with Professors A. Dalgarno, K. L. Joshipura, K. C. Mathur, W. J. Meath, R. Srivastava, A. N. Tripathi, and J. M. Wadhera and Drs. A. K. Bhatia and A. Temkin are gratefully acknowledged. Invaluable help has been provided to me by my children, Vandana, Seema, Arun, and Jaydeep and my student Manoj in the preparation of this manuscript. I heartily thank all of them. Thanks are also due to Mr. Ravi Jain and Mr. R. R. Verma for their typing services and to Mr. Chandra Prakash Rastogi for preparing all the diagrams. Above all, I would like to acknowledge the help received from my wife, Pushpa, to whom this volume is dedicated.

I sincerely hope that this book will be useful to students and young workers in the field of atomic collisions. Suggestions and criticisms are welcome.

S. P. Khare
*Meerut*

# Contents

# Introduction to the Theory
# of Collisions of Electrons
# with Atoms and Molecules

# Basics of Collisions

## 1.1  Introduction

A collision is basically an interaction between two or more systems, each containing one or more particles. In a collision process, two particles or systems approach each other from a big distance, interact (collide) for a short time, and then separate again. The interaction time is very short in comparison with the time for which the system (formed by the colliding partners) can be observed. During the collision a large force acts between the colliding partners and the post-collision state is different than the precollision state. From a study of these two states it should be possible to learn about the nature of interaction between the colliding partners. Collision techniques have been successfully employed to learn about the internal structure of the microparticles and the nature of interactions between different microparticles. For collisions involving macroparticles, the associated de Broglie wavelengths are very small in comparison to the size of the particles, so classical mechanics can be employed to describe such collisions. However, for the microscopic objects, the de Broglie wavelengths are large, so classical mechanics becomes inadequate and must be replaced by quantum mechanics. In the collision of microscopic objects, the concept of physical contact between the colliding partners becomes irrelevant.

## 1.2  Collision Cross Section

One of the most important parameters in collision physics is the collision cross section. A schematic diagram for the measurement of the collision cross section is shown in Fig. 1.1. A well-collimated beam *A* of monoenergetic particles falls on a thin scattering chamber *C* that contains *n* number of *B* targets. Due to collisions, some of the projectiles are scattered in all possible directions around *C* and a few of them are detected by the detector *D*. The distances *SC* and *CD*

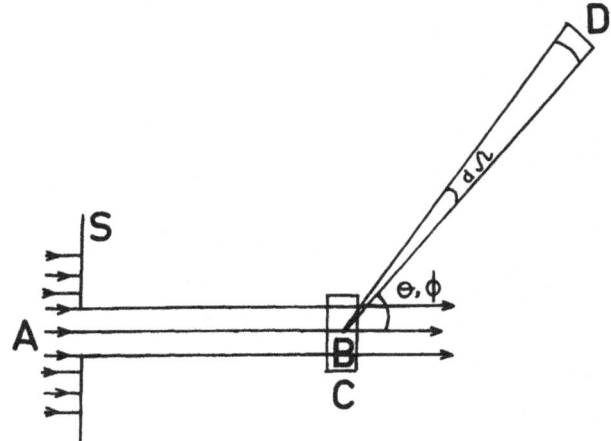

**FIGURE 1.1** Schematic diagram for the scattering of projectiles $A$ by the targets $B$.

are several orders of magnitude greater than the sizes of the projectiles and the targets. The experiment is carried out under steady state conditions; i.e., the flux $F$ of the projectiles and the number of scattered particles detected by $D$ per unit time are independent of time.

Let the detector $D$ be in the direction $(\theta, \phi)$ with respect to the direction of the projectiles and at a distance $r$ from the chamber $C$. If $D$ makes a solid angle $d\Omega$ with $C$, then the number of particles $\Delta N$ reaching the detector per unit time is proportional to $F$, $n$, and $d\Omega$. Denoting the constant of proportionality by $I(\theta, \phi)$, we have

$$\Delta N = I(\theta, \phi) F n d\Omega \qquad (1.2.1)$$

The quantity $I(\theta, \phi)$ is known as the differential cross section and is expressed in terms of area/steradian. $I(\theta, \phi) d\Omega$, is equal to the number of particles scattered in the direction $(\theta, \phi)$, in the solid angle $d\Omega$ per unit time per unit incident flux per target. An integration of the differential cross section over the solid angle yields the integrated (also known as total) cross section $\sigma$. Hence

$$\sigma = \int_0^\pi \int_0^{2\pi} I(\theta, \phi) \sin \theta \, d\theta \, d\phi \qquad (1.2.2)$$

This is equal to the total number of particles scattered in all possible directions per unit flux per unit time per target. In many cases, owing to cylindrical symmetry $I(\theta, \phi)$ is independent of $\phi$ and we have

$$\sigma = 2\pi \int_0^\pi I(\theta) \sin\theta \, d\theta \qquad (1.2.3)$$

The momentum transfer cross section $\sigma_m$ is also obtained by integrating $I(\theta,\phi)$ but with a weight factor $(1 - \cos\theta)$. Hence, with axial symmetry,

$$\sigma_m = 2\pi \int_0^\pi I(\theta)(1 - \cos\theta) \sin\theta \, d\theta \qquad (1.2.4)$$

Now, unit flux means one projectile per unit area per unit time; hence $\sigma$ is the cross-sectional area that a target presents to the direction of the incident beam. Similarly, $I(\theta,\phi)d\Omega$ is the effective area of the target, which deflects the projectiles in the solid angle $d\Omega(\theta,\phi)$. Equation (1.2.1) assumes that the projectiles and targets do not interact among themselves and that one projectile collides with only one target. This is possible only for small values of $F$ and $n$, however, and for more accuracy in the measurements of $\Delta N$, the values of $F$ and $n$ should be large. Hence, one is required to choose optimum values of $F$ and $n$.

## 1.3   Types of Collisions

Since we are examining collisions under steady state conditions, the total energy $E_T$ of the whole system remains conserved. Broadly speaking we have two types of collisions: (i) elastic and (ii) inelastic.

### 1.3.1   Elastic Collisions

In an elastic collision between a projectile $A$ and a target $B$, the internal structures (the potential energies) of $A$ and $B$ do not change. We represent it by

$$A + B \rightarrow A + B \qquad (1.3.1)$$

with

$$E_T = K_A + V_A + K_B + V_B = K_A' + V_A' + K_B' + V_B' \qquad (1.3.2)$$

$$V_A = V_A' \quad \text{and} \quad V_B = V_B' \qquad (1.3.3)$$

where $K_A$ and $V_A$ are the kinetic and the potential energies, respectively, of the projectile $A$ before the collision. Due to the collision they change to $K_A'$ and $V_A'$, respectively. $K_B$, $V_B$, $K_B'$, and $V_B'$ are similar quantities for the target $B$.

Use of Eq. (1.3.3) in (1.3.2) yields

$$K_T = K_A + K_B = K_A' + K_B' \tag{1.3.4}$$

i.e., along with $E_T$ in an elastic collision, the total kinetic energy is also conserved. If in such a collision, $A$ loses $\Delta K$ kinetic energy then $B$ gains the same amount of kinetic energy. The reaction

$$e + H(1s) \rightarrow e + H(1s) \tag{1.3.5}$$

where $H(1s)$ represents a hydrogen atom in its ground state and $e$ is an electron, is an example of an elastic collision. The hydrogen atom continues to be in the ground state after the collision. If the relative energy of the electron with respect to the atom is $E$, then in the collision the electron loses approximately $2mE/M$ of its kinetic energy, where $m$ and $M$ are the masses of the electron and the hydrogen atom, respectively (see Problem 1.9). This energy is taken up by the hydrogen atom and its kinetic energy increases by that amount. The well-known Rayleigh scattering is another example of an elastic collision between photons and atoms.

### 1.3.2   Inelastic Collisions

In an inelastic collision the internal energy of at least one of the colliding partners changes. The collision

$$A + B \rightarrow A + B^* \tag{1.3.6}$$

is an example of an inelastic collision. The asterisk on $B$ indicates that it is an excited atom, and $V_B'$ is greater than $V_B$. The reverse of (1.3.6), in which $B^*$ is de-excited to $B$, is also an inelastic collision. However, in this case the potential energy is converted into kinetic energy. Such collisions are also known as super-elastic collisions. In the inelastic collision

$$e + H(1s) \rightarrow e + H(2p) \tag{1.3.7}$$

the electron loses kinetic energy equal to $\varepsilon_{2p} - \varepsilon_{1s}$, where $\varepsilon_n$ is the eigenenergy of the $n$th state of the hydrogen atom. A hydrogen atom can also be excited from its ground state to its $2p$ state by a photon. However, the energy of the photon must be equal to $\varepsilon_{2p} - \varepsilon_{1s}$, which means that photoexcitation, unlike electron impact excitation, is a resonant process.

Raman scattering,

$$h\nu_1 + M \rightarrow h\nu_2 + M^* \tag{1.3.8}$$

where $h$ is Planck's constant, is also a photoexcitation, but it is not a resonant process; here the photon of the frequency $\nu_1$ is completely absorbed and a new photon of frequency $\nu_2$ is produced. Furthermore, the postcollision de-excitation of $M^*$ to $M$ produces a line of frequency $\nu_1 - \nu_2$. This line is a characteristic line of the molecule and does not depend upon $\nu_1$. A change in $\nu_1$ also changes $\nu_2$, such that $\nu_1 - \nu_2$ remains the same. The collision

$$A^* + B \rightarrow A + B^* \qquad (1.3.9)$$

is also an inelastic collision. However, here $A^*$ loses its potential energy, which is utilized to excite $B$ to $B^*$. The law of conservation of energy requires that

$$\varepsilon_{A^*} + \varepsilon_B = \varepsilon_A + \varepsilon_{B^*} \qquad (1.3.10)$$

which is an example of the resonant transfer of energy.

In all the inelastic collisions discussed so far, the atomic particles make transitions from one bound state to another bound state. These are called bound–bound transitions. However, in the collision

$$A + e^+ \rightarrow A^+ + e^- + e^+ \qquad (1.3.11)$$

the initially bound atomic electron goes to a continuum state. Hence, such an inelastic collision represents a bound–free transition. Here the atom $A$ is ionized by a positron and a free electron is produced. For this reaction to proceed, the initial kinetic energy of the positron must be greater than the ionization potential $I$ of the target $A$. If in the collision, the positron loses energy $W$, then

$$W = I + \varepsilon_e \qquad (1.3.12)$$

where $\varepsilon_e$ is the kinetic energy of the ejected electron. $W$ can vary continuously from $I$ to $E$. In the photoionization

$$h\nu + A \rightarrow A^+ + e \qquad (1.3.13)$$

and

$$h\nu = I + \varepsilon_e \qquad (1.3.14)$$

Thus knowledge of $h\nu$ and $\varepsilon_e$ can be utilized to determine the ionization potential of the target atom. This is the basic principle of photoelectron spectroscopy.

It is also possible that the energy required to ionize $B$ may come from the excited atom $A*$:

$$A* + B \rightarrow A + B^+ + e \qquad (1.3.15)$$

For example, a metastable helium atom on collision with a sodium atom may ionize the latter with simultaneous de-excitation of itself:

$$He*(2^3S_1) + Na \rightarrow He(1^1S_0) + Na^+ + e \qquad (1.3.16)$$

Two metastable helium atoms $He(2^3S_1)$, one with $M_J = 1$ and another $M_J = 0$, may collide with one another to produce a ground state helium atom $He(2^1S_0)$, a helium ion, and a free electron:

$$He*(2^3S_1, M_J = 1) + He*(2^3S_1, M_J = 0) \rightarrow He(1^1S_0) + He^+ + e \quad (1.3.17)$$

which is known as the Penning ionization. The reaction

$$A + B^+ \rightarrow A^+ + B* \qquad (1.3.18)$$

involving electron (charge) transfer from $A$ to $B$ is another example of an inelastic collision. The atom $B*$ may be in the ground or excited state. Such collisions are known as rearrangement collisions.

## 1.4   The Total Cross Section

When an incident beam $A$ collides with targets $B$, in general, there are both elastic and inelastic collisions. Every excited state of the target, having excitation energy $\varepsilon_{ex} \leq E$, constitutes an open channel. The elastic channel is always an open one. Since there are an infinite number of eigenstates in an atomic target there are infinite number of channels (closed + open). Each channel has its own differential $I(\theta,\phi)$ and integrated $\sigma_C$ cross sections. The total collision cross section is defined by

$$\sigma_T = \underset{C}{S}\,\sigma_C = \underset{C}{S} \int I_C(\theta,\phi)d\Omega \qquad (1.4.1)$$

The symbol $S$ signifies that we sum over the open discrete channels and integrate over the open continuum (say, ionizing) channels. In a collision the momentum of the projectile changes from $\hbar k_i$ to $\hbar k_f$ and the projectile is removed from its initial channel $k_i$. Hence, the total collision cross section $\sigma_T$ can be determined

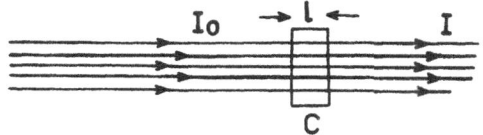

**FIGURE 1.2** Measurement of the total cross section $\sigma_T$ by a transmission experiment.

by carrying out a transmission experiment. In such an experiment, shown schematically in Fig. 1.2, a beam of monoenergetic particles is allowed to pass through a collision chamber $C$. The intensities of the beam before and after the passage through $C$ are measured. Let these intensities be $I_0$ and $I$; then according to Beart's law,

$$I = I_0 \exp(-n_0 l \sigma_T) \tag{1.4.2}$$

where $n_0$ is the number density of the targets in the chamber $C$ whose length is $l$. From the above relation

$$\sigma_T = \frac{1}{n_0 l} \ln\left(\frac{I_0}{I}\right) \tag{1.4.3}$$

## 1.5   Applications of Collision Cross Sections

Cross sections for collisions between projectiles such as photons, electrons, and protons with various atoms and molecules find their applications in a number of fields, including astrophysics, space physics, plasma physics, fusion, lasers, radiation physics, mass spectrometry, chemical reactions, and biological science. Due to collisions the atoms are excited. After a short interval (equal to their lifetime), these excited atoms decay to their low-lying states and emit characteristic electromagnetic radiation. Most of astrophysics is based on interpretation of the spectra of radiation reaching earth from outer space. In our own upper atmosphere, this radiation produces fluorescence, known as day or night air glow. During magnetic storms, a large number of charged particles reach the earth's magnetic poles and excite atmospheric gases. Their de-excitation produces a bright glow called aurora. Measurements of the collision cross sections and their theoretical evaluation have greatly increased our knowledge of microparticles. For example, in the famous Rutherford scattering experiment, an analysis of the differential cross sections led Rutherford to conclude that all the positive charges of an atom are concentrated in a very small space, which we now know as the nucleus, and that a large empty space in the atom is available for the movement of the electrons. Similarly, the study of inelastic collisions of photons with molecules (Raman scattering) has enabled us to determine rotational and vibrational structures even of homonuclear molecules.

In this energy-hungry world there is a concentrated effort to harness the fusion process for the purpose of obtaining a virtually limitless and relatively noncontaminating energy source. Research on controlled thermonuclear reactions also requires knowledge of thousands of atomic collision cross sections. Collision cross sections are also required in the monitoring of energy deposition by incident particles in medical applications. Photoionization cross sections control the temperature of corona. The formation of the ionic layers in our upper atmosphere is also due to collision processes. Thus atomic collisions play a very important role in our day-to-day life. For more information on the application of collision cross sections see Massey et al. (1969), Christophorou (1971), Fliescher et al. (1975), Dalgarno (1979), and Lindinger and Howorka (1985).

## 1.6   *Laboratory and Center-of-Mass Systems*

Collision cross sections are measured in laboratories, where the targets are at rest and the projectiles move. However, it is more convenient to calculate the cross sections in the center-of-mass frame of reference, in which the center of mass of the system (projectile plus target) is always at rest. In such a frame, before the collision both the projectile and the target move toward center of mass (CM) and after the collision, both move away from the CM in such a way that the CM is always at rest. As expected, the value of the differential cross section $I_L(\theta_L,\phi_L)$, measured in the laboratory frame of reference, is different from $I_C(\theta_C,\phi_C)$, calculated in the CM frame. In order to compare experiment with theory it is necessary to find a relation between $I_L(\theta_L,\phi_L)$ and $I_C(\theta_C,\phi_C)$.

Figure 1.3(a) shows a collision between two particles of masses $m_A$ and $m_B$ in the laboratory frame of reference (LF). Before the collision, the projectile $A$ of mass $m_A$, moves toward the target $B$ (mass $m_B$) with velocity $C_A$. Since $B$ is at rest, the CM of the system also moves toward $B$ with velocity $V_{CM} = m_A C_A/ (m_A + m_B)$ . After the collision, the projectile is scattered in the direction $(\theta_L,\phi_L)$ with velocity $C_A'$ and the target recoils in some other direction. The same collision as seen in the CM frame of reference is shown in Fig. 1.3(b). Before the collision $m_A$ and $m_B$ move toward each other with velocities and $V_A$ and $V_B$, respectively. After the collision the projectile is scattered in the direction $(\theta_C,\phi_C)$ with velocity $V_A'$. Since in this frame CM is always at rest, after the collision the target moves in a direction opposite to that of the projectile. The vector relationship between the different velocities is shown in Fig. 1.3(c).

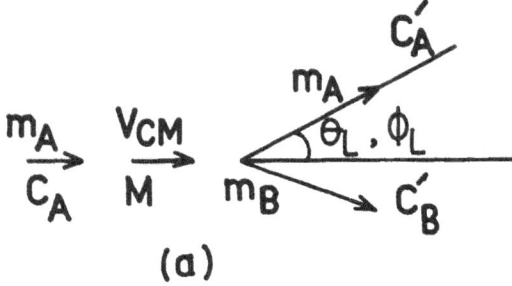

(a)

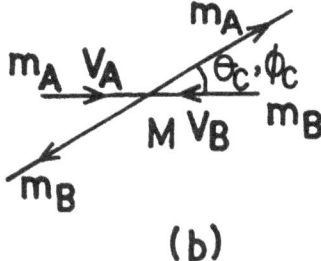

(b)

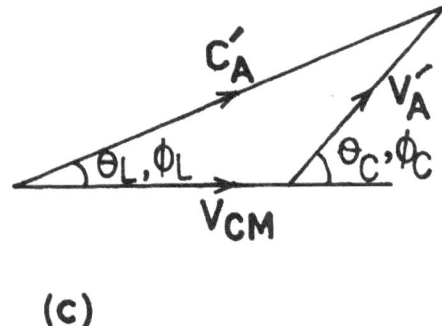

(c)

**FIGURE 1.3** Collision of two particles A and B: (a) in the laboratory frame, (b) in the CM frame, (c) vector relationship between $C'_A$ and $V_{CM}$, and $V'_A$.

Now

$$C_A' = V_{\mathrm{CM}} + V_A' \tag{1.6.1}$$

Equating their $z$ components we get

$$C_A' \cos\theta_L = V_{\mathrm{CM}} + V_A' \cos\theta_C \tag{1.6.2}$$

Similarly by equating the $x$ and $y$ components we get

$$V_A' \sin\theta_C = C_A' \sin\theta_L \tag{1.6.3}$$

and

$$\phi_C = \phi_L \tag{1.6.4}$$

Thus we obtain

$$\tan\theta_L = \frac{\sin\theta_C}{\cos\theta_C + \alpha} \tag{1.6.5}$$

where

$$\alpha = V_{\mathrm{CM}}/V_A' \tag{1.6.6}$$

In the laboratory frame, the number of projectiles scattered in the direction $(\theta_L, \phi_L)$, in solid angle $d\Omega(\theta_L, \phi_L)$ is proportional to $I_L(\theta_L, \phi_L)d\Omega(\theta_L, \phi_L)$. The same quantity in the CM frame is proportional to $I_C(\theta_C, \phi_C)d\Omega(\theta_C, \phi_C)$. Hence, by definition

$$I_L(\theta_L, \phi_L)d\Omega_L = I_C(\theta_C, \phi_C)d\Omega_C$$

or

$$I_L(\theta_L, \phi_L)d(\cos\theta_L) = I_C(\theta_C, \phi_C)d(\cos\theta_C) \tag{1.6.7}$$

because $\phi_L = \phi_C$. From (1.6.5),

$$\cos\theta_L = \frac{\alpha + \mu}{\left(1 + 2\mu\alpha + \alpha^2\right)^{1/2}} \tag{1.6.8}$$

where $\mu = \cos\theta_C$, so

$$d(\cos\theta_L) = \frac{1+\mu\alpha}{\left(1+2\mu\alpha+\alpha^2\right)^{3/2}}\,d\mu \qquad (1.6.9)$$

and

$$I_L(\theta_L,\phi) = \frac{\left(1+2\mu\alpha+\alpha^2\right)^{3/2}}{|1+\mu\alpha|}\,I_C(\theta_C,\phi) \qquad (1.6.10)$$

In the above equation the mod value of $1 + \alpha\cos\theta_C$ is taken to keep $I_L(\theta_L,\phi_L)$ always positive. Equation (1.6.5) shows that for $\alpha < 1$, $\theta_L$ increases from 0 to $\pi$ as $\theta_C$ increases from 0 to $\pi$. However, for $\alpha = 1$, we obtain $\theta_L = \theta_C/2$. Thus as $\theta_C$ increases from 0 to $\pi$ the laboratory angle $\theta_L$ increases only from 0 to $\pi/2$, i.e., no particle is scattered in the backward hemisphere in the laboratory, and in this case (1.6.10) reduces to

$$I_L(\theta_L,\phi) = 4\cos(\theta_C/2)I_C(\theta_C,\phi) \qquad (1.6.11)$$

with

$$\theta_L = \theta_C/2 \qquad (1.6.12)$$

For $\alpha > 1$, $\theta_L = 0$ at $\theta_C = 0$ and increases with $\theta_C$, but reaches a maximum value of $\sin^{-1}(1/\alpha)$ for $\theta_C = \cos^{-1}(-1/\alpha)$. A further increase in $\theta_C$ decreases $\theta_L$ and at $\theta_C = \pi$, $\theta_L = 0$. As far as the total cross sections $\sigma_L$ and $\sigma_C$ are concerned, they depend on the total number of particles scattered in the whole space. These numbers are the same in both frames of reference, and we have $\sigma_L = \sigma_C$.

The initial kinetic energy of the system in the $L$ frame is

$$(K_i)_L = \tfrac{1}{2}m_A C_A^2 \qquad (1.6.13)$$

The initial and the final kinetic energies in the frame are given by

$$(K_i)_C = \tfrac{1}{2}m_A V_A^2 + \tfrac{1}{2}m_B V_B^2 \qquad (1.6.14)$$

and

$$(K_f)_C = \tfrac{1}{2}m_A V_A'^2 + \tfrac{1}{2}m_B V_B'^2 \qquad (1.6.15)$$

respectively. We also have

$$m_A V_A + m_B V_B = m_A V_A' + m_B V_B' = 0 \qquad (1.6.16)$$

Hence, (1.6.14) changes to

$$(K_i)_C = \tfrac{1}{2}(m_A/m_B)M V_A^2 \qquad (1.6.17)$$

where $M = m_A + m_B$. Similarly,

$$(K_f)_C = \tfrac{1}{2}(m_A/m_B)M V_A'^2 \qquad (1.6.18)$$

We also have

$$(K_i)_C = \tfrac{1}{2}m_A C_A^2 - \tfrac{1}{2}M V_{CM}^2 \qquad (1.6.19)$$

and

$$m_A C_A = M V_{CM} \qquad (1.6.20)$$

So that (1.6.19) reduces to

$$(K_i)_C = \tfrac{1}{2}(m_B/m_A)M V_{CM}^2 \qquad (1.6.21)$$

Equating (1.6.17) and (1.6.21), we get

$$V_{CM} = m_A/m_B V_A \qquad (1.6.22)$$

Suppose owing to collision the kinetic energy of the system changes by $\Delta E$; then

$$(K_f)_C = (K_i)_C + \Delta E$$

or

$$V_A' = V_A\left[1 + \frac{\Delta E}{(K_i)_C}\right]^{1/2} \qquad (1.6.23)$$

where we have used (1.6.17) and (1.6.18). Putting (1.6.22) and (1.6.23) into (1.6.6), we obtain

$$\alpha = \frac{m_A}{m_B}[1 + \Delta E/(K_i)_c]^{-1/2} \tag{1.6.24}$$

If $\Delta E$ is positive, i.e., if the reaction is exothermic, then a collision is always possible. However, for endothermic reactions (negative values of $\Delta E$) the collision is physically possible only of $(K_i)_c \geq |\Delta E|$. Hence, $|\Delta E|$ is the threshold of the reaction in the CM frame. To obtain the value of the $\Delta E$ in the $L$ frame we use (1.6.20) and (1.6.21) and replace $(K_i)_c$ by $|\Delta E|$ in (1.6.21). Thus we get

$$|\Delta E|_L = \frac{m_A + m_B}{m_B}|\Delta E|_C \tag{1.6.25}$$

where $|\Delta E|_L = \frac{1}{2}m_A C_A^2$ is the threshold energy in the $L$ frame with $C_A$ as the threshold velocity of the projectile in the same frame. Equation (1.6.24) shows that for $m_B \gg m_A$, $\alpha$ is very small. So that from (1.6.5) we get $\theta_L = \theta_C$ and $I(\theta_C, \phi) = I(\theta_C, \phi)$, i.e., there is hardly any difference in the results obtained in the CM and $L$ frames.

For elastic scattering, $\Delta E = 0$ and $(K_i)_C = (K_f)_C$ and from (1.6.24), $\alpha = m_A/m_B$. Further, from (1.6.14) to (1.6.18) $V_A' = V_A$ and $V_B' = V_B$, i.e., in elastic collisions the speeds of the particles in the CM frame do not change.

## Questions and Problems

1.1 Distinguish among elastic, inelastic, and superelastic collisions. Describe briefly various types of inelastic collisions.

1.2 An electron with an energy of 12.5 eV is scattered by a ground state hydrogen atom. What is the highest possible principal quantum number $n$ of the atom after the collision? If the energy of the incident electron is raised to 50 eV what are the possible values of the energy of the electron ejected from the atom?

1.3 What is Penning ionization? A metastable helium atom in the $2^3S_1$ state, having thermal energy, collides with a ground state lithium atom and an electron is ejected. If the ionization potential of the lithium atom is 5.37 eV, find the energy of the ejected electron in eV. The energy of the excitation of the metastable helium atom is $5.985 \times 10^6 \, m^{-1}$

1.4 In a pure rotational Raman scattering a hydrogen molecule is excited from the $j = 0$ to the $j = 2$ rotational level by light of $6 \times 10^{-7} \, m$ wavelength. If the internuclear distance of the hydrogen molecule is $0.741 \times 10^{-10} \, m$, find the wavelength of the scattered radiation.

1.5  A deuteron is elastically scattered by a stationary helium atom. The angle of scattering in the CM frame is 60°. Find the scattering angle in the laboratory frame.

1.6  Calculate the maximum angle of scattering in the laboratory frame for elastic scattering of deuterons by stationary hydrogen atoms.

1.7  Prove that for $m_A > m_B$ the maximum value of $\theta_L$ is $\sin^{-1}(1/\alpha)$ and that it occurs for $\theta_C = \cos^{-1}(-1/\alpha)$.

1.8  The threshold of excitation of the hydrogen atoms is $\frac{3}{4}$ Rydberg. What should be the minimum energy and velocity of a proton so that it can excite the atom?

1.9  A structureless particle $A$ of mass $m_A$ and energy $E$ collides with a stationary particle $B$ of mass $m_B$. Show that the energy transferred from $A$ to $B$ in the CM frame is zero but that in the $L$ frame it is given by

$$\Delta E = \frac{2m_A m_B}{(m_A + m_B)^2}(1 - \cos\theta_C)E$$

where $\theta_C$ is the angle of scattering in the CM frame.

1.10  If in the above problem the particle $A$ recoils in the backward direction in the laboratory frame, i.e., $\theta_L = \pi$, show that the energy of the recoiled particle is

$$E_A' = \left(\frac{m_A - m_B}{m_A + m_B}\right)^2 E$$

# Motion of a Free Microparticle

## 2.1 Introduction

In a two-body collision, two free particles approach each other from a big distance, interact with one another for a brief period of time, and then separate again, moving far away from one another. For a proper understanding of this collision process, we start with a study of the motion of a free particle. The changes in the characteristics of a free particle as a result of an interaction (collision) are considered in the next chapter. Since microparticles are involved, in atomic collisions, we need quantum mechanics for our study.

In this chapter we shall also discuss a number of special mathematical functions that are employed to represent a free particle.

## 2.2 Energy and Linear Momentum of a Free Particle

According to classical mechanics, a free particle of mass $m$ moves on a straight line with a constant linear momentum $p$ and kinetic energy $E$ $(=p^2/2m)$, the potential energy being zero. For well-defined $p$ and $E$ the uncertainties $\Delta p$ and $\Delta E$ are zero. Hence, according to the Heisenberg uncertainty principle,

$$\Delta x \Delta p = \Delta E \Delta t = \hbar \qquad (2.2.1)$$

where $\hbar$ is the Planck's constant divided by $2\pi$, the uncertainties in the position $x$ and the time $t$ will be infinite, i.e., such a particle cannot be localized in space and time. Thus in quantum mechanics a straight-line trajectory for a free particle does not make any sense and definite values of $p$ and $E$ cannot be assigned.

Fortunately, in atomic collision experiments (see Fig.1.1) the uncertainty in the position of the projectile is equal to the width of the slit $S$, which is larger than the de Broglie wavelength of the projectile by many orders of magnitude. Hence, a very small value of $\Delta p$ can satisfy (2.2.1). Further, as the experiment is carried out under steady state conditions, a very long time becomes available to measure the energy of the projectiles, and as $\Delta E$ can also be quite small, single values of $p$ and $E$ may be assigned to a microparticle.

## 2.3    Wave Function of a Free Particle

Let $\psi_k(r, t)$ be the wave function of a free particle having linear momentum $\hbar k$ and mass $m$. The magnitude of $k$ is equal to $2\pi/\lambda$, where $\lambda$ is the de Broglie wavelength of the object. In the nonrelativistic domain, $\psi_k$ is the solution of the following time-dependent Schrödinger equation:

$$i\hbar \frac{\partial \Psi_k(r,t)}{\partial t} = H\Psi_k(r,t)$$  (2.3.1)

where the Hamiltonian operator $H$ is equal to $-\hbar^2/2m$ times the Laplacian operator $\nabla^2$. Under steady state conditions the energy $E$ is a constant of motion and $\Psi_k(r, t)$ can be factored as

$$\Psi_k(r,t) = \psi_k(r)\exp(-iEt/\hbar)$$  (2.3.2)

Putting (2.3.2) into (2.3.1) we find that the space wave function $\psi_k(r)$ satisfies the following time-independent Schrödinger equation:

$$-\frac{\hbar^2}{2m}\nabla^2\psi_k(r) = E\psi_k(r)$$  (2.3.3)

or

$$(\nabla^2 + k^2)\psi_k(r) = 0$$  (2.3.4)

It is easy to see that the solution of the above equation is

$$\psi_k(r) = Ae^{ik\cdot r}$$  (2.3.5)

where $A$ is the normalization constant. A solution of (2.3.3) is possible for all real values of the eigenenergy $E$, which varies in a continuous manner, and $\psi_k(r)$ are continuum eigenfunctions. The wave given by (2.3.5) propagates in the direction of $k$ and its wave fronts are the planes perpendicular to $k$, so these waves are known as plane waves.

According to the probabilistic concept of Born, the probability of finding the object at $(r, t)$ in the elementary volume $dr$ is given by

$$dP = \Psi_k^*(r,t)\Psi_k(r,t)dr \qquad (2.3.6)$$

Using (2.3.2) in the above equation gives

$$dP = \psi_k^*(r)\psi_k(r)dr \qquad (2.3.7)$$

Thus $dP$ is independent of time. This is expected because the state represented by (2.3.5) is a steady state. To obtain the value of the normalization constant $A$ one is required to integrate (2.3.7) over the whole space and equate the result to 1, because the probability of finding the particle in the whole space is unity. However, it is evident from (2.3.5) and (2.3.7) that $\int dP$ is infinite. Hence, the continuum wave functions cannot be normalized in this way. They are un-normalizable wave functions, but for the quantitative calculation, we have to use normalized wave functions.

## 2.4  Normalization of Plane Waves

The following two types of normalization of plane waves are employed:

(a)  Box Normalization

In box normalization it is assumed that the particle is confined to a cubical box of length $L$. Due to this confinement, $\Psi_k(r)$ vanishes at the edges of the box. Under such a condition, $k$ does not vary in a continuous manner but takes discrete values given by

$$k_x = \frac{2\pi}{L}n_x, \quad k_y = \frac{2\pi}{L}n_y, \quad \text{and} \quad k_z = \frac{2\pi}{L}n_z \qquad (2.4.1)$$

where $n_x$, $n_y$, and $n_z$ are integers (positive as well as negative), but the three are not zero simultaneously. The above procedure yields $A = L^{-3/2}$. Using this normalization we carry out the calculations and finally obtain the limit of the calculated quantity as $L \to \infty$.

(b)  The Dirac Delta Function Normalization

The one-dimensional Dirac delta function $\delta$ is defined by

$$\delta(z - z') = 0, \quad \text{if} \quad z \neq z' \tag{2.4.2a}$$

$$= \infty, \quad \text{if} \quad z = z' \tag{2.4.2b}$$

and

$$\int\limits_{-\infty}^{+\infty} \delta(z - z')dz' = 1$$

or

$$\int\limits_{-\infty}^{+\infty} f(z')\delta(z - z')dz' = f(z) \tag{2.4.2c}$$

provided that $z$ lies in the range of $z'$. Because of (2.4.2b) it should be understood that the Dirac delta function has significance only as part of an integrand and never as an end result. It finds its application in the form of (2.4.2c).

One of the useful representations of the delta function is

$$\delta(k - k') = \frac{1}{2\pi} \int\limits_{-\infty}^{+\infty} e^{i(k-k')x}dx$$

$$= \lim_{a \to \infty} \frac{\sin a(k - k')}{\pi(k - k')} \tag{2.4.3}$$

The function given by the right-hand side of the above equation satisfies (2.4.2). Now

$$\int \psi_k^*(r)\psi_{k'}(r)dr = AA^* \int e^{i(k'-k)\cdot r}dr \tag{2.4.4}$$

Using (2.4.3) in (2.4.4) we get

$$\int \psi_k^*(r)\psi_{k'}(r)dr = AA^*(2\pi)^3 \delta(k - k') \tag{2.4.5}$$

where $\delta(k - k')$ is the three-dimensional Dirac delta function. Hence, a Dirac delta function normalized plane wave is represented by

$$\psi_k(r) = \frac{1}{(2\pi)^{3/2}} e^{ik\cdot r} \tag{2.4.6}$$

where $A$ is assumed to be real, which satisfies the orthogonality relation (2.4.5). The above wave function is subjected to $k$ (wave vector) normalization. We can also have plane waves subjected to $p$ (momentum) normalization. This is given by

$$\psi_p(r) = \frac{1}{(2\pi\hbar)^{3/2}} e^{(ip\cdot r)/\hbar} \tag{2.4.7}$$

For a one-dimensional motion, the energy-normalized plane wave is given by

$$\psi_E(r) = \left(\frac{m}{2h^2 E}\right)^{1/4} e^{\pm ikz} \tag{2.4.8}$$

We note that whereas $\psi_k(r)$ is dimensionless, the wave functions $\psi_p(r)$ and $\psi_E(r)$ have dimensions. We shall continue to represent a plane wave by (2.3.5) but shall take $A = (2\pi)^{-3/2}$ whenever required. We also have

$$\int \psi_k^*(r')\psi_k(r)dk = \frac{1}{(2\pi)^3} \int e^{ik\cdot(r-r')}dk$$

$$= \delta(r - r') \tag{2.4.9}$$

Thus the wave function $\psi_k(r)$ also satisfies closure relation (2.4.9). These functions are eigenfunctions of the Hermitian operator $-\nabla^2$ with eigenvalue $k^2$ and form a complete set. In quantum mechanics, a member of a complete set is regarded as a basis vector in a multidimensional Hilbert space. Like the three basis vectors $\hat{i}$, $\hat{j}$, and $\hat{k}$, the basis vectors $\psi_k(r)$, $\psi_{k'}(r)$, etc. are orthonormal. Furthermore, just as we can write

$$f(r) = a_i\hat{i} + a_j\hat{j} + a_k\hat{k} \tag{2.4.10}$$

where $a_i$ etc are expansion coefficients, similarly we have

$$\chi(r) = \int a_k\psi_k(r)dk \tag{2.4.11}$$

The expansion coefficient $a_k$ is the projection of the vector $\chi(r)$ over $\psi(r)$ in an infinite dimensional Hilbert space and is given by

$$a_k = \int \psi_k^*(r)\chi(r)dr \tag{2.4.12}$$

In the derivation of the above equation, (2.4.5) has been utilized. We further obtain

$$\int \chi^*(r)\chi(r)dr = \int\int\int a_k a_{k'}^* \psi_k(r)\psi_{k'}^*(r)dkdk'dr$$

$$= \int\int a_k a_{k'}^* \delta(k-k')dkdk'$$

or

$$\int |\chi(r)|^2 dr = \int |a_k|^2 dk \qquad (2.4.13)$$

It is evident from the above equation that just as $\chi(r)$ is the wave function of the object in $r$ space, similarly $a_k$ is the wave function of the same object in $k$ space. Equation (2.4.12) shows that $a_k$ is simply the Fourier transform of $\chi(r)$. Thus an object has different wave functions (representations) in different spaces.

## 2.5   Dirac's Bra and Ket Notation

Dirac invented an extremely compact notation to represent state functions and state vectors. In this notation the function $\chi_a(r)$ is represented by a ket $|a\rangle$ and its Hermitian adjoint state $\chi_a^+(r)$ by a bra $\langle a|$. This notation also gives rise to compact representation of the integrals. For example, the following equality

$$\int \psi_a^*(r)\psi_b(r)dr = \left[\int \psi_b^*(r)\psi_a(r)dr\right]^* \qquad (2.5.1)$$

in Dirac's notation is written as

$$\langle a \,|\, b\rangle = \langle b \,|\, a\rangle^* \qquad (2.5.2)$$

The bra $\langle a|$ and the ket $|b\rangle$ are abstract "bra" and "ket" state vectors in the bra and ket spaces, respectively. The names bra and ket come from the word bracket. Although bra and ket spaces are different they are related by (2.5.1). These abstract vectors can be utilized to obtain wave functions in different representations. For example, if an object is represented by the abstract vector $|\psi\rangle$, its wave function in the position representation is given by $\langle r|\psi\rangle \equiv \psi(r)$. The complex conjugate of $\psi$, i.e., $\psi^*(r)$ is equal to $\langle\psi|r\rangle$. Further, if $|a\rangle$ are eigenkets of an Hermitian operator they form a complete set.

Thus,

$$|\psi\rangle = \sum_a c_a |a\rangle \qquad (2.5.3)$$

and

$$\langle\psi| = \sum_a \langle a|c_a^* \qquad (2.5.4)$$

Assuming $|a\rangle$ to be discrete, we get

$$\langle a'|a\rangle = \delta_{aa'} \tag{2.5.5}$$

where $\delta_{aa'}$ is the Kronecker delta and is 1 for $a = a'$ and 0 for $a \neq a'$. Hence, from (2.5.3) $c_a = \langle a|\psi\rangle$ and equation (2.5.3) takes the form

$$|\psi\rangle = \sum_a |a\rangle\langle a|\psi\rangle \tag{2.5.6}$$

The operator $|a\rangle\langle a|$ is known as the projection operator because it projects out of $|\psi\rangle$, the eigenket $|a\rangle$. The above equation also shows that

$$\sum_a |a\rangle\langle a| = 1 \tag{2.5.7}$$

If the eigenkets $|r\rangle$ vary in a continuous manner, then (2.5.5) and (2.5.7) change to

$$\langle r'|r\rangle = \delta(r - r') \tag{2.5.8}$$

and

$$\int |r\rangle\langle r|dr = 1 \tag{2.5.9}$$

respectively. In the bra and ket notation

$$\int \psi_k^*(r')\psi_k(r)dk = \int \langle r|k\rangle\langle k|r'\rangle dk \tag{2.5.10}$$

but

$$\int |k\rangle\langle k|dk = 1$$

Hence,

$$\int \psi_k^*(r')\psi_k(r)dk = \langle r|r'\rangle = \delta(r - r') \tag{2.511a}$$

Thus we again obtain (2.4.9). If $k$ varies in a discrete manner the integration of (2.5.11a) changes to a summation. Replacing $k$ by a summation index $n$, we have from the above equation

$$\sum_n \psi_n^*(r')\psi_n(r) = \delta(r - r') \tag{2.511b}$$

where $\Psi_n(r)$ are the members of a complete set. Furthermore, if in a complete set $n$ varies in a discrete manner up to a certain term and beyond that it varies in a continuous manner (e.g., the complete sets formed by the eigenstates of atoms), then (2.5.11a) modifies to

$$\sum_n \psi_n^*(r')\psi_n(r) + \int \psi_k^*(r')\psi_k(r)dk = S \psi_n^*(r')\psi_n(r) = \delta(r - r') \quad (2.5.11c)$$

For atoms, $\psi_n(r)$ and $\psi_k(r)$ represent their bound and ionized states, respectively. The kinetic energy of the ejected electron due to ionization is $Rk^2a_0^2$, where $R$ is the Rydberg energy and $a_0$ is the first Bohr radius. The symbol $S$ represents summation over the discrete states and integration over the continuum states. As Dirac's notation is so compact we shall use it quite often.

## 2.6   Partial Wave Expansion of Plane Waves

For the plane wave given by (2.3.5), $k \cdot r = kr\cos\theta$. Hence, the wave function $\psi_k(r)$ is independent of the angle $\phi$ and can be expanded in terms of a complete set having the polar angle $\theta$ as a variable. The Legendre polynomial $P_l(\cos\theta)$ with the positive integer $l$ (including 0) represents such a complete set, which satisfies the following differential equation (Arfken, 1968):

$$\frac{d}{d\mu}\left[(1-\mu^2)\frac{d}{d\mu}\right]P_l(\mu) = -l(l+1)P_l(\mu) \quad (2.6.1)$$

where $\mu = \cos\theta$. Hence, $P_l(\mu)$ are eigenfunctions of the Hermitian operator $d/d\mu[(1-\mu^2)d/d\mu]$ with the eigenvalues $-l(l+1)$. Their orthogonality and closure relation are given by

$$\int_{-1}^{+1} P_l(\mu)P_{l'}(\mu)d\mu = \frac{2}{(2l+1)}\delta_{ll'} \quad (2.6.2)$$

and

$$\sum_{l=0}^{\infty} \frac{2l+1}{2}P_l(\mu)P_l(\mu') = \delta(\mu - \mu') = \delta(\theta - \theta')/\sin\theta \quad (2.6.3)$$

respectively. The first few $P_l(\mu)$ for small values of $l$ are given by

$$\begin{aligned}
P_0(\mu) &= 1 & P_0(\mu) &= \mu \\
P_2(\mu) &= \tfrac{1}{2}(3\mu^2 - 1) & P_3(\mu) &= \tfrac{1}{2}(5\mu^2 - 3\mu) \\
P_4(\mu) &= \tfrac{1}{8}(35\mu^4 - 30\mu^2 + 3) & P_5(\mu) &= \tfrac{1}{8}(63\mu^5 - 70\mu^3 + 15\mu)
\end{aligned} \quad (2.6.4)$$

A partial wave expansion of the plane wave in terms of $P_l(\mu)$ is taken as

$$\psi_k(r) = Ae^{ikr\mu} = A\sum_{l=0}^{\infty} A_l R_{l_0}(k,r)P_l(\mu) \qquad (2.6.5)$$

where $A_l$ are the coefficients of expansion and $R_{l_0}(k,r)$ is the radial wave function of the $l^{th}$ component of the plane wave. The subscript 0 indicates that the particle is free. In the above equation $k$ is taken as the axis of reference and $\theta(=\cos^{-1}\mu)$ is the angle between $k$ and $r$.

In spherical polar coordinates

$$\nabla^2 = \frac{1}{r^2}\frac{\partial}{\partial r}\left(r^2\frac{\partial}{\partial r}\right) - \frac{1}{\hbar^2 r^2}L^2 \qquad (2.6.6)$$

where $L^2$, the square of the angular momentum operator $L$, is

$$L^2 = -\hbar^2\left\{\frac{\partial}{\partial\mu}\left[(1-\mu^2)\frac{\partial}{\partial\mu}\right] + \frac{1}{1-\mu^2}\frac{\partial^2}{\partial\phi^2}\right\} \qquad (2.6.7)$$

From Eqs. (2.6.7) and (2.6.1) we get

$$L^2 P_l(\mu) = l(l+1)\hbar^2 P_l(\mu) \qquad (2.6.8)$$

which shows that the Legendre polynomials are the eigenfunctions of the operator $L^2$ with eigenvalues $l(l+1)\hbar^2$. Using (2.6.5), (2.6.6), and (2.6.8) in (2.3.4) gives the following differential equation for the radial wave function:

$$\left[\frac{1}{r^2}\frac{d}{dr}\left(r^2\frac{d}{dr}\right) + k^2 - \frac{l(l+1)}{r^2}\right]R_{l_0}(k,r) = 0 \qquad (2.6.9)$$

Let us define $Z(kr) = (kr)^{1/2}R_{l_0}(kr)$; then (2.6.9) changes to

$$\left[r^2\frac{d^2}{dr^2} + r\frac{d}{dr} + k^2r^2 - (l+\tfrac{1}{2})^2\right]Z(k,r) = 0 \qquad (2.6.10)$$

The above equation is the differential equation for the Bessel function $J_{l+1/2}(kr)$ of the order $l+\frac{1}{2}$ (Arfken, 1968). This Bessel function is employed to define the following two linearly independent spherical Bessel functions:

$$j_l(x) = \left(\frac{\pi}{2x}\right)^{1/2}J_{l+1/2}(x) \qquad (2.6.11)$$

and

$$n_l(x) = (-1)^{l+1} \left(\frac{\pi}{2x}\right)^{1/2} J_{-l-1/2}(x) \qquad (2.6.12)$$

The latter is known as the Neumann function. At small values of $x$ we have

$$j_l(x) \underset{x \to 0}{\to} \frac{x^l}{(2l+1)!!} \qquad (2.6.13)$$

and

$$n_l(x) \underset{x \to 0}{\to} -\frac{(2l-1)!!}{x^{l+1}} \qquad (2.6.14)$$

where $(2l + 1)!! = 1 \cdot 3 \cdot 5 \ldots (2l - 1) \cdot (2l + 1)$. The above equations show that $j_l(x)$ are regular but $n_l(x)$ are irregular at the origin. For large values of $x$, the above functions are given by

$$j_l(x) \underset{x \to \infty}{\to} \frac{1}{x} \sin(x - l\pi/2) \qquad (2.6.15)$$

$$n_l(x) \underset{x \to \infty}{\to} -\frac{1}{x} \cos(x - l\pi/2) \qquad (2.6.16)$$

The first three $j_l(x)$ and $n_l(x)$ at all the values of $x$ are given by:

$$j_0(x) = \frac{\sin x}{x} \qquad\qquad n_0(x) = -\frac{\cos x}{x}$$

$$j_1(x) = \frac{\sin x}{x^2} - \frac{\cos x}{x} \qquad n_1(x) = -\left(\frac{\cos x}{x^2} + \frac{\sin x}{x}\right)$$

$$j_2(x) = \left(\frac{3}{x^3} - \frac{1}{x}\right)\sin x - \frac{3}{x^2}\cos x \qquad n_2(x) = -\left(\frac{3}{x^3} - \frac{1}{x}\right)\cos x - \frac{3}{x^2}\sin x$$

$$\qquad\qquad\qquad\qquad\qquad\qquad\qquad\qquad\qquad (2.6.17)$$

In general $R_{l_0}(x)$ is a linear combination of $j_l(x)$ and $n_l(x)$. Hence,

$$R_{l_0}(x) = a_l j_l(x) + b_l n_l(x) \qquad (2.6.18)$$

Since $R_{l_0}(x)$ is the radial wave function of an object, it has to be finite everywhere, but $n_l(x)$ diverges at the origin, so $b_l$ in (2.6.18) has to be zero at $x = 0$. Since the particle is free, the form of $R_{l_o}(x)$ will not change with $x$, and $b_l$ is zero for all values of $x$.

Putting (2.6.18) in (2.6.5) and absorbing $a_l$ in $A_l$, we get

$$\psi_k(r) = A \sum_l A_l j_l(kr) P_l(\mu) \qquad (2.6.19)$$

To evaluate the value of the expansion coefficient $A_l$ we multiply (2.6.19) by $P_{l'}(\mu)$ and integrate over $\mu$. This gives

$$\int_{-1}^{+1} e^{ikr\mu} P_{l'}(\mu) d\mu = \sum_l A_l j_l(kr) \int_{-1}^{+1} P_{l'}(\mu) P_l(\mu) d\mu \qquad (2.6.20)$$

Integration of the left-hand side by parts and the use of the orthogonality relation for $P_l(\mu)$ gives

$$\frac{1}{ikr} \left[ e^{ikr\mu} P_{l'}(\mu) \right]_{-1}^{+1} - \frac{1}{ikr} \int_{-1}^{+1} e^{ikr\mu} \left[ \frac{dP_{l'}(\mu)}{d\mu} \right] d\mu = \frac{2}{2l'+1} A_{l'} j_{l'}(kr) \quad (2.6.21)$$

As the second term on the left-hand side is of the order of $r^{-2}$, at large $r$ we get

$$\frac{1}{ikr} \left[ e^{ikr} - (-1)^l e^{-ikr} \right] = \frac{A_l}{2l+1} \frac{1}{ikr} \left( e^{i(kr-l\pi/2)} - e^{-i(kr-l\pi/2)} \right) \qquad (2.6.22)$$

Equating the coefficients of $e^{ikr}$ on both the sides yields

$$A_l = (2l+1) e^{il\pi/2} = i^l (2l+1) \qquad (2.6.23)$$

Hence, the partial wave expansion of a plane wave is given by

$$\psi_k(r) = A \sum_{l=0}^{\infty} i^l (2l+1) j_l(kr) P_l(\mu) \qquad (2.6.24)$$

$$\xrightarrow[r \to \infty]{} A \sum_{l=0}^{\infty} i^l (2l+1) \frac{1}{kr} \sin(kr - l\pi/2) P_l(\mu) \qquad (2.6.25)$$

Each partial wave has a well-defined angular momentum, characterized by the quantum number $l$.

## 2.7   Spherical Harmonics

In the expansion given by (2.6.24) the reference axis is along the $k$ axis. If we remove this condition the plane wave also becomes a function of the polar angle $\phi$. Thus instead of $P_l(\mu)$ we require a complete set that depends on $\theta$ as well as on $\phi$. Such a complete set is provided by the spherical harmonics $Y_{lm}(\theta,\phi)$, where the magnetic quantum number $m$ is an integer and varies from $-l$ to $+l$ in steps of 1. Hence, for a given $l$ there are $(2l + 1)$ spherical harmonics. Like the Legendre polynomials, spherical harmonics are also eigenfunctions of the operator $L^2$ [given by (2.6.7)] with the same eigenvalues, i.e., $l(l+1)\hbar^2$. They are also eigenfunctions of $L_z$ with eigenvalues $m\hbar$. Their orthogonality and closure relations are given by

$$\int_0^\pi \int_0^{2\pi} Y_{lm}^*(\theta,\phi)Y_{l'm'}(\theta,\phi)\sin d\theta \, d\phi = \delta_{ll'}\delta_{mm'} \tag{2.7.1}$$

and

$$\sum_l \sum_{m=-l}^{+l} Y_{lm}^*(\theta,\phi)Y_{lm}(\theta',\phi') = \frac{\delta(\theta-\theta')\delta(\phi-\phi')}{\sin\theta} \tag{2.7.2}$$

For a given $l$ we also have

$$\sum_{m=-l}^{+l} Y_{lm}^*(\theta,\phi)Y_{lm}(\theta',\phi') = \delta(\phi=\phi') \tag{2.7.3}$$

Normalized $Y_{lm}(\theta,\phi)$ are given by

$$Y_{lm}(\theta,\phi) = K\left[\frac{(2l+1)}{4\pi}\frac{(l-|m|)!}{(l+|m|)!}\right]^{1/2} P_l^{|m|}(\cos\theta)e^{im\phi} \tag{2.7.4}$$

where $K = (-1)^m$ for $m > 0$ and $K = 1$ for $m \le 0$. The associated Legendre polynomials $P_l^{|m|}(\cos\theta)$ are the solutions of

$$\left\{\frac{d}{d\mu}\left[(1-\mu)^2\frac{d}{d\mu}\right] + l(l+1) - \frac{m^2}{1-\mu^2}\right\}P_l^{|m|}(\mu) = 0 \tag{2.7.5}$$

It is easy to see that $P_l^{|m|}(\mu)$ reduces to $P_l(\mu)$ for $m = 0$.

The first few $Y_{lm}(\theta,\phi)$ are given by:

$$Y_{00} = \left(\frac{1}{4\pi}\right)^{1/2} \qquad\qquad Y_{20} = \left(\frac{5}{16\pi}\right)^{1/2}(3\cos^2\theta - 1)$$

$$Y_{10} = \left(\frac{3}{4\pi}\right)^{1/2}\cos\theta \qquad\qquad Y_{2\pm1} = \mp\left(\frac{15}{8\pi}\right)^{1/2}\sin\theta\cos\theta e^{\pm i\phi}$$

$$Y_{1\pm1} = \mp\left(\frac{3}{8\pi}\right)^{1/2}\sin\theta e^{\pm i\phi} \qquad Y_{2\pm2} = \left(\frac{15}{32\pi}\right)^{1/2}\sin^2\theta e^{\pm 2i\phi} \tag{2.7.6}$$

According to the addition theorem, if $\theta$ is the angle between $r$ and $k$, then

$$P_l(\cos\theta) = \frac{4\pi}{2l+1} \sum_{m=-l}^{+l} Y_{lm}^*(\hat{r})Y_{lm}(\hat{k}) \tag{2.7.7}$$

Putting this equation into (2.6.24), we get

$$\psi_k(r) = A4\pi \sum_l \sum_m i^l j_l(kr)Y_{lm}^*(\hat{r})Y_{lm}(\hat{k}) \tag{2.7.8}$$

In the next chapter we shall examine the changes in $\Psi_k(r)$ and its partial waves given by (2.6.25) when a microparticle collides (is scattered) with (by) a potential field. Those changes will be used to obtain the differential and integrated cross sections.

## Questions and Problems

2.1 Define the Dirac delta function and show that

$$\delta(x-a) = \operatorname*{Lim}_{l\to 0} \frac{1}{l\sqrt{2\pi}} \exp\left[-\frac{(x-a)^2}{2l^2}\right]$$

2.2 Prove the following relations for the Dirac delta function

(a) $x\delta(x) = 0$
(b) $\delta(ax) = \delta(x)/|a|$ for $a \neq 0$
(c) $x\delta'(x) = -\delta(x)$

where $\delta'$ is the differential of $\delta$

(d) $\delta(x^2 - a^2) = \frac{1}{|2a|}[\delta(x-a) + \delta(x+a)]$ for $a \neq 0$

2.3 Verify the relations (2.4.7) and (2.4.8).

2.4 (a) Show that for a projection operator $P$

$$P^2 = P$$

(b) A $3 \times 3$ matrix is given by $\begin{pmatrix} a & 0 & 0 \\ 0 & 0 & 0 \\ 0 & 0 & b \end{pmatrix}$

Find the value of $a$ and $b$ so that this matrix represents a projector operator. Show that with the proper values of $a$ and $b$ this operator projects a three-dimensional

vector $\begin{pmatrix} x \\ y \\ z \end{pmatrix}$ on a two-dimensional subspace to give $\begin{pmatrix} x \\ 0 \\ z \end{pmatrix}$.

2.5  Write the matrix element $H_{kl}$ in the Dirac bra and ket notation. Show that $H$ is a diagonal matrix if $|k\rangle$ and $|l\rangle$ are eigenkets of $H$. On the other hand, if $|n\rangle$ are eigenkets of $H$ instead of $|k\rangle$ and $|l\rangle$, then

$$H_{kl} = \underset{n}{S}\varepsilon_n \langle k|n\rangle\langle n|l\rangle$$

where $\varepsilon_n$ are the eigenvalues of the kets $|n\rangle$. Give a physical interpretation of the term $\langle n|k\rangle$.

2.6  Use the expressions for $P_4(\mu)$ and $P_5(\mu)$ given by (2.6.4) and evaluate the required integrals to show that

$$\langle P_5(\mu)|P_5(\mu)\rangle = \tfrac{2}{11} \qquad \text{and} \qquad \langle P_5(\mu)|P_4(\mu)\rangle = 0$$

2.7  A function $\psi(r)$ when expanded in the complete set of $P_l(\cos\theta)$ is given by

$$\psi(r) = \sum_{l=0}^{\infty} A_l R_l(r) P_l(\cos\theta)$$

Asymptotically,

$$\psi(r) \underset{r\to\infty}{\longrightarrow} e^{ik\cdot r} + \frac{f(\theta)}{r}e^{ikr} \qquad \text{and} \qquad R_l(r) \underset{r\to\infty}{\longrightarrow} \cos\eta_l j_l(kr) - \sin\eta_l n_l(kr)$$

where $\theta$ is the angle between $k$ and $r$. Show that $A_l = i^l (2l + 1) e^{i\eta_l}$.

2.8  Prove the following recurrence relation for the Legendre polynomials:

$$lP_l(\mu) = (2l-1)\mu P_{l-1}(\mu) - (l-1)P_{l-2}(\mu)$$

2.9  (a) If $(\theta_1,\phi_1)$ and $(\theta_2,\phi_2)$ are two different directions in spherical polar coordinates and $\theta$ is the angle between these two directions, prove the addition theorem

$$P_l(\cos\theta) = \frac{4\pi}{2l+1}\sum_{m=-l}^{l} Y_{lm}(\theta_1,\phi_1)Y_{lm}^*(\theta_2,\phi_2)$$

(b) Use the addition theorem to show that

$$\frac{1}{|r_1 - r_2|} = \sum_{l=0}^{\infty} \sum_{m=-l}^{l} \frac{4\pi}{2l+1} \frac{r_<^l}{r_>^{l+1}} Y_{lm}^*(\hat{r}_1) Y_{lm}(\hat{r}_2)$$

where $r_<$ is smaller than $r_1$ and $r_2$ and $r_>$ is greater than $r_1$ and $r_2$.

2.10 Evaluate the transition amplitude $\langle \psi_1(r) | z | \psi_2(r) \rangle$ where

$$\psi_1(r) = 2\left(\frac{z}{a_0}\right)^{3/2} \exp(-zr/a_0) Y_{00}(\hat{r}) \qquad \text{and} \qquad \psi_2(r) = \sum_n A_n \exp(ik_n \cdot r)$$

# Collision of a Free Particle with a Potential Field

## 3.1 Introduction

In the previous chapter we discussed the motion of a free particle. Now we consider the collision (scattering) of a free particle with (by) a potential field. In the presence of a potential, Eq. (2.3.3) changes to (Burke, 1977)

$$\left[-\frac{\hbar^2}{2m}\nabla^2 + V(r)\right]\psi(r) = E\psi(r) \tag{3.1.1}$$

where $V(r)$ is the potential energy of the particle. In the asymptotic region, where $V(r) = 0$, (3.1.1) admits two solutions. One is the plane wave given by (2.3.5) and the other is either a spherically outgoing wave $\exp(ikr)$ or a spherically incoming wave $\exp(-ikr)$ having $f(\theta, \varphi)/r$ as its amplitude. The polar coordinates of the scattered particle measured from the center of the field are given by $(r, \theta, \varphi)$, and $f(\theta, \varphi)$ is the scattering amplitude. Taking a linear combination of both solutions, the wave function of the scattered particle in the asymptotic region is given by

$$\psi_E^\pm(r) \underset{r \to \infty}{\sim} A\left[e^{ik \cdot r} + \frac{f(\theta, \varphi)}{r} e^{\pm ikr}\right] \tag{3.1.2}$$

where + and − denote the outgoing and incoming solutions, respectively. We shall consider only the outgoing solution and drop the superscript ±. It is easy to verify that up to the order of $1/r$ for any arbitrary form of $f(\theta, \varphi)$, $Af(\theta, \varphi)\exp(ikr)/r$ is a solution of (3.1.1) in the region where $V(r) = 0$. Hence, for (3.1.2) to be valid, $V(r)$ should fall faster than $r^{-2}$ in the asymptotic region. Now we proceed to derive a relation between the differential cross section $I(\theta, \varphi)$, a quantity measured by the experimentalists, and the scattering amplitude $f(\theta, \varphi)$, which is calculated by the theoreticians. To achieve this we first consider the continuity equation and its relationship with the collision cross section.

## 3.2   Continuity Equation and Cross Section

The differential equation for $\psi^*(r)$, the complex conjugate of $\psi(r)$, as obtained from (3.1.1) is given by

$$\left[-\frac{\hbar^2}{2m}\nabla^2 + V^*(r)\right]\psi^*(r) = E\psi^*(r) \tag{3.2.1}$$

Now we multiply (3.1.1) by $\psi^*(r)$ and (3.2.1) by $\psi(r)$ from the left and subtract the former from the latter to get

$$-\frac{\hbar^2}{2m}\left(\psi\nabla^2\psi^* - \psi^*\nabla^2\psi\right) + \psi V^*\psi^* - \psi^*V\psi = 0 \tag{3.2.2}$$

We take $V = V_R - iV_I$, where $V_R$ and $V_I$ are the real and imaginary parts of the complex potential. This substitution reduces (3.2.2) to

$$\frac{\hbar^2}{2m}\nabla\cdot\left(\psi^*\nabla\psi - \psi\nabla\psi^*\right) + 2iV_I\psi^*\psi = 0 \tag{3.2.3}$$

or

$$\nabla\cdot c(r) + \frac{2}{\hbar}V_I|\psi|^2 = 0 \tag{3.2.4}$$

where the probability current density $c(r)$ is equal to

$$c(r) = \mathrm{Re}\left(\frac{\hbar}{mi}\psi^*\nabla\psi\right) \tag{3.2.5}$$

Hence, $c(r)$ is a flux vector and its radial component $c(r)\cdot\hat{r}$ is given by

$$c(r)\cdot\hat{r} = \mathrm{Re}\left(\frac{\hbar}{mi}\psi^*\frac{\partial}{\partial r}\psi\right) \tag{3.2.6}$$

To compute $c(r) \cdot \hat{r}$ we use $\psi$ as given by (3.1.2). Thus $c(r) \cdot \hat{r}$ is the sum of three terms. The first term, $c_i(r) \cdot \hat{r}$, is due to the plane wave, the second, $c_0 \cdot \hat{r}$, is due to the outgoing spherical wave, and the third, $c_{in}(r) \cdot \hat{r}$, arises due to the interference between the plane and spherical waves. These terms are given by

$$c_i \cdot \hat{r} = vAA^* \cos\theta \tag{3.2.7}$$

$$c_0 \cdot \hat{r} = vAA^* \frac{|f(\Omega)|^2}{r^2} + 0\left(\frac{1}{r^3}\right) \tag{3.2.8}$$

$$c_{in} \cdot \hat{r} = \text{Re}\left[AA^* v\left(\frac{f(\Omega)}{r} e^{ikr(1-\cos\theta)} + \frac{f^*(\Omega)}{r} e^{-ikr(1-\cos\theta)} \cos\theta\right)\right] + 0\left(\frac{1}{r^2}\right) \tag{3.2.9}$$

Equation (3.2.9) shows that at oblique angles ($\theta \neq 0$) and large $r$, $c_{in} \cdot \hat{r}$ oscillates very rapidly as a function of $r$. Furthermore, due to collimating slits in any experimental arrangement (see Fig. 1.1), the contribution of the incident beam to $c(r) \cdot \hat{r}$ in the oblique direction is also negligibly small. Thus for $\theta \neq 0$, we take the outgoing flux equal to $c_0 \cdot \hat{r}$.

Suppose the detector, which is at a distance $r$ from the scattering center, makes a solid angle $d\Omega$ with the center. Then $\Delta N$, the number of particles entering into the detector $D$ per unit time is given by

$$\Delta N = c_0 \cdot \hat{r} r^2 d\Omega \tag{3.2.10}$$

The incidence flux $F$ is

$$F = c_i \cdot \hat{z} = \text{Re}\left[\frac{\hbar}{mi} AA^* e^{-ikz} \frac{\partial}{\partial z}(e^{ikz})\right]$$

or

$$F = AA^* v$$

Hence, from (1.2.1) with $n = 1$ and (3.2.8) and (3.2.10), we get

$$I(\theta, \varphi)d\Omega = |f(\Omega)|^2 d\Omega \tag{3.2.11}$$

The integrated cross section is obtained by integrating (3.2.11) over the angles $\theta$ and $\varphi$. Since for atomic collisions it yields the elastic cross section, we shall denote it by $\sigma_{el}$, where

$$\sigma_{el} = \int |f(\Omega)|^2 d\Omega \tag{3.2.12}$$

Let us now integrate (3.2.4) over the volume and use Green's theorem on the first term. Thus we get

$$r^2 \int c(r) \cdot \hat{r} \, d\Omega + \frac{2}{\hbar} \int V_I |\psi|^2 dr = 0 \tag{3.2.13}$$

The first term of the above equation is the net number of particles leaving the surface of a sphere of radius $r$ per unit time. It is not equal to zero but is equal to the negative of the second term. Hence, in the scattering of a beam by a complex potential, a certain number of particles are absorbed, i.e., there is a sink, owing to the fact that we have taken $V = V_R - iV_I$, whereas $V = V_R + iV_I$ would have produced a source. The incident beam provides the particles, which are scattered and absorbed. For a real potential, the second term is zero and the particles are conserved. In this case there are no sources or sinks for the particles, i.e., there is neither creation nor absorption of particles but the incident beam provides the scattered particles. Now in (3.2.13) we replace $c(r) \cdot \hat{r}$ by its three components and obtain

$$r^2 \int c_I \hat{r} d\Omega + r^2 \int c_0 \hat{r} d\Omega + r^2 \int c_{in} \cdot \hat{r} d\Omega + \frac{2}{\hbar} \int V_I |\psi|^2 dr = 0 \tag{3.2.14}$$

It can be shown (Joachain, 1987) that

$$r^2 \int_{\delta\Omega} c_{in} \cdot \hat{r} d\Omega = -4\pi AA^* \frac{\hbar}{m} \text{Im} \, f(\theta = 0) \tag{3.2.15}$$

where $\delta\Omega$ is an infinitesimal solid angle around the forward direction. As noted earlier, $c_{in} \cdot \hat{r}$ does not contribute in other directions. Using (3.2.7), (3.2.8), and (3.2.15) in (3.2.14), we get

$$vAA^* \int |f(\Omega)|^2 d\Omega - 4\pi AA^* \frac{\hbar}{m} \text{Im} \, f(\theta = 0) + \frac{2}{\hbar} \int V_I |\psi|^2 dr = 0 \tag{3.2.16}$$

or

$$\sigma_{el} + \frac{2}{\hbar} \frac{1}{vAA^*} \int V_I |\psi|^2 dr = \frac{4\pi}{k} \text{Im} \, f(\theta = 0) \tag{3.2.17}$$

Since the second term is due to absorption, we denote it by $\sigma_{ab}$, the absorption cross section. Then the total cross section $\sigma_T$ is given by

$$\sigma_T \equiv \sigma_{el} + \sigma_{ab} = \frac{4\pi}{k} \operatorname{Im} f(\theta = 0) \qquad (3.2.18)$$

The above equation, known as the optical theorem, is a direct consequence of the conservation of the particles. For a real potential it reduces to

$$\sigma_{el} = \frac{4\pi}{k} \operatorname{Im} f(\theta = 0) \qquad (3.2.19)$$

Obviously, $\sigma_{el}$ as obtained from (3.2.18) is different from that given by (3.2.19) even for the same $V_R$. As a matter of fact, the difference between the two represents the effect of absorption on the elastic scattering. In atomic collisions, we have a number of inelastic channels apart from the elastic channel and, in general, elastic and inelastic collisions take place simultaneously. Hence, to take into account the effect of the inelastic processes on the elastic cross section, we use a model complex interaction potential. The complex potential should be such that $\sigma_{ab}$ is equal to the sum of all the inelastic cross sections.

## 3.3   Relationship between the Scattering Amplitude and the Scattered Wave Function

Equation (3.2.11) shows that we need $f(\theta, \varphi)$ to calculate $I(\theta, \varphi)$. In this section we derive a relationship between $f(\theta, \varphi)$ and $\psi_k^+(r)$. We rewrite (3.1.1) as

$$(\nabla^2 + k^2)\psi(r) = U(r)\psi(r) \qquad (3.3.1)$$

where the reduced interaction potential $U(r)$ is equal to $(2m/\hbar^2)V(r)$ and has the dimension $L^{-2}$. The above equation is an inhomogeneous differential equation. The corresponding integral equation for the outgoing scattered wave $\psi_k^+(r)$ is given by

$$\psi_k^+(r) = \varphi_k(r) + \int G_0^+(r, r')U(r')\psi_k^+(r')dr' \qquad (3.3.2)$$

where $\varphi_k(r)$ is the solution of (3.3.1) when $U(r) = 0$; i.e., it is identical to the plane wave $Ae^{ik\cdot r}$, given by (2.3.5). The above integral equation is known as the Lippmann–Schwinger equation and in symbolic form is written as

$$\psi^+ = \varphi + G_0^+ U\psi^+ \qquad (3.3.3)$$

The free-particle Green's function $G_0^+$ satisfies the following differential equation:

$$(\nabla^2 + k^2)G_0^+(r, r') = \delta(r - r') \qquad (3.3.4)$$

and has the solution

$$G_0^+(r, r') = -\frac{1}{4\pi} \frac{e^{ik|r-r'|}}{|r - r'|} \qquad (3.3.5)$$

Its integral representation is given by

$$G_0^+(r, r') = -\frac{1}{(2\pi)^3} \lim_{\varepsilon \to 0^+} \int \frac{e^{ik' \cdot (r-r')}}{k'^2 - k^2 - i\varepsilon} dk' \qquad (3.3.6)$$

$$= -\lim_{\varepsilon \to 0^+} \int \frac{|k'\rangle\langle k'|}{k'^2 - k^2 - i\varepsilon} dk' \qquad (3.3.7)$$

where, in Dirac's notation,

$$\langle r | k \rangle = \frac{1}{(2\pi)^{3/2}} e^{ik \cdot r} = \varphi_k(r) \qquad (3.3.8)$$

Now,

$$|r - r'| = (r^2 - 2r \cdot r' + r'^2)^{1/2} \approx r\left(1 - \frac{2r \cdot r'}{r^2}\right)^{1/2} \text{ for large } r.$$

Hence,

$$|r - r'| \approx r - \hat{r} \cdot r' + \cdots \qquad (3.3.9)$$

where $\hat{r}$ is a unit vector in the direction of the scattered particle. To evaluate the phase term of $G_0^+$ at large $r$, we take the first two terms of (3.3.9), but for the amplitude, $|r - r'|$ is taken as $r$. Then (3.3.5) reduces to

$$G_0^+(r, r') \underset{r \to \infty}{\sim} -\frac{1}{4\pi} \frac{e^{-ik\hat{r} \cdot r'}}{r} e^{ikr} \qquad (3.3.10)$$

The final momentum vector $k_f$ is equal to $k\hat{r}$, whereas the initial momentum vector is $k_i$. Putting (3.3.10) into (3.3.2), we get

$$\psi_{k_i}^+(r) \underset{r \to \infty}{\sim} \varphi_{k_i}(r) - \frac{e^{ikr}}{4\pi r} \int e^{-ik_f \cdot r'} U(r') \psi_{k_i}^+(r') dr' \qquad (3.3.11)$$

Comparing (3.3.11) with (3.1.2), we obtain

$$f(\theta, \varphi) = -\frac{1}{4\pi A} \int e^{-ik_f \cdot r} U(r) \psi_{k_i}^+(r) dr$$

or, in bra and ket notation,

$$f_{fi}(\theta, \varphi) = -2\pi^2 \langle \varphi_{k_f} | U | \psi_{k_i}^+ \rangle \tag{3.3.12}$$

where we have taken $\langle \varphi_{k_f} | r \rangle = A^* e^{-ik_f \cdot r}$ and $A = (2\pi)^{-3/2}$. The term $f_{fi}(\theta, \varphi)$ represents the scattering amplitude for the scattering of the particle from the initial state $i$ to the final state $f$. Equation (3.3.12) may be rewritten as

$$f_{fi} = -\frac{4\pi^2 m}{\hbar^2} T_{fi} \tag{3.3.13}$$

where the transition matrix element from the initial state $|k_i\rangle$ to the final state $|k_f\rangle$ is given by

$$T_{fi} = \langle \varphi_{k_f} | V | \psi_{k_i}^+ \rangle \tag{3.3.14}$$

Equation (3.3.12) shows that to calculate $f(\theta, \varphi)$ and, hence, $I(\theta, \varphi)$ and $\sigma$, we must have $\psi_{k_i}^+(r)$ in the region where $U(r)$ is nonzero. On the other hand, (3.1.2) shows that to calculate $f(\theta, \varphi)$ we need the asymptotic value of $\psi_{k_i}^+(r)$. These two equations have given rise to two different approaches, namely the integral and differential approaches, to evaluate $f(\theta, \varphi)$.

## 3.4 The Integral Approach

In the integral approach, the Lippmann–Schwinger equation, given by (3.3.2) or (3.3.3), is solved for $\psi_{k_i}^+(r)$ by the iterative method. In these equations, the second term represents the distortion of the initial wave function $\psi_{k_i}(r)$ by $U(r)$. For convenience we represent the initial wave by $\psi_0(r)$. The distorted part of the wave function is given by

$$\psi_d(r) = \int G_0^+(r, r') U(r') \psi_{k_i}^+(r') dr' \tag{3.4.1}$$

The above equation is solved by iteration. We replace $\psi_{k_i}^+(r')$ in this equation by the initial wave function $\psi_0(r')$, which is the zeroth-order solution of the Lippmann–Schwinger integral equation. This gives the first-order correction to $\psi_0(r)$ and is equal to

$$\psi_1(r) = \int G_0^+(r, r')U(r')\psi_0(r')dr' \tag{3.4.2}$$

A replacement of $\psi_{k_i}^+(r')$ by $\psi_1(r')$ in (3.4.1) yields the second-order correction to $\psi_0(r)$:

$$\psi_2(r) = \int G_0^+(r, r')U(r')\psi_1(r')dr' \tag{3.4.3}$$

Use of (3.4.2) in (3.4.3) gives

$$\psi_2(r) = \int\int G_0^+(r, r')U(r')G_0^+(r', r'')U(r'')\psi_0(r'')dr'dr'' \tag{3.4.4}$$

Symbolically,

$$\psi_2 = G_0^+ U G_0^+ U \psi_0 = \left(G_0^+ U\right)^2 \psi_0 \tag{3.4.5}$$

Hence, in general,

$$\psi_n = \left(G_0^+ U\right)^n \psi_0 = G_0^+ U \psi_{n-1} \tag{3.4.6}$$

where $n$ is a positive integer. Adding all the corrections, we obtain the following series:

$$\psi_{k_i}^+ = \psi_0 + \psi_1 + \psi_2 + \cdots + \psi_n + \cdots$$
$$= \sum_{n=1}^{\infty} \left(G_0^+ U\right)^{n-1} \psi_0 \tag{3.4.7}$$

We may also write

$$\psi_{k_i}^+ = \psi_0 + G^+ U \psi_0 \tag{3.4.8}$$

where

$$G^+ = G_0^+ + G_0^+ U G_0^+ + G_0^+ U G_0^+ U G_0^+ + \cdots \tag{3.4.9}$$

is the full Green's function. The series given by (3.4.7) is known as the Born series for the scattered wave function $\psi_{k_i}^+(r)$. The use of this series in (3.3.12) gives

$$f_{fi}(\theta, \varphi) = -2\pi^2 \left\langle \varphi_{k_f} |U| \sum_{n=1}^{\infty} \left(G_0^+ U\right)^{n-1} \psi_0 \right\rangle$$
$$= f_{B1} + \bar{f}_{B2} + \bar{f}_{B3} + \cdots + \bar{f}_{Bn} + \cdots \tag{3.4.10}$$

where the $n$th Born term is given by

$$\bar{f}_{Bn} = -2\pi^2 \langle \varphi_{k_f} | U | (G_0^+ U)^{n-1} \psi_0 \rangle \qquad (3.4.11)$$

The $n$th Born scattering amplitude is the sum of the first $n$ terms of the Born series given by (3.4.10). Hence,

$$f_{Bn} = \sum_{p=1}^{n} \bar{f}_{Bp} \qquad (3.4.12)$$

By our definition the first Born term and the first Born scattering amplitude are the same. We also note that except for $f_{B1}$, all the Born terms involve the reduced interaction energy more than once and hence represent multiple scattering terms. For example, the second Born term

$$\bar{f}_{B2}(k_f, k_i) = -2\pi^2 \int \varphi_{k_f}^*(r) U(r) G_0^+(r, r') U(r') \psi_0(r') dr dr' \qquad (3.4.13)$$

involves $U$ twice, and hence is a double scattering term. It is interesting to visualize $\bar{f}_{B2}$ due to the following processes: The incident wave $\psi_0(r')$ interacts with the potential at $r'$ and is converted into a new wave given by $U(r')\psi_0(r')$. This wave is propagated to $r$ by the Green's function propagator $G_0^+(r, r')$. Since $r'$ is any point in the space, the wave function of the object at $r$ is given by $\int G_0^+(r, r') U(r') \psi_0(r') dr'$. This object at $r$ interacts again with $U$ to become

$$U(r) \int G_0^+(r, r') U(r') \psi_0(r') dr'$$

Now, to obtain the probability amplitude of finding the object in the final state, $f$, we take the overlap of $\psi_{k_f}(r)$ with the above wave function. Again $r$ can be anywhere in space; hence, the resultant expression is integrated over $r$ and we get (3.4.13). Using (3.3.7) in (3.4.13), we get

$$\bar{f}_{B2}(k_f, k_i) = 2\pi^2 \lim_{\varepsilon \to 0} \int \langle \varphi_f | U | k \rangle \frac{1}{k^2 - k_i^2 - i\varepsilon} \langle k | U | \psi_0 \rangle dk \qquad (3.4.14)$$

The above equation can be represented by a simple Feynman diagram, shown in Fig. 3.1, which can be interpreted as follows. The object in the initial state $|k_i\rangle$ collides with $U$ and goes to an intermediate state $|k\rangle$. This intermediate state again collides with $U$ and goes to the final state $|k_f\rangle$. Although in the steady state the final energy of the system must be equal to its initial energy, in the intermediate processes such as $|k_i\rangle \to |k\rangle$ or $|k\rangle \to |k_f\rangle$, the energy need not be conserved. Hence, there are an infinite number of intermediate states. To include contribu-

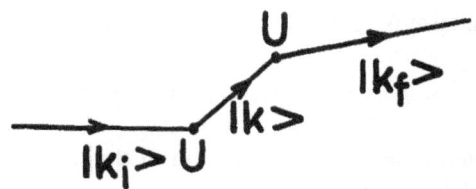

**FIGURE 3.1** Feynman diagram for the second Born scattering term.

tions from all the intermediate states, equation (3.4.14) involves integration over $k$.

Let us now put $\psi_{ki}^+$ as given by (3.3.2) into (3.3.12). This gives

$$f(k_f, k_i) = f_{B1}(k_f, k_i) - 2\pi^2 \langle \varphi_{k_f} | UG_0^+ U | \psi_{k_i}^+ \rangle \qquad (3.4.15)$$

Putting the integral expression of $G_0^+$ from (3.3.7) into the above equation, we get

$$f(k_f, k_i) = f_{B1}(k_f, k_i) - 2\pi^2 \int \langle \varphi_{k_f} | U | k \rangle \frac{1}{k_i^2 - k^2 + i\varepsilon} \langle k | U | \psi_{k_i}^+ \rangle \, dk \quad (3.4.16)$$

or

$$f(k_f, k_i) = f_{B1}(k_f, k_i) - \frac{1}{2\pi^2} \int f_{B1}(k_f, k) \frac{1}{k_i^2 - k^2 + i\varepsilon} f(k, k_i) \, dk \quad (3.4.17)$$

The above equation is known as the Fredholm integral equation, and can also be employed to generate the Born series. It can be shown that the Born series converges for a repulsive potential. It also converges for an attractive potential provided that the potential field does not support any bound state.

In principle, $f(k_f, k_i)$ can be calculated correct to any order. However, with an increase in $n$ the difficulties in the evaluation of $\bar{f}_{Bn}$ increase rapidly. In most cases calculations are limited to $f_{B1}$ and $\bar{f}_{B2}$, which we consider now.

## 3.5  The First Born Approximation

The *first Born approximation* (FBA) is the simplest but one of the most celebrated approximations of collision theory. Almost all the investigations start from the FBA, which is given by

$$f_{B1} = -2\pi^2 \langle \varphi_{k_f} | U | \varphi_{k_i} \rangle \qquad (3.5.1)$$

Hence, it completely neglects the distortion of the incident wave by the interacting potential. Let us represent the change in the momentum vector of the incident particle due to collision by $\boldsymbol{K}$; then

$$\boldsymbol{K} = \boldsymbol{k}_i - \boldsymbol{k}_f \qquad (3.5.2)$$

For elastic scattering $|\boldsymbol{k}_i| = |\boldsymbol{k}_f| = k$; hence,

$$K = 2k\sin(\theta/2) \qquad (3.5.3)$$

where $\theta$ is the scattering angle. Using plane waves for the initial and final states and (3.5.2) in (3.5.1), we get

$$f_{B1}(K) = -\frac{1}{4\pi}\int e^{i\boldsymbol{K}\cdot\boldsymbol{r}}U(r)d\boldsymbol{r} \qquad (3.5.4)$$

where we have taken $A = (2\pi)^{-3/2}$. It is useful to note some of the characteristics of the FBA. First of all, (3.5.4) shows that $f_{B1}$ depends only on $K(= \boldsymbol{k}_i - \boldsymbol{k}_f)$ and not on $\boldsymbol{k}_i$ and $\boldsymbol{k}_f$ individually. Secondly, $f_{B1}(K)$ is simply the Fourier transform of the reduced interaction energy $U(r)$. Furthermore, due to the oscillation of the phase term $\exp(i\boldsymbol{K}\cdot\boldsymbol{r})$ with $\boldsymbol{r}$, the contributions of the integrand to the integral from the different regions of $\boldsymbol{r}$ are positive as well as negative. However, in the forward direction $K = 0$. Hence, the contributions from all the different regions of $\boldsymbol{r}$ are in phase and add up. Thus $f_{B1}(0)$ is a maximum. With an increase in $K$ the cancellation starts and $f_{B1}(K)$ falls with $K$.

To proceed further, let us assume that the interaction potential is spherically symmetric (central); then $U(\boldsymbol{r})$ is equal to $U(r)$ and it does not depend upon the polar coordinates of $\boldsymbol{r}$. Taking $\boldsymbol{K}$ as the reference axis and integrating over the polar coordinate $\varphi$, we get

$$f_{B1}(K) = -\frac{1}{2}\int_0^\infty\int_{-1}^{+1} e^{iKr\mu}U(r)r^2dr\,d\mu \qquad (3.5.5)$$

where $\mu$ is $\cos\theta$, $\theta$ being the angle between $\boldsymbol{K}$ and $\boldsymbol{r}$. Integration over $\mu$ yields

$$f_{B1}(K) = -\frac{1}{K}\int \sin(Kr)U(r)r\,dr \qquad (3.5.6)$$

Any further evaluation of $f_{B1}$ requires knowledge of $U(r)$. Let us take the interaction potential to be a screened Coulomb (Yukawa) potential given by

$$U(r) = \frac{2m}{\hbar^2}\left(-\frac{Ze^2}{r}\right)e^{-\lambda r} \tag{3.5.7}$$

where $\lambda$ is the screening parameter and $e$ is the electronic charge. Then

$$f_{B1}(K) = \frac{2}{Ka}\int \sin(Kr)e^{-\lambda r}dr \tag{3.5.8}$$

where $a = \hbar^2/Zme^2$. To evaluate (3.5.8) we note that

$$\int_0^\infty \sin(Kr)e^{-\lambda r}r^n dr = \text{Im}\int_0^\infty e^{-r(\lambda - iK)}r^n dr = \text{Im}\frac{n!}{(\lambda - iK)^{n+1}} \tag{3.5.9}$$

where Im $F(x)$ is the imaginary part of $F(x)$.

Hence,

$$f_{B1}(K) = \frac{2}{a}\frac{1}{K^2 + \lambda^2} \tag{3.5.10}$$

and

$$I_{B1}(K) = \frac{4a^2}{\left[(Ka)^2 + (\lambda a)^2\right]^2} \tag{3.5.11}$$

Figure 3.2 shows $I_{B1}(Ka)$ as a function of $Ka$ for $\lambda a = 1$. It is evident that $I_{B1}(Ka)$ falls monotonically with an increase in $Ka$. We also note that for a real interaction potential, $f_{B1}$ is purely real. Now, since from (3.5.3)

$$KdK = k^2 \sin\theta d\theta \tag{3.5.12}$$

We get from (1.2.2)

$$\sigma_{el}^{B1} = \frac{2\pi}{k^2}\int_0^{2k} I_{B1}(K)KdK \tag{3.5.13}$$

With the help of (3.5.11) we get

$$\sigma_{el}^{B1} = \frac{16\pi}{\lambda^2 a^2(4k^2 + \lambda^2)} \tag{3.5.14}$$

If we consider a pure Coulomb interaction, then $\lambda = 0$ and

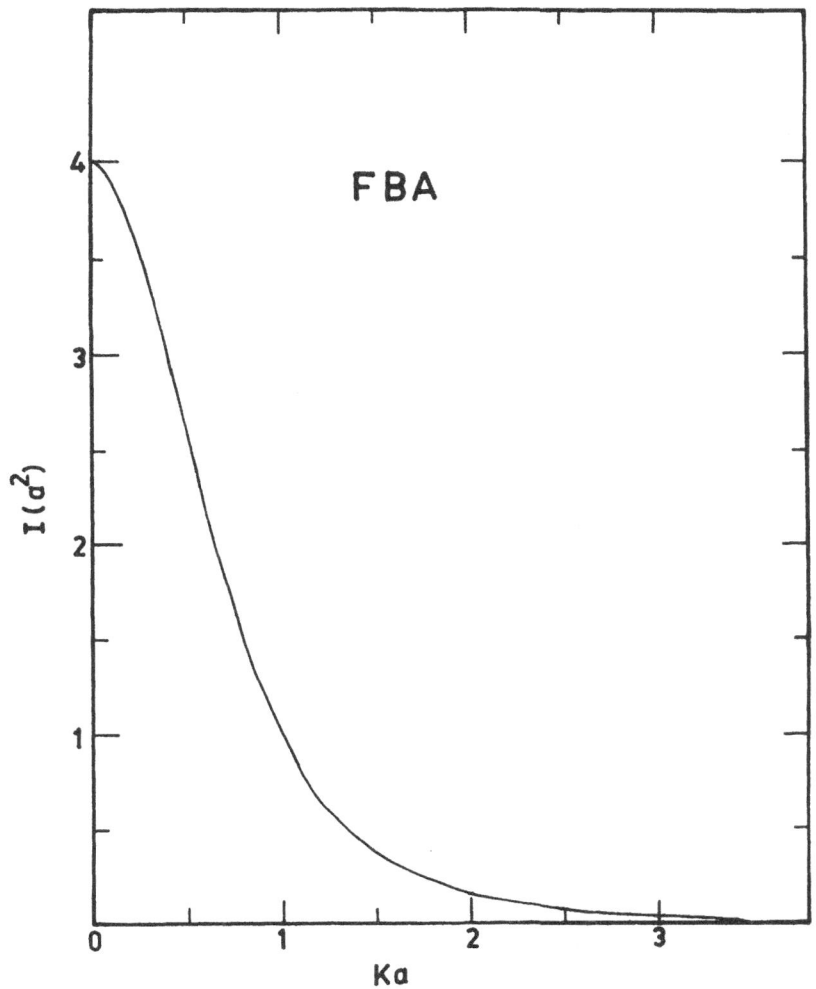

**FIGURE 3.2** Variation of the differential cross section with $Ka$ in the first Born approximation for the scattering of a particle by a screened Coulomb potential with $\lambda a = 1$.

$$I_{B1}(K) = \frac{4}{K^4 a^2}$$

or

$$I_{B1}(\theta) = \left(\frac{Zme^2}{2\hbar^2}\right)^2 \frac{\operatorname{cosec}^4(\theta/2)}{k^4} \qquad (3.5.15)$$

Thus the differential cross section diverges in the forward direction and $\sigma_{el}^{B1}$ also tends to infinity. It should be noted that (3.5.15) is the Rutherford scattering formula for the scattering of electrons by a nucleus of charge $Ze$.

*Validity of the First Born Approximation.* Since the FBA neglects distortion of the plane wave, it is expected to be valid for $|\psi_1(0)|/|\psi_0(0)| \langle\langle 1$. We have taken $r = 0$ because the correction to $\psi_0(r)$ is expected to be largest at the origin. Now

$$\frac{\psi_1(r)}{\psi_0(r)} = \int G_0^+(r, r')U(r')\psi_0(r')dr'/\psi_0(r) \qquad (3.5.16)$$

Using (3.3.5) at $r = 0$ and taking $U(r')$ to be spherically symmetric, we get

$$\frac{\psi_1(0)}{\psi_0(0)} = -\frac{1}{2}\int_0^\infty\int_{-1}^{+1} \exp(ik\,r')U(r')r'\exp(ik\,r'\mu)d\mu\,dr'$$

or

$$\frac{|\psi_1(0)|}{|\psi_0(0)|} = \frac{1}{2k}\left|\int_0^\infty U(r)(e^{2ikr} - 1)dr\right| \qquad (3.5.17)$$

Hence, for the FBA to be valid we should have

$$\frac{1}{2k}\left|\int_0^\infty U(r)(e^{2ikr} - 1)dr\right| \langle\langle 1 \qquad (3.5.18)$$

Thus the FBA, given by (3.5.1), is a weak potential approximation whose validity in the nonrelativistic domain increases with an increase in the projectile's energy. Since for a real potential $f_{B1}$ is purely real, it does not satisfy the optical theorem.

## 3.6   The Second Born Approximation

From (3.4.17) we get for the second Born term

$$\bar{f}_{B2}(k_f,\ k_i) = -\frac{1}{2\pi^2}\int f_{B1}(k_f, k')\frac{1}{k_i^2 - k'^2 + i\varepsilon}f_{B1}(k', k_i)dk' \qquad (3.6.1)$$

Again taking $U(r)$ as given by (3.5.7), i.e., Yukawa potential, we get

$$\bar{f}_{B2}(k_f,\ k_i) = -\frac{1}{\pi^2}\frac{2}{a^2}\int\frac{1}{K_1^2+\lambda^2}\frac{1}{k_i^2-k'^2+i\varepsilon}\frac{1}{K_2^2+\lambda^2}dk' \qquad (3.6.2)$$

where $K_1 = k' - k_f$, $K_2 = k_i - k'$, and the expression for $f_{B1}$ is taken from (3.5.10). The above integral is evaluated using Dalitz's technique (Joachain, 1987) and we get

$$\bar{f}_{B2}(k_f, k_i) = \operatorname{Re}\bar{f}_{B2}(k_f, k_i) + i\operatorname{Im}\bar{f}_{B2}(k_f, k_i)$$

where the real part is

$$\operatorname{Re}\bar{f}_{B2}(k_f, k_i) = \frac{4}{Ka^2}\frac{1}{[\lambda^4+k^2(4\lambda^2+K^2)]^{1/2}} \times \tan^{-1}\frac{\lambda K}{2[\lambda^4+k^2(4\lambda^2+K^2)]^{1/2}}$$

$$(3.6.3)$$

and the imaginary part is given by

$$\operatorname{Im}\bar{f}_{B2}(k_f,\ k_i) = \frac{2}{Ka^2}\ln\left[\frac{[\lambda^4+k^2(4\lambda^2+K^2)]^{1/2}+kK}{[\lambda^4+k^2(4\lambda^2+K^2)]^{1/2}-kK}\right]$$
$$\times\frac{1}{[\lambda^4+k^2(4\lambda^2+K^2)]^{1/2}} \qquad (3.6.4)$$

The above equation shows that $\bar{f}_{B2}$ is complex. This is true for all the higher Born terms. All of them partially include the effect of the distortion of the plane wave. Furthermore, using the optical theorem with $\bar{f}_{B2}$, we get

$$\sigma = \frac{4\pi}{k}\operatorname{Im}\bar{f}_{B2}(k_i, k_i) = \frac{16\pi}{\lambda^2 a^2(\lambda^2+4k^2)} \qquad (3.6.5)$$

which is equal to $\sigma_{el}^{B1}$. Hence, the second Born term satisfies the optical theorem. Similarly, it can be shown that

$$\sigma_{el}^{B2}(E) = \frac{4\pi}{k}\operatorname{Im}[\bar{f}_{B2}(k_i, k_i) + \bar{f}_{B3}(k_i, k_i)]$$

or

$$\sigma_{el}^{B2}(E) - \sigma_{el}^{B1}(E) = \frac{4\pi}{k}\operatorname{Im}\bar{f}_{B3}(\theta = 0) \quad \text{etc.} \qquad (3.6.6)$$

## 3.7  The Schwinger Variational Principle

Let us go back to (3.3.12). If instead of the outgoing waves we consider the incoming waves, then (3.3.12) changes to

$$f_{fi}(\theta, \varphi) = -2\pi^2 \langle \psi_{k_f}^- | U | \varphi_{k_i} \rangle \tag{3.7.1}$$

where the incoming scattered wave in the final channel, as given by the Lippmann–Schwinger equation, is

$$\langle \psi_{k_f}^- | = \langle \phi_{k_f} | + \langle \psi_{k_f}^- | U G_0^+ \tag{3.7.2}$$

The use of the above equation in (3.3.12) yields a third expression for $f(\theta, \varphi)$, which given by

$$f_{fi}(\theta, \varphi) = -2\pi^2 \langle \psi_{k_f}^- | U - U G_0^+ U | \psi_{k_i}^+ \rangle \tag{3.7.3}$$

All the above three forms of $f$ given by (3.3.12), (3.7.1), and (3.7.3) combine to yield

$$[f] = -2\pi^2 (\langle \psi_{k_f}^- | U | \varphi_{k_i} \rangle + \langle \phi_{k_f} | U | \psi_{k_i}^+ \rangle) + 2\pi^2 \langle \psi_{k_f}^- | U - U G_0^+ U | \psi_{k_i}^+ \rangle \tag{3.7.4}$$

The scattering amplitude given by Eq. (3.7.4) is still exact but, as shown below, in addition $[f]$ is also stationary with respect to any arbitrary variation of either $|\psi_{k_i}^+\rangle$ or $\langle \psi_{k_f}^- |$. The above equation is known as the bilinear form of the *Schwinger variational principle*. A variation of $|\psi_{k_i}^+\rangle$ yields

$$\delta[f] = -2\pi^2 [\langle \varphi_{k_f} | U | \delta \psi_{k_i}^+ \rangle - \langle \psi_{k_f}^- | U - U G_0^+ U | \delta \psi_{k_i}^+ \rangle] \tag{3.7.5}$$

Now from (3.7.2)

$$\langle \varphi_{k_f} | U = \langle \psi_{k_f}^- | (U - U G_0^+ U)$$

Hence, $\delta[f]$ is equal to zero. Similarly we can show that $[f]$ is stationary with respect to any arbitrary variation of $\langle \psi_{k_f}^- |$.

Let us now replace $|\psi_{k_i}^+\rangle$ and $\langle \psi_{k_f}^- |$ by the trial wave functions $|\psi_{k_i}^+\rangle_t = a|\psi_{k_i}^+\rangle$ and $\langle \psi_{k_f}^- |_t = b \langle \psi_{k_f}^- |$, where $a$ and $b$ are variational parameters. Then equating $\partial [f]/\partial a$ and $\partial [f]/\partial b$ to zero, we obtain

$$b = \frac{\langle \varphi_{k_f} | U | \psi_{k_i}^+ \rangle}{\langle \psi_{k_f}^- | U - U G_0^+ U | \psi_{k_i}^+ \rangle} \tag{3.7.6a}$$

and

$$a = \frac{\langle \psi_{k_f}^- | U | \varphi_{k_i} \rangle}{\langle \psi_{k_f}^- | U - U G_0^+ U | \psi_{k_i}^+ \rangle} \tag{3.7.6b}$$

Hence,

$$[f] = -2\pi^2 \frac{\langle \psi_{k_f}^- | U | \varphi_{k_i} \rangle \langle \varphi_{k_f} | U | \psi_{k_i}^+ \rangle}{\langle \psi_{k_f}^- | U - U G_0^+ U | \psi_{k_i}^+ \rangle} \tag{3.7.7}$$

The above equation gives the exact scattering amplitude as a fraction. We now approximate $|\psi_{k_i}^+\rangle$ and $\langle \psi_{k_f}^- |$ by

$$|\psi_{k_i}^+\rangle = \sum_{l=1}^{n} (G_0^+ U)^{l-1} |\varphi_{k_i}\rangle \quad \text{and} \quad \langle \psi_{k_f}^- | = \sum_{m=1}^{p} \langle \varphi_{k_f} | (G_0^+ U)^{m-1}$$

Then,

$$[f_{pn}] = \frac{f_{Bp} f_{Bn}}{f_{B1} + \bar{f}_{B2} + \cdots + \bar{f}_{Bp} - (\bar{f}_{\overline{Bn+1}} + \bar{f}_{\overline{Bn+2}} + \cdots + \bar{f}_{\overline{Bn+p}})}$$

or

$$[f_{pn}] = \frac{f_{Bp} f_{Bn}}{f_{Bp} + f_{Bn} - f_{\overline{Bn+p}}} \tag{3.7.8}$$

where, $l$, $m$, $n$, and $p$ are integers and $f_{Bi}$ is the $i$th Born scattering amplitude. For $n = p = 1$, we get

$$[f_{11}] = \frac{f_{B1} f_{B1}}{f_{B1} - \bar{f}_{B2}} \tag{3.7.9}$$

## 3.8   The Eikonal Approximation

In the previous section we obtained a Born series for $\psi_{k_i}^+(r)$. However, evaluation of the higher Born terms is very difficult. In this section we discuss the *eikonal approximation*, which gives $\psi_k^+(r)$ in a closed form involving a one-dimensional integral with the reduced interaction potential energy $U$ as the integrand. The eikonal approximation assumes that the potential energy changes very slowly and, hence, the local momentum $\{2m[E - V(r)]\}^{1/2}$ is practically constant over many de Broglie wavelengths of the projectile. Under the above condition, it is justified to take

$$\psi_{k_i}^+(r) = \phi_{k_i}(r)\varphi(r) \tag{3.8.1}$$

Now we rewrite (3.3.6) as

$$G_0^+(R) = -(2\pi)^{-3} \int \frac{e^{ik\cdot R} e^{-ik_i\cdot R} e^{ik_i\cdot R}}{k^2 - k_i^2 - i\varepsilon} dk \tag{3.8.2}$$

where

$$R = r - r' \tag{3.8.3}$$

and $\varepsilon \to 0$. We put $p = k - k_i$ in Eq. (3.8.2) to get

$$G_0^+(R) = -(2\pi)^{-3} e^{ik_i\cdot R} \int \frac{e^{ip\cdot R}}{p^2 + 2p\cdot k_i - i\varepsilon} dp \tag{3.8.4}$$

Putting (3.8.1), (3.8.3), and (3.8.4) into (3.3.2), we obtain the following integral equation

$$\varphi(r) = 1 - (2\pi)^{-3} \int \frac{e^{ip\cdot R}}{p^2 + 2p\cdot k_i - i\varepsilon} U(r - R)\varphi(r - R) dp dR \tag{3.8.5}$$

Since $U\varphi$ is a slowly varying function and $e^{ip\cdot R}$ is an oscillating function, the major contribution to the above integral comes from small values of $p$. Hence, we neglect $p^2$ in comparison to $2p \cdot k_i$ and take

$$\varphi(r) = 1 - (2\pi)^{-3} \int \frac{e^{ip\cdot R}}{2p \cdot k_i - i\varepsilon} U(r - R)\varphi(r - R) dp dR \tag{3.8.6}$$

or

$$\varphi(r) = 1 + \int G_{oL}^+(R) e^{-ik_i\cdot R} U(r - R)\varphi(r - R) dR \tag{3.8.7}$$

where the linearized Green's function is given by

$$G_{0L}^+(R) = -\frac{e^{ik_i\cdot R}}{(2\pi)^3} \int \frac{e^{ip\cdot R}}{2p \cdot k_i - i\varepsilon} dp$$

$$= -e^{ik_i\cdot R} \frac{1}{2\pi} \int_{-\infty}^{+\infty} e^{ip_x X} dp_x \frac{1}{2\pi} \int_{-\infty}^{+\infty} e^{iP_y Y} dp_Y \times \frac{1}{2\pi} \int_{-\infty}^{+\infty} \frac{e^{ip_z Z}}{2p_z k_i - i\varepsilon} dp_z \tag{3.8.8}$$

with $k_i$ as the $z$-axis. Integration over $p_X$ and $p_Y$ yields

$$G_{0L}^+(R) = -e^{ik_i \cdot R}\delta(X)\delta(Y)\frac{1}{2\pi}\int_{-\infty}^{+\infty}\frac{e^{ip_z z}}{2p_z k_i - i\varepsilon}dp_z$$

The $p_z$ integral has a pole on the imaginary axis at $p_z = i\varepsilon/2k_i$. Hence, by the Cauchy theorem,

$$G_{0L}^+(R) = -\frac{i}{2k_i}e^{ik_i \cdot R}\delta(X)\delta(Y)\theta(Z) \tag{3.8.9}$$

where $\theta(Z)$ is the step function. Thus $G_{0L}^+(\mathbf{R})$ propagates only along the forward direction of the $z$-axis. Putting (3.8.9) into (3.8.7), we obtain

$$\varphi(r) = 1 - \frac{i}{2k_i}\int_0^\infty U(x, y, z - Z)\varphi(x, y, z - Z)dZ$$

A change of the variable to $z'$ yields

$$\varphi(r) = 1 - \frac{i}{2k_i}\int_{-\infty}^z U(x, y, z')\varphi(x, y, z')dz' \tag{3.8.10}$$

The solution to (3.8.10) is given by

$$\varphi(r) = \exp\left[-\frac{i}{2k_i}\int_{-\infty}^z U(x, y, z')dz'\right]$$

To verify it we note that

$$\frac{d\varphi(r)}{dz} = -\frac{i}{2k_i}\varphi(r)\left[\frac{d}{dz}\int_{-\infty}^z U(x, y, z')dz'\right]$$

$$= -\frac{i}{2k_i}\varphi(r)\left[\int_{-\infty}^z \frac{dU(x, y, z')}{dz}dz' + U(x, y, z)\frac{d(z)}{dz} - U(x, y, -\infty)\frac{d(-\infty)}{dz}\right]$$

$$= -\frac{i}{2k_i}\varphi(r)U(r)$$

Now integration over $z$ from $-\infty$ to $z''$ yields

$$\varphi(z'') = \phi(-\infty) - \frac{i}{2k_i}\int_{-\infty}^{z''} U(x, y, z)\varphi(x, y, z)dz$$

Since $\phi(-\infty) = 1$, the above equation gives (3.8.10). Hence, in the eikonal approximation, the outgoing scattered wave function is given by

$$\psi_E^+(r) = A\exp\left[ik_i \cdot r - \frac{i}{2k_i}\int\limits_{-\infty}^{z} U(x, y, z')dz'\right] \qquad (3.8.11)$$

It is evident from the above equation that $\psi_E^+(r)$ differs from the plane wave only by a phase term and like the plane wave this also does not satisfy the proper asymptotic condition (3.1.2). The eikonal approximation assumes that propagation is in the forward direction.

Now we use $\psi_E^+(r)$ in (3.3.12) to obtain the scattering amplitude in the eikonal approximation. With $A = (2\pi)^{-3/2}$ we get

$$f_E(\theta, \varphi) = -\frac{1}{4\pi}\int dr\, e^{iK \cdot r} U(r)\exp\left[-\frac{i}{2k_i}\int\limits_{-\infty}^{z} U(x, y, z')dz'\right] \qquad (3.8.12)$$

This differs from $f_{B1}$ given by (3.5.4) only by a phase term, which shows that the FBA takes the exponential term of (3.8.12) equal to unity. This can be satisfied only by assuming the potential to be weak and/or $E$ to be high. The phase integral of (3.8.12) is to be evaluated along the $z(k_i)$-axis. However, we know that initially the projectile moves along $z$ direction but after the scattering it moves along the direction of $k_f$. Hence, to be more realistic, we integrate along $OD$ (see Fig. 3.3), which is the bisector of the angle between $k_i$ and $k_f$. For the potential (elastic) scattering, $OD$ is perpendicular to $K$.

Using the cylindrical coordinate system $r = b + z\hat{z}$, where $b$ is a two-dimensional vector on the $x$–$y$ plane and $\hat{z}$ is a unit vector along $OD$, in (3.8.12), we get

$$f_E(k_f, k_i) = -\frac{1}{4\pi}\int db\int\limits_{-\infty}^{+\infty} dz\, e^{iK \cdot b} U(b, z)\exp\left[-\frac{i}{2k}\int\limits_{-\infty}^{z} U(b, z')dz'\right] \qquad (3.8.13)$$

because $K \cdot \hat{z} = 0$. Now let us take

$$I(z) = \exp\left[-\frac{i}{2k}\int\limits_{-\infty}^{z} U(b, z')dz'\right]$$

Then

$$\frac{dI(z)}{dz} = -\frac{i}{2k}I(z)\left[\int\limits_{-\infty}^{z}\frac{d}{dz}U(b, z')dz' + U(b, z)\frac{d(z)}{dz} - U(b, -\infty)\frac{d(-\infty)}{dz}\right]$$

$$= -\frac{i}{2k}I(z)U(b, z)$$

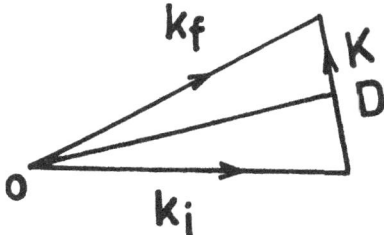

**FIGURE 3.3** In the eikonal approximation the path of integration is taken along *OD*, which is perpendicular to *K*.

Hence,

$$f_E(\mathbf{k}_f, \mathbf{k}_i) = -\frac{ik}{2\pi} \int d\mathbf{b}\, e^{i\mathbf{K}\cdot\mathbf{b}} \frac{dI(z)}{dz} dz$$

$$= -\frac{ik}{2\pi} \int d\mathbf{b}\, e^{i\mathbf{K}\cdot\mathbf{b}} \{\exp[i\xi(\mathbf{b}, k)] - 1\} \qquad (3.8.14)$$

where the phase $\xi$ is given by

$$\xi(\mathbf{b}, k) = -\frac{1}{2k} \int_{-\infty}^{+\infty} U(\mathbf{b}, z) dz \qquad (3.8.15)$$

For a cylindrically symmetric potential $U(\mathbf{b}, z) = U(b, z)$; i.e., $U$ is independent of the polar angle $\varphi$ (angle between $\mathbf{K}$ and $\mathbf{b}$). Under the above condition (3.8.14) reduces

$$f_E(\mathbf{k}_i, \mathbf{k}_f) = -\frac{ik}{2\pi} \int_0^\infty \int_0^{2\pi} b\, db\, d\varphi\, e^{iKb\cos\varphi} \{\exp[i\xi(b, k)] - 1\}$$

Hence, finally,

$$f_E(\mathbf{k}_f, \mathbf{k}_i) = -ik \int_0^\infty J_0(Kb)[\exp(i\xi) - 1] b\, db \qquad (3.8.16)$$

where $J_0(x)$ is the zeroth-order Bessel function. We note that for a spherically symmetrical potential, the expression for $f_{B1}$ given by (3.5.6) involves only a one-dimensional integral. But for a cylindrically symmetrical potential, $f_E$ given by (3.8.16) involves a two-dimensional integral, one over $z$, to evaluate the phase $\xi$, and the other over $b$. However, $f_{B1}$ is correct only up to first order in the interaction, whereas $f_E$ includes all orders of interaction. Hence, the eikonal approximation should be regarded as superior to the FBA. It may also be noted that $f_{B1}$ depends only upon $K$, whereas $f_E$ depends upon $K$ and $k_i$.

Let us now examine whether the eikonal approximation satisfies the optical theorem. In the forward direction $K = 0$ and $J_0(0) = 1$. Hence,

$$f_E(\theta = 0) = -ik \int (e^{i\xi} - 1) b \, db$$

Therefore, according to the optical theorem,

$$\sigma_{el} = \frac{4\pi}{k} \operatorname{Im} f_E(\theta = 0)$$
$$= -4\pi \int b \, db (\cos \xi - 1) \tag{3.8.17}$$

From (3.2.12), (3.5.13), and (3.8.16), we also have

$$\sigma_{el} = 2\pi \iint (-ik) J_0(Kb)(e^{i\xi} - 1) b \, db \int (ik) J_0(Kb')(e^{-i\xi'} - 1) b' \, db' \frac{K \, dK}{k^2} \tag{3.8.18}$$

Now using a closure relation for the Bessel functions we get

$$\int J_0(Kb) J_0(Kb') K \, dK = \frac{\delta(b - b')}{b} \tag{3.8.19}$$

Putting (3.8.19) into (3.8.18) and taking $\xi$ to be real, we obtain

$$\sigma_{el} = 2\pi \int b \, db (e^{i\xi} - 1)(e^{-i\xi} - 1)$$

or

$$\sigma_{el} = 4\pi \int b \, db (1 - \cos \xi) \tag{3.8.20}$$

The above equation is identical to (3.8.17). Thus the eikonal approximation satisfies the optical theorem. In this respect also it is superior to the FBA, which does not satisfy optical theorem.

An expansion of $e^{i\xi}$ in the powers of $i\xi$ gives the eikonal series

$$f_E = \sum_{n=1}^{\infty} \bar{f}_{En} \tag{3.8.21}$$

where the $n$th eikonal term from (3.8.14) is given by

$$\bar{f}_{En} = i^{n-1} \frac{k}{2\pi} \frac{1}{n!} \int db \, e^{iK \cdot b} \xi^n \qquad (3.8.22)$$

The first eikonal term (same as the first eikonal amplitude) is given by

$$f_{E1} = \frac{k}{2\pi} \int db \, e^{iK \cdot b} \int (-1) \frac{U(b, z)}{2k} dz$$

or

$$f_{E1}(K) = -\frac{1}{4\pi} \int e^{iK \cdot r} U(r) dr \qquad (3.8.23)$$

because $K$ is perpendicular to $\hat{z}$. A comparison of (3.8.23) and (3.5.4) shows that for $\hat{z}$ perpendicular to $K$, the $f_{E1}$ is identically equal to $f_{B1}$. However, the higher eikonal terms as given by (3.8.22) are alternately imaginary and real. For example, for real interaction potential, $\bar{f}_{E2}$ is imaginary, whereas $\bar{f}_{E3}$ is real. On the other hand, all the higher Born terms are complex.

## 3.9  The Differential Approach

In the differential approach, the differential equation (3.1.1) is solved subject to two boundary conditions. Since it is a three-dimensional differential equation, in the general case its solution is quite difficult. However, for a central potential $V(r)$ the angular momentum of the each partial wave is a constant of motion and, like a plane wave, $\psi_{ki}^{+}(r)$ can also be expanded in terms of $P_l(\mu)$. Such a partial wave expansion is given by

$$\psi_{ki}^{+}(r) = A \sum_l A_l R_l(r) P_l(\mu) \qquad (3.9.1)$$

where $A_l$ are the coefficients of expansion and $R_l(r)$ satisfies the following one-dimensional differential equation:

$$\left[ \frac{1}{r^2} \frac{d}{dr} \left( r^2 \frac{d}{dr} \right) + k^2 - U(r) - \frac{l(l+1)}{r^2} \right] R_l(kr) = 0 \qquad (3.9.2)$$

As expected, for $U(r) = 0$ Eq. (3.9.2) reduces to Eq. (2.6.9). Hence, in the asymptotic region, where $U(r)$ is zero, $R_l(r)$ is given by

$$R_l(r) \underset{r \to \infty}{\sim} a_l j_l(kr) + b_l n_l(kr) \qquad (3.9.3)$$

provided $U(r)$ falls faster than $r^{-2}$ at large values of $r$. Due to the presence of $U(r)$ in (3.9.2) the coefficient $b_l$ is not zero, and the above equation constitutes one of the boundary conditions imposed on $R_l(r)$. Near the origin the terms $k^2$ and $U(r)$ are small in comparison to $l(l+1)/r^2$. Hence, (3.9.2) is satisfied with $R_l = r^l$. Thus the second boundary condition is

$$R_l(r) \underset{r \to 0}{\sim} r^l \tag{3.9.4}$$

Now we choose $a_l = \cos \eta_l$ and $b_l = -\sin \eta_l$, which yields

$$R_l(r) \underset{r \to \infty}{\sim} \frac{1}{kr} \sin(kr - l\pi/2 + \eta_l) \tag{3.9.5}$$

The above choice of $a_l$ and $b_l$ ensures that for $U(r) = 0$, as required, (3.9.5) reduces to (2.6.15). We have taken $b_l$ with a negative sign so that the phase $\eta_l$ will be positive for an attractive potential field. The use of (3.9.5) in (3.9.1) gives

$$\psi_{k_i}^+(r) \underset{r \to \infty}{\sim} A \sum_{l=0}^{\infty} A_l \frac{1}{kr} \sin(kr - l\pi/2 + \eta_l) P_l(\mu) \tag{3.9.6}$$

or

$$\psi_{k_i}^+(r) \underset{r \to \infty}{\sim} A \sum_{l=0}^{\infty} \frac{A_l}{2ikr} (e^{i(kr - l\pi/2 + \eta_l)} - e^{-i(kr - l\pi/2 + \eta_l)}) P_l(\mu) \tag{3.9.7}$$

An alternative expression for $\psi_{k_i}^+(r)$ at large $r$ is given by (3.1.2). In this equation we take

$$f(\theta, \varphi) = \sum_{l=0}^{\infty} c_l P_l(\mu) \tag{3.9.8}$$

and use (2.6.25) and (3.9.8) in (3.1.2) to get

$$\psi_{k_i}^+(r) \underset{r \to \infty}{\sim} A \sum_{l=0}^{\infty} \frac{i^l(2l+1)}{2ikr} (e^{ikr} e^{-il\pi/2} - e^{-ikr} e^{il\pi/2}) P_l(\mu) + A \sum_{l=0}^{\infty} c_l P_l(\mu) \frac{e^{ikr}}{r} \tag{3.9.9}$$

Equating the coefficients of $e^{-ikr}$ in (3.9.7) and (3.9.9), we get

$$A_l = i^l(2l+1) e^{i\eta_l} \tag{3.9.10}$$

Similarly, a comparison of the coefficients of $e^{ikr}$ in the same two equations with (3.9.10) yields

$$c_l = \frac{2l+1}{2ik}(e^{2i\eta_l} - 1) \tag{3.9.11}$$

Hence,

$$f(\theta) = \sum_{l=0}^{\infty} \frac{2l+1}{2ik}(e^{2i\eta_l} - 1)P_l(\mu) \tag{3.9.12}$$

$$= \sum_l \frac{1}{k}(2l+1)e^{i\eta_l}\sin\eta_l P_l(\cos\theta) \tag{3.9.13}$$

Furthermore, from (3.9.1) and (3.9.10)

$$\psi_{k_i}^+(\mathbf{r}) = A\sum_l i^l(2l+1)e^{i\eta_l}R_l(kr)P_l(\cos\theta) \tag{3.9.14}$$

The above equation constitutes the partial wave expansion of the scattered wave in terms of the Legendre polynomials and the radial wave function. Such an expansion is possible only for the central potentials. It is easy to see that in the asymptotic region, (3.9.14) goes to (2.6.24) for $\eta_l = 0$. Hence, we conclude that the effect of the potential scattering is to shift the phase of the $l$th incident partial wave by an angle $\eta_l$. Therefore, $\eta_l$ is known as the phas shift. The scattering by a central potential does not change the value of $l$. For $\eta_l = 0$, the scattering amplitude $f(\theta)$ reduces to zero; i.e., there is no scattering. Hence, $\eta_l$ carries the signature of the collision.

To evaluate $\eta_l$, we consider $R_l(kr)$ at two large values of $r$, and divide one by the other to get

$$\frac{R_l(kr_1)}{R_l(kr_2)} = \frac{j_l(kr_1) - \tan\eta_l n_l(kr_1)}{j_l(kr_2) - \tan\eta_l n_l(kr_2)}$$

The above equation yields

$$\tan\eta_l = \frac{R_l(kr_2)j_l(kr_1) - R_l(kr_1)j_l(kr_2)}{R_l(kr_2)n_l(kr_1) - R_l(kr_1)n_l(kr_2)} \tag{3.9.15}$$

To eliminate the first-order differential from (3.9.2), we take

$$R_l(kr) = \frac{f_{lk}(r)}{r} \tag{3.9.16}$$

which yields

$$\left[\frac{d^2}{dr^2} + k^2 - U(r) - \frac{l(l+1)}{r^2}\right] f_{lk}(r) = 0 \tag{3.9.17}$$

The boundary conditions (3.9.4) and (3.9.5) now change to

$$f_l(r) \underset{r \to 0}{\sim} r^{l+1}$$

and

$$f_l(r) \underset{r \to \infty}{\sim} \frac{1}{k} \sin(kr - l\pi/2 + \eta_l) \tag{3.9.18}$$

In most cases numerical methods are employed to solve the differential equation (3.9.17) and obtain the value of $f_l(r)$ at large $r$, and thus $\eta_l$ are calculated with the help of (3.9.16) and (3.9.15). Due to the presence of the centrifugal term $l(l+1)/r^2$ in (3.9.17), in general, the value of $\eta_l$ falls with an increase in $l$. At low impact energies the zeroth-order phase shift $\eta_0$ dominates.

From (3.2.12) and (3.9.12) we get

$$\sigma_{el} = \frac{\pi}{k^2} \sum_l (2l+1)(1 - S_l)(1 - S_l^*)$$

$$= \frac{\pi}{k^2} \sum_l (2l+1)|(1 - S_l)|^2 \tag{3.9.19}$$

where the scattering matrix element $S_l$ is defined by

$$S_l = e^{2i\eta_l} \tag{3.9.20}$$

Hence,

$$\tan \eta_l = i\frac{1 - S_l}{1 + S_l} \tag{3.9.21}$$

To obtain an expression for the absorption cross section in terms of the scattering matrix element $S_l$ we note that according to (3.2.13), (3.2.17), and (3.2.18), for large $r$,

$$\sigma_{ab} = -r^2 \int c(r) \cdot \hat{r} \, d\Omega / (vAA^*) \qquad (3.9.22)$$

and from (3.2.6)

$$c(r) \cdot \hat{r} = \frac{\hbar}{2mi}\left(\psi^* \frac{\partial}{\partial r}\psi - \psi \frac{\partial}{\partial r}\psi^*\right)$$

Using (3.9.14) for $\psi$, we get

$$\int \psi^* \frac{\partial}{\partial r}\psi \, d\Omega = \frac{4\pi AA^*}{kr^2}\sum_l (2l+1)e^{i(\eta_l - \eta_l^*)}\sin\left(kr - l\pi/2 + \eta_l^*\right)$$
$$\times \cos(kr - l\pi/2 + \eta_l) + O\!\left(\frac{1}{r^3}\right)$$

Similarly,

$$\int \psi \frac{\partial}{\partial r}\psi^* \, d\Omega = \frac{4\pi AA^*}{kr^2}\sum_l (2l+1)e^{i\left(\eta_l - \eta_l^*\right)}\sin(kr - l\pi/2 + \eta_l)$$
$$\times \cos\left(kr - l\pi/2 + \eta_l^*\right) + O\!\left(\frac{1}{r^3}\right)$$

Hence,

$$\sigma_{ab} = \frac{\pi}{k^2}\sum_l (2l+1)\left(1 - |S_l|^2\right) \qquad (3.9.23)$$

Using (3.9.12) and the optical theorem, we obtain

$$\sigma_T = \frac{2\pi}{k^2}\sum_l (2l+1)[1 - \mathrm{Re}(S_l)] \qquad (3.9.24)$$

It is easy to verify that the sum of $\sigma_{el}$ and $\sigma_{ab}$ is equal to $\sigma_T$.

The scattering matrix $S$, whose elements are $S_{ll'}$, is a diagonal matrix. Hence, $SS^\dagger$ (where $S^\dagger$ is the adjoint of $S$) is also a diagonal matrix with the elements $e^{2i(\eta_l - \eta_l^*)}$. Now, for a real potential $\eta_l = \eta_l^*$ and $\sigma_{ab} = 0$. For such a potential $SS^\dagger$ is a unit matrix, which confirms the conservation of particles. We also get

$$\sigma_T = \sigma_{el} = \sum_l \sigma_l \qquad (3.9.25)$$

where the partial cross section $\sigma_l$ is given by

$$\sigma_l = \frac{4\pi}{k^2}(2l+1)\sin^2 \eta_l \tag{3.9.26}$$

Hence,

$$\sigma_l \leq \frac{4\pi}{k^2}(2l+1) \tag{3.9.27}$$

The above equation is a statement of the theorem of maximum cross section. Equations (3.2.11) and (3.9.13) show that the structure in the differential cross section is due to $l > 0$. In general, the stronger the potential field, the more structure in the curve of $I(\theta)$ vs. $\theta$. Furthermore,

$$I(\theta) = \frac{1}{k^2}\sum_l \sum_{l'} (2l+1)(2l'+1)e^{i\eta_l}e^{-i\eta_{l'}} \sin\eta_l \sin\eta_{l'} P_l(\cos\theta)P_{l'}(\cos\theta)$$

Hence, the cross terms for the different values of $l$ and $l'$ contribute to $I(\theta)$. But (3.9.25) and (3.9.26) show that the cross terms do not contribute to $\sigma_{el}$. This is a consequence of the orthogonality of $P_l(\mu)$.

At low incident energies the scattering is dominated by the lower partial waves. At very low energies we may assume that only the $l = 0$ partial wave is of significance. Then, from (3.9.25) and (3.9.26),

$$\sigma_{el} = \frac{4\pi}{k^2}\sin^2\eta_0 \tag{3.9.28}$$

In general, the ratio $\eta_0/k$ as $k \to 0$ tends to a finite limit equal to $-a_S$, where $a_S$ is known as the scattering length. Hence,

$$a_S = -\lim_{k\to 0}\left(\frac{\tan\eta_0}{k}\right) \tag{3.9.29}$$

Thus, from (3.9.28),

$$\sigma_{el} = \frac{4\pi}{k^2}\frac{a_S^2 k^2}{\sqrt{1+a_S^2 k^2}} \approx 4\pi a_S^2 \tag{3.9.30}$$

For the repulsive potential, $\eta_0$ tends to zero, as $k$ tends to zero. However, for the attractive potential, according to Levinson's theorem,

$$\eta_0 = \underset{k \to 0}{\longrightarrow} n\pi \qquad\qquad (3.9.31)$$

where the integer $n$ represents the number of bound states that the potential can support. A study of the square well potential (see Sec. 3.10.2) shows that the number of bound states it can support depends upon its depth. For a shallow well, $\eta_0 \to 0$ as $k \to 0$. With an increase in its depth, a situation will arise when it is able to support one bound state. For that depth, $\eta_0 \to \pi$ as $k \to 0$. In between we encounter a depth for which $\eta_0 = \pi/2$. For this value of the phase shift, $a_s$ will tend toward infinity and so will the cross section. Such a phenomenon is known as resonance.

## 3.10 Scattering by a Hard Sphere and a Three-Dimensional Potential Well

### 3.10.1 Hard Sphere

Let us now apply the method of partial waves to the collision of a particle with a hard sphere of radius $R$ and infinite mass. Since outside the sphere $U(r) = 0$, the radial wave function is given by

$$R_l(r) = j_l(kr) - \tan\eta_l n_l(kr), \quad \text{for} \quad r \geq R$$

Furthermore,

$$R_l(r) = 0, \quad \text{for} \quad r \leq R$$

Hence, at $r = R$ we get

$$\tan\eta_l = j_l(kR)\big/n_l(kR) \qquad\qquad (3.10.1)$$

In the limit of zero energy only the $l = 0$ partial wave contributes, and from the above equation

$$\tan\eta_0 = -\tan(kR)$$

or

$$\eta_0 = -kR$$

Hence, we get

$$\sigma_l = 4\pi R^2 \tag{3.10.2}$$

The above value is four times the geometrical area of the sphere. In the high-energy limit, where a large number of the partial waves contribute, $\sigma_l$ reduces to $2\pi R^2$.


### 3.10.2 Three-Dimensional Potential Well

Let us now consider the scattering of a particle of mass $m$, energy $E$, and momentum $\hbar k$ by a three-dimensional potential well of depth $-V_0$ and width $b$. The Schrödinger equations describing the system are

$$(\nabla^2 + \beta^2)\psi_{in}(r) = 0 \quad \text{for} \quad 0 \le r \le b \tag{3.10.3}$$

and

$$(\nabla^2 + k^2)\psi_{out}(r) = 0 \quad \text{for} \quad r \ge b \tag{3.10.4}$$

where

$$\beta^2 = \frac{2m}{\hbar^2}(E + V_0) \tag{3.10.5}$$

In general the radial wave function of the $l$th partial wave is a linear combination of $j_l(r)$ and $n_l(r)$. Further, it has to be finite everywhere. Hence, $R_{in}^l(r)$ cannot contain $n_l(r)$ because the Neumann function diverges at $r = 0$. Therefore, we take

$$R_{in}^l(r) = A_l j_l(\beta r) \tag{3.10.6}$$

and

$$R_{out}^l(r) = B_l[j_l(kr) - \tan\eta_l n_l(kr)] \tag{3.10.7}$$

where $\eta_l$ is the phase shift for the $l$th partial wave. Now $R_{in}^l(r)$ and $R_{out}^l(r)$ and their first derivatives are to be continuous $r = b$. Hence,

$$\lambda_l = \frac{1}{R_{in}^l(\beta r)} \left.\frac{dR_{in}^l(\beta r)}{dr}\right|_{r=b} = \frac{1}{R_{out}^l(kr)} \left.\frac{dR_{out}^l(kr)}{dr}\right|_{r=b} \tag{3.10.8}$$

or

$$\lambda_l = \frac{\beta j_l'(\beta b)}{j_l(\beta b)} = \frac{k[j_l'(kb) - \tan \eta_l n_l'(kb)]}{j_l(kb) - \tan \eta_l n_l(kb)} \tag{3.10.9}$$

where the prime denotes differentiation with respect to $x = \beta r$ or $kr$. From the above equation we get

$$\tan \eta_l = \frac{kj_l'(kb) - \lambda_l j_l(kb)}{kn_l'(kb) - \lambda_l n_l(kb)} \tag{3.10.10}$$

For small $k$, with the help of (2.6.13) and (2.6.14), we get

$$\tan \eta_l = \frac{(kb)^{2l+1}}{(2l+1)!!(2l-1)!!} \frac{l - \lambda_l b}{l + 1 + \lambda_l b} \tag{3.10.11}$$

Thus for $l \geq 1$, $\tan \eta_l$ goes to zero faster than $k^2$; hence $\sigma_l$ as obtained from (3.9.26) for $l \geq 1$ is zero at $k = 0$. However, for $l = 0$,

$$\tan \eta_0 = \frac{-kb^2 \lambda_0}{1 + \lambda_0 b} \tag{3.10.12}$$

Hence, for $k \to 0$, $\tan \eta_0$ also tends to zero provided $\lambda_0 b \neq -1$. Now from (3.10.9) and (2.6.17)

$$\lambda_0 = \beta \cot \beta b - 1/b \tag{3.10.13}$$

Hence,

$$\sin \eta_0 \approx \tan \eta_0 = -kb(1 - \tan \beta b / \beta b)$$

and

$$\sigma_0 = 4\pi b^2 (1 - \tan \beta b / \beta b)^2 \tag{3.10.14}$$

which is finite even at $k = 0$.

Equation (3.10.11) shows that at $l + 1 + \lambda_l b = 0$, $\eta_l$ goes to $(2n + 1)\pi/2$, where $n$ is an integer. Hence, the cross section $\sigma_l$ assumes its maximum value. In such a situation, the $l$th partial wave is said to be in resonance with the scattering well. The value of the resonance energy $E_r^l$ depends upon $l$, $m$, $b$, and $V_0$. If $E$ is close to $E_r^l$ then the total elastic cross section $\sigma_{el} = \sum_l \sigma_l$ is controlled only by the $l$th partial wave, and we have

$$\sigma_{el} \approx \sigma_l = \frac{4\pi}{k^2}(2l+1)\frac{\Gamma^2}{4(E-E_r^l)^2+\Gamma^2} \qquad (3.10.15)$$

The above equation is known as the one-level Breit–Wigner formula. The energy $E_r^l$ and $\Gamma$ are known as are the position and the width of the resonance.

For $l = 0$ the resonance occurs at

$$\beta b = (2n+1)\pi/2$$

For $E \lll V_0$ the above equation gives

$$V_0 b^2 = \frac{(2n+1)^2\pi^2\hbar^2}{8m} \qquad (3.10.16)$$

As expected, each value of $n$ corresponds to a resonance and the appearance of a new bound state. At each resonance the potential produces a large distortion in the wave function of the incident particle and so a large amount of scattering.

## 3.11  Integral Equation for $R_l(r)$ and $\tan\eta_l$

We expand $\psi_{k_i}^+(r)$ and $\phi_{k_i}(r)$ in the complete set of spherical harmonics and put the expansion into the Lippmann–Schwinger equation given by (3.3.2) to get

$$\sum_{lm} C_{lm} R_l(r) Y_{lm}^*(\hat{r}) Y_{lm}(\hat{k}_i) = 4\pi A \sum_{lm} i^l j_l(kr) Y_{lm}^*(\hat{r}) Y_{lm}(\hat{k}_i)$$

$$+ \int G_0^+(r,r') U(r') \psi_{k_i}^+(r') dr' \qquad (3.11.1)$$

Now

$$G_0^+(r,r') = \sum_{lm} g_l^+(r,r') Y_{lm}(\hat{r}') Y_{lm}^*(\hat{r})$$

with

$$g_l^+(r,r') = -ikj_l(kr_<)h_l^{(1)}(kr_>)$$

where $h_l^{(1)}$ is a first-order Hankel function and is equal to $j_l + in_l$.

The second term on the right hand side of (3.11.1) takes the form

$$\sum_{lm} \int g_l^+(r,r')Y_{lm}(\hat{r}')Y_{lm}^*(\hat{r})U(r')\sum_{l'm'}C_{l'm'}R_l(r') \times Y_{l'm'}^*(\hat{r}')Y_{l'm'}(\hat{k}_i)r'^2dr'd\Omega'$$

$$= \sum_{lm} \int g_l^+(r,r')Y_{lm}^*(\hat{r})U(r')C_{lm}R_l(r')Y_{lm}(\hat{k}_i)r'^2dr' \tag{3.11.2}$$

Use of (3.11.2) in (3.11.1) yields

$$R_l(r) = \frac{i^l j_l(kr)}{C_l}4\pi A + \int g_l^+(r,r')U(r')R_l(r')r'^2dr' \tag{3.11.3}$$

where $C_{lm}$, being independent of $m$, is replaced by $C_l$. Expressing $g_l^+(r,r')$ in terms of $j_l$ and $h_l^{(1)}$ in the second term of (3.11.3), reduces it to

$$-ik\int_0^\infty j_l(kr_<)h_l^{(1)}(kr_>)U(r')R_l(r')r'^2dr'$$

$$= -ik\int_0^r j_l(kr')h_l^{(1)}(kr)U(r')R_l(r')r'^2dr' - ik\int_r^\infty j_l(kr)h_l^{(1)}(kr')U(r')R_l(r')r'^2dr'$$

In the limit of large $r$ the second term of the above equation goes to zero and the upper limit of integration in the first term goes to $\infty$. Hence, we obtain

$$R_l(r) \underset{r\to\infty}{\longrightarrow} j_l(kr)\left[4\pi A\frac{i^l}{C_l} - ik\int_0^\infty j_l(kr')U(r')R_l(r')r'^2dr'\right]$$

$$+ n_l(kr)\left[k\int_0^\infty j_l(kr')U(r')R_l(r')r'^2dr'\right] \tag{3.11.4}$$

A comparison of the above equation with

$$R_l(r) \underset{r\to\infty}{\longrightarrow} j_l(kr) - \tan\eta_l n_l(kr)$$

gives

$$\tan\eta_l = -k\int_0^\infty j_l(kr)U(r)R_l(r)r^2dr \tag{3.11.5}$$

and

$$C_l = \frac{i^l 4\pi A}{1 - i\tan\eta_l} \tag{3.11.6}$$

Putting (3.11.6) into (3.11.3) we get

$$R_l(r) = j_l(kr)(1 - i\tan\eta_l) - ik\int_0^\infty j_l(kr_<)[j_l(kr_<) + in_l(kr_<)] \times U(r')R_l(r')r'^2 dr'$$

Finally, with the use of (3.11.5), we obtain

$$R_l(r) = j_l(kr) + \int G_l(r, r')U(r')R_l(r')r'^2 dr' \tag{3.11.7}$$

where $G_l(r, r') = kj_l(kr_<)n_l(kr_>)$.

Equations (3.11.5) and (3.11.7) are the integral representations of the phase shift $\eta_l$ and the radial wave function $R_l(r)$, respectively. Both of these also depend upon $k$.

In the first Born approximation $R_l(r) = j_l(kr)$. Hence, in the FBA (3.11.5) and (3.11.7) reduce to, respectively,

$$\tan\eta_l^{B1} = -k\int_0^\infty [j_l(kr)]^2 U(r)r^2 dr \tag{3.11.8}$$

and

$$R_l^{B1}(r) = j_l(kr) + \int G_l(r, r')U(r')j_l(kr')r'^2 dr' \tag{3.11.9}$$

## 3.12 The Distorted Wave Born Approximation

Sometimes it is convenient to break $U$ into two parts and take $U = U_1 + U_2$. This procedure is quite useful if the scattered wave function due to $U_1$ can be obtained exactly and $U_2$ can be treated as a perturbation. The Lippmann–Schwinger equation in the bra form due to $U_1$ alone is given by

$$\langle\psi_{1k_f}^-| = \langle\varphi_{k_f}| + \langle\psi_{1k_f}^-|U_1 G_0^+$$

or

$$\langle\varphi_{k_f}| = \langle\psi_{1k_f}^-| - \langle\psi_{1k_f}^-|U_1 G_0^+ \tag{3.12.1}$$

Putting the above equation into (3.3.12), we get

$$f_{fi}(\theta, \varphi) = -2\pi^2\langle\psi_{1k_f}^-|U|\psi_{k_i}^+\rangle + 2\pi^2\langle\psi_{1k_f}^-|U_1 G_0^+ U|\psi_{k_i}^+\rangle$$

Now from (3.3.2)

$$G_0^+ U |\psi_{k_i}^+\rangle = |\psi_{k_i}^+\rangle - |\varphi_{k_i}\rangle$$

Hence,

$$
\begin{aligned}
f_{fi}(\theta, \varphi) &= -2\pi^2 \langle \psi_{1k_f}^- |U| \psi_{k_i}^+ \rangle + 2\pi^2 \langle \psi_{1k_f}^- |U_1| \psi_{k_i}^+ \rangle - 2\pi^2 \langle \psi_{1k_f}^- |U_1| \varphi_{k_i} \rangle \\
&= -2\pi^2 \langle \psi_{1k_f}^- |U_1| \varphi_{k_i} \rangle - 2\pi^2 \langle \psi_{1k_f}^- |U_2| \psi_{k_i}^+ \rangle
\end{aligned}
\tag{3.12.2}
$$

This equation is still exact. However, $\psi_{k_i}^+(r)$ cannot be determined exactly for $U_1 + U_2$. Hence, in the *distorted wave Born approximation* $\psi_{k_i}^+$ is replaced by $\psi_{1k_i}^+(r)$, and we get

$$f_{fi}^{\text{DWBA}}(\theta, \varphi) = -2\pi^2 \langle \psi_{1k_f}^- |U_1| \varphi_{k_i} \rangle - 2\pi^2 \langle \psi_{1k_f}^- |U_2| \psi_{1k_i}^+ \rangle \tag{3.12.3}$$

The first term on the right-hand side is the exact scattering amplitude for a particle due to potential $V_1$ and the second term is the matrix element of $U_2$ due to distorted outgoing scattered wave $\psi_{1k_i}^+$ and distorted incoming scattering wave $\psi_{1k_f}^-$, both distorted by $U_1$. Asymptotically, $\psi_{1k_i}^+$ is the sum of the plane wave $\phi_{k_i}$ and the spherically outgoing wave $e^{ikr}/r$. On the other hand, asymptotically, $\psi_{1k_f}^-$ is the sum of the plane wave $\phi_{k_f}$ and the spherically incoming wave $e^{-ikr}/r$.

If both $U_1$ and $U_2$ are spherically symmetric then $f_{fi}^{\text{DWBA}}(\theta)$ can be expanded in the partial waves and we finally obtain (Schiff, 1968)

$$
\begin{aligned}
f_{fi}^{\text{DWBA}}(\theta) &= f_{fi}^{\text{ex}}(\theta) + f_{fi}^{\text{B1}}(\theta) \\
&\quad - \sum_{l=0}^{\infty} (2l+1) P_l(\cos\theta) \int_0^{\infty} U_2 [R_l^2(r) - j_l^2(kr)] r^2 dr
\end{aligned}
\tag{3.12.4}
$$

where $f^{\text{ex}}$ is the exact scattering amplitude due to $U_1$ and $f^{B1}$ is the first Born scattering amplitude due to $U_2$. $R_l(r)$ is the radial wave function due to $U_1$. The above equation is valid only if $U_2$ falls faster than $1/r$ at large $r$.

## 3.13  The Critical Points

It is found that for a strong interaction potential the differential cross sections (DCS) possess deep minima. Such minima exist at one or more impact energy and scattering angle. A small change in either causes an increase in the DCS. Such impact energies and scattering angles are known as critical energies ($E_c$) and critical angles ($\theta_c$), respectively, and we have

$$\frac{d^2 I(E, \theta)}{dE d\theta} \bigg|_{E=E_c, \theta=\theta_c} = 0 \tag{3.13.1}$$

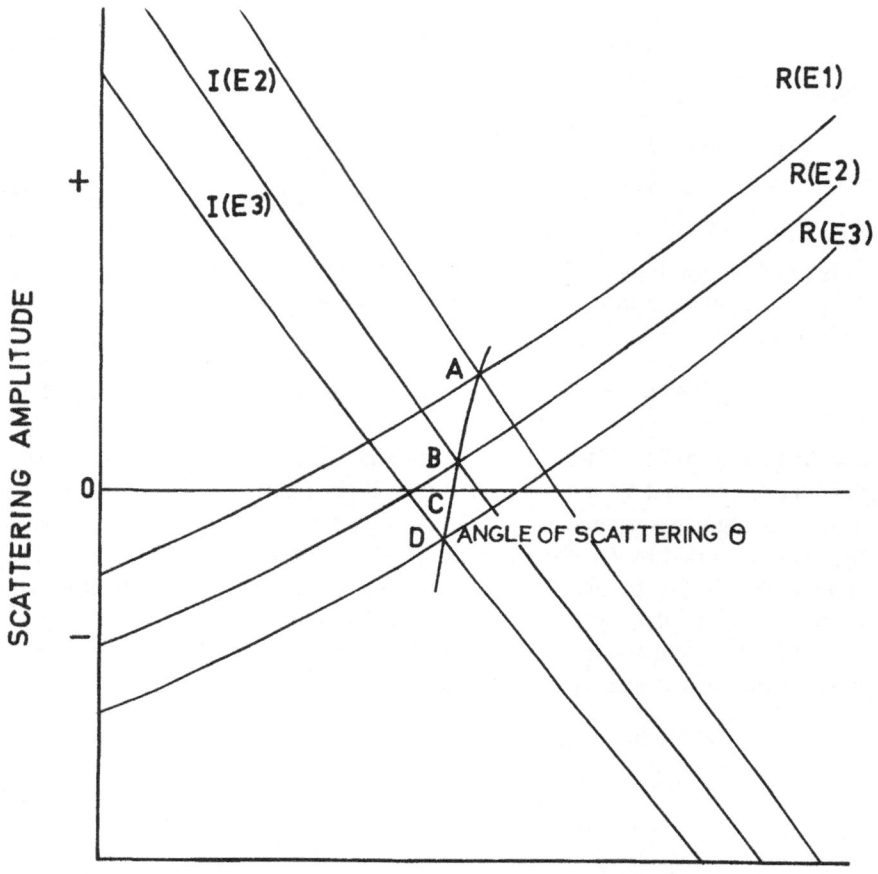

**FIGURE 3.4** The variation of the real $R(E_i)$ and imaginary $I(E_i)$ parts of the scattering amplitudes for the scattering of electrons by a potential at different energies $E_i$.

For a theoretical determination of the critical points, Khare and Raj (1980) have suggested the following simple method. Since $I(E, \theta)$ is very small at $(E_c, \theta_c)$ we take $I(E_c, \theta_c) = 0$ without introducing any significant error. Now,

$$I = |f_R|^2 + |f_I|^2 \tag{3.13.2}$$

where $f_R$ and $f_I$ are the real and imaginary parts of the scattering amplitude. Hence, at $(E_c, \theta_c)$ we have $f_R = f_I = 0$. To determine $E_c$ and $\theta_c$, the values of $f_R(E, \theta)$ and $f_I(E, \theta)$ are generated by the numerical solution of the differential equation (3.9.17). In Fig. 3.4 they are represented by $R(E_i)$ and $I(E_i)$, respectively, and are plotted as functions of $\theta$. The interaction curve $ABCD$ of $R(E_i)$ and $I(E_i)$ is obtained, as shown in the figure. The intersection of the curve $ABCD$ with $\theta$-axis yields the critical angle $\theta_c$.

Near the critical point the variation of $\theta$ with $f_i(= f_R = f_I)$ can be given by

$$\theta = af_i + bf_i^2 + c \qquad (3.13.3)$$

Hence, we require $f_i$ at three impact energies to determine $\theta_c$. Similarly, $E_c$ is determined from the equation

$$E = ef_i + gf_i^2 + h \qquad (3.13.4)$$

## 3.14 The Hulthen–Kohn Variational Principle

We have already discussed [in Sec. (3.7)] the Schwinger variational method, which employs an integral approach. With trial wave functions this method gives variationally correct scattering amplitudes. Variational methods based on the differential approach have also been developed. As we have seen, in this approach one is required to solve the differential equation (3.9.2) with proper boundary conditions to obtain exact phase shifts $\eta_l^e$ and thus the exact scattering amplitude. However, we may start with a trial radial function $f_l^t(r)$ and a trial phase shift $\eta_l^t$ and employ a variational technique to obtain a better phase shift $\eta_l^b$ and a better radial function $f_l^b(r)$.

In the *Hulthen-Kohn variational method* the exact radial function $f_l^e(r)$ and trial radial function $f_l^t(r)$ are subjected to following boundary conditions:

$$f_l^e(0) = f_l^t(0) = 0 \qquad (3.14.1)$$

$$f_l^e(r) \underset{r\to\infty}{\longrightarrow} \sin(kr - l\pi/2 + \eta_l^e) \qquad (3.14.2)$$

and

$$f_l^t(r) \underset{r\to\infty}{\longrightarrow} \sin(kr - l\pi/2 + \eta_l^t) \qquad (3.14.3)$$

The boundary condition (3.14.1) is consistent with (3.9.4) but (3.14.2) and (3.14.3) are slightly different from (3.9.5). The latter are obtained by adopting a different normalization. Let us now assume that

$$f_l^t(r) = f_l^e(r) + \delta[f_l(r)] \qquad (3.14.4)$$

and

$$\eta_l^i = \eta_l^e + \delta(\eta_l) \tag{3.14.5}$$

where $\delta[f_l(r)]$ and $\delta(\eta_l)$ are infinitesimal quantities. Hence, from (3.14.1) and (3.14.4),

$$\delta[f_l(0)] = 0 \tag{3.14.6}$$

With the help of (3.14.2) to (3.14.5), we get

$$\delta[f_l(r)] \underset{r\to\infty}{\longrightarrow} \cos(kr - l\pi/2 + \eta_l^e)\delta(\eta_l) \tag{3.14.7}$$

Now we define a functional:

$$Q_l = \int_0^\infty f_l(r) P_l f_l(r) dr \tag{3.14.8}$$

where the operator $P_l$ is given by

$$P_l = \frac{d^2}{dr^2} + k^2 - \frac{l(l+1)}{r^2} - U(r) \tag{3.14.9}$$

Hence,

$$P_l f_l^e = Q_l^e = 0 \tag{3.14.10}$$

and

$$Q_l^i = \int_0^\infty \{f_l^e(r) + \delta[f_l(r)]\} P_l \{f_l^e(r) + \delta[f_l(r)]\} dr$$

$$= Q_l^e + \int_0^\infty \delta[f_l(r)] P_l f_l^e dr + \int_0^\infty \delta[f_l(r)] P_l \delta[f_l(r)] dr + \int_0^\infty f_l^e P_l \delta[f_l(r)] dr \tag{3.14.11}$$

The first two terms are zero and the third term is of second order. Hence, up to first order,

$$\delta(Q_l) = Q_l^i - Q_l^e = \int_0^\infty f_l^e(r) P_l \delta[f(r)] dr \tag{3.14.12}$$

Using Green's theorem or carrying out partial integration, we get

$$\int_0^\infty f_l^e(r) \frac{d^2}{dr^2} \delta[f_l(r)] dr = \int_0^\infty \delta[f_l(r)] \frac{d^2}{dr^2} f_l^e(r) dr - k\delta(\eta_l) \tag{3.14.13}$$

Then, (3.14.12) and (3.14.13) give

$$\delta(Q_l) = -k\delta(\eta_l) \tag{3.14.14}$$

or

$$\delta(Q_l + k\eta_l) = 0 \tag{3.14.15}$$

Thus up to first order $Q_l + k\eta_l$ is a stationary quantity. According to this variational principle, which was propounded by Hulthen (1944, 1948), up to first order,

$$Q_l^e + k\eta_l^e - Q_l^i - k\eta_l^i = 0 \tag{3.14.16}$$

Since $Q_l^e = 0$, we start from a trial radial wave function $f_l^i(r)$, which obeys (3.14.1) and (3.14.2), and a trial phase shift $\eta_l^i$, and get a better phase shift $\eta_l^b$ from (3.14.16):

$$\eta_l^b = \eta_l^i + \frac{1}{k}Q_l^i \tag{3.14.17}$$

To obtain $\eta_l^b$, we take $f_l^i(C_1, C_2, \ldots, C_n, \eta_l^i)$, which depends upon $(n + 1)$ parameters given by $C_i$ and $\eta_l^i$. This is used to calculate $Q_l$ from (3.14.8). The calculated $Q_l$ is made stationary with respect to $C_i$ and $\eta_l^i$ by imposing the conditions

$$\frac{\partial Q_l}{\partial C_i} = 0 \qquad (i = 1, 2, \ldots, n) \tag{3.14.18}$$

and

$$\frac{\partial Q_l}{\partial C_l^i} = 0 \tag{3.14.19}$$

The resulting $(n + 1)$ equations are solved to obtain variationally correct $C_i$ and $\eta_l^i$ and thus variationally correct $f_l(r)$, $Q_l$, and phase shift $\eta_l^b$ are calculated. This method was developed by Kohn (1948).

Instead of determining $\eta_l^b$ we can obtain $(\tan \eta_l)^b$ by changing (3.14.2) and (3.14.3) to

$$f_l^{e,t} \underset{r \to \infty}{\to} \sin(kr - l\pi/2) + (\tan \eta_l)^{e,t} \cos(kr - l\pi/2) \tag{3.14.20}$$

Proceeding as before, we now get

$$(\tan \eta_l)^b = (\tan \eta_l)^t + \frac{1}{k} Q_l \tag{3.14.21}$$

Partial differentiation of $\theta_l$ with respect to the variational parameter $C_i$ gives (3.14.18), but instead of (3.14.19) we get

$$\frac{\partial Q_l}{\partial (\tan \eta_l)} = -k \tag{3.14.22}$$

Solutions of (3.14.19) and (3.14.22) give variationally correct $Q_l$ and $(\tan \eta_l)^t$ and thus $(\tan \eta_l)^b$ [from (3.14.21)].

Equation (3.14.21) can be extended to obtain a variational principle for the scattering length $a_S$. For the zeroth partial wave this equation gives

$$-\frac{\tan \eta_o^b}{k} = -\frac{\tan \eta_o^t}{k} - \int_0^\infty \frac{f_o^t(r)}{k} P_0 \frac{f_o^t(r)}{k} dr \tag{3.14.23}$$

We define

$$f_0^o(r) = \lim_{k \to 0} \left[ -\frac{1}{k} f_0^t(k, r) \right] \tag{3.14.24}$$

Taking the limit of (3.14.23) as $k \to 0$, we get

$$a_S^b = a_S^t - \int_0^\infty f_0^o(r) P_0 f_0^o(r) dr \tag{3.14.25}$$

Now $f_0^o$ is expressed as a function of variational parameters $C_i$ and $a_S^t$. This gives

$$\frac{\partial I}{\partial C_i} = 0 \tag{3.14.26}$$

and

$$\frac{\partial I}{\partial (a_S^t)} = 0 \tag{3.14.27}$$

where

$$I = a_S^t - \int_0^\infty f_0^o(r) P_0 f_0^o(r) dr \tag{3.14.28}$$

The value of the variational parameters $C_i$ and $a_s^i$ are determined from (3.14.26) and (3.14.27), and thus a better value of the scattering length is obtained from (3.14.25).

It should be noted that although $\eta_l^b$ differs from the exact phase shift $\eta_l^e$ by a second-order quantity, it is not possible to make a definite statement as to whether the difference $(\eta_l^b - \eta_l^e)$ is negative or positive. Thus $\eta_l^b$ does not provide a bound to the phase shifts, not does $(\tan \eta_l)^b$. Hence, in this respect the variational principles of scattering theory are inferior to the *Rayleigh–Ritz variational principle*, which provides an upper bound to the eigenenergy of bound states. However, it has been shown by Rosenberg et al. (1960) that $a_s^b$ provides an upper bound to the scattering length, provided that the interaction potential is too weak to support negative energy states.

## 3.15  The Atomic Units

In atomic physics quite often atomic units are employed. Here the length is expressed in the units of $a_0$ and the unit of energy is taken to be 1 Hartree, which is equal to $e^2/a_0$, i.e., 2 Rydbergs. In these units $\hbar = m = e = a_0 = 1$. Equation (3.1.1) written in atomic units becomes

$$[\nabla^2 + k^2 - 2V(r)]\psi(r) = 0 \qquad (3.15.1)$$

Sometimes the unit of energy is taken to be 1 Rydberg, instead of 2 Rydbergs. Then the above equation changes to

$$[\nabla^2 + k^2 - V(r)]\psi(r) = 0 \qquad (3.15.2)$$

In equations written in atomic units all the quantities are dimensionless. However, we shall continue to write equations in terms of $\hbar$, $m$, $e$, and $a_0$.

## Questions and Problems

3.1  Show that for the interaction potentials that fall faster than $r^{-2}$ in the asymptotic region, $f(\theta, \varphi)e^{ikr}/r$ satisfies Eq. (3.1.1).

3.2  An electron of energy $54\,\text{eV}$ is scattered by an absorptive complex potential. The forward scattering amplitude is equal to $(2 + i0.5)a_0$, where $a_0$ is the first Bohr radius. Calculate the differential cross section in the forward direction and the total collision cross section.

3.3 A particle of mass $m$ is scattered by a potential field, which is given by $-Ze^2[(1/r) + (Z/a_0)]e^{-\alpha Zr/a_0}$. Obtain the scattering amplitude in the first Born approximation. Also obtain the ratio of the differential cross sections in the forward and backward directions.

3.4 Use the first Born approximation to show that the integrated cross section for the scattering of a particle by a Coulomb field is infinite.

3.5 According to Simpson's rule for numerical integration

$$\int_{x_0}^{x_0+nh} f(x)dx = \frac{h}{3}\left\{f(x_0) + 4\left[f(x_0+h)+f(x_0+3h)+\dots+f\left(x_0+\overline{n-1}h\right)\right]\right.$$
$$\left. + 2\left[f(x_0+2h)+f(x_0+4h)+\dots+f\left(x_0+\overline{n-2}h\right)\right]+f(x_0+nh)\right\}$$

where $n$ is an even positive integer. Use the above equation along with (3.5.11) and (3.5.13) to evaluate $\sigma_{el}^{B1}$ in the units of $\pi a^2$ for $\lambda a = 1$. Take $k = 2.5$ and $n = 20$. Compare your result with that obtained from (3.5.14).

3.6 According to the trapezoidal rule for numerical integration

$$\int_{x_0}^{x_0+nh} f(x)dx = \frac{h}{2}\left\{f(x_0+h)+2[f(x_0+h)+f(x_0+2h)+\dots\right.$$
$$\left.+f\left(x_0+\overline{n-2}h\right)+f\left(x_0+\overline{n-1}h\right)\right]+f(x_0+nh)\right\}$$

where $n$ is a positive integer. Use this rule also to evaluate the above cross section with the same values of $ka$ and $n$. Comment on the accuracy of this rule vis-à-vis Simpson's rule.

3.7 On the both sides of (3.4.17) the exact scattering amplitude $f$ is approximated by $\lambda f_{B1}$, where $\lambda$ is a complex number. Obtain the real and imaginary parts of $\lambda$ in terms of $f_{B1}$ and the real and imaginary parts of $\bar{f}_{B2}$. Further, show that in the above approximation

$$[f] = \frac{f_{B1}^2}{f_{B1} - \bar{f}_{B2}}$$

Compare the above expression with the Born series for $f_{B1} \gg \bar{f}_{B2}$.

3.8 Point out the main differences between the Born and the eikonal series. Out of the first Born and the eikonal approximations, which one is superior and why?

3.9 Discuss the concept of phase shift. Why, in general, does the value of the phase shift decrease with an increase in $l$?

3.10 For a particle scattered by a central potential, the values of the first three phase shifts, in radians, are $\eta_0 = 1.960$, $\eta_1 = 0.453$, and $\eta_2 = 0.112$. Obtain the values of $I(\theta)$ and plot a graph of $I(\theta)$ vs. $\theta$. Also calculate the partial and integrated cross sections.

*4*

# Collision of Electrons with a Potential

## 4.1  Introduction

   The methods developed in the previous chapter are applicable to the collision (scattering) of electrons with (by) a potential provided we assume that an electron is a spinless particle. However, an electron is a spin-$\frac{1}{2}$ particle and possesses magnetic moment $\boldsymbol{\mu}$. The coupling of the spin angular momentum $S$ with the orbital angular momentum $L$ of the electron produces a new term in the Hamiltonian of the system. The potential energy of the electron changes from $V(r)$ to $V_{\text{eff}}(r) = V(r) + V_{SO}(r)$, where $V_{SO}(r)$ is the additional potential energy due to spin–orbit coupling. Hence, the phase shift, $\eta_l$ as obtained from (3.9.15), changes. This additional term depends upon three quantum numbers $j$, $l$, and $s$. However, $s$ is always $\frac{1}{2}$. Hence, we represent the phase shift by $\eta_{j,l}$. Not only are the $\eta_{j,l}$ different from the $\eta_l$ (obtained with $V_{SO} = 0$), but one value of $l$ gives two phase shifts corresponding to $j = l \pm \frac{1}{2}$. In such collisions $m_l$ and $m_s$ need not be separately conserved, but their sum, i.e., $m_j$, is a constant of motion. Hence,

$$\Delta m_j = \Delta m_l + \Delta m_s = 0 \tag{4.1.1}$$

Since $m_s = \pm\frac{1}{2}$, it is possible that in the collision an incident electron with $m_s = +\frac{1}{2}$ may flip its spin and appear with $m_s = -\frac{1}{2}$ after the collision. Thus, due to the spin–orbit interaction, we have following two types of collisions:

$$e\uparrow + V_{\text{eff}}(r) \rightarrow e\uparrow + V_{\text{eff}}(r) \tag{4.1.2}$$

and

$$e\!\uparrow + V_{\text{eff}}(r) \rightarrow e\!\downarrow + V_{\text{eff}}(r) \tag{4.1.3}$$

In (4.1.2) $\Delta m_s = \Delta m_l = 0$ but in (4.1.3) $\Delta m_s = 1$; hence, $\Delta m_l = -1$. The former is the direct collision while the latter is known as the spin-flip collision.

## 4.2   Spin–Orbit Interaction Potential

Suppose an electron is moving in a potential field with a velocity $v$ and the potential produces an electrical field $E$ at the position of electron. Then, in the rest frame of the electron, a magnetic field $B$ equal to $(v \times E)/c$ is produced. This magnetic field $B$ interacts with the magnetic moment $\mu[= -eS/(mc)]$ of the electron to produce a new term in the Hamiltonian. This term is given by

$$H_{\text{SO}} = -\mu \cdot B = -\frac{e}{m^2c^2} S \cdot (E \times p) \tag{4.2.1}$$

where $p$ is the linear momentum of the electron. For a central potential

$$E = -\frac{1}{er}\frac{dV(r)}{dr}r \tag{4.2.2}$$

Hence, in a frame in which the electron is at rest,

$$V_{\text{SO}}(r) = H_{\text{SO}} = \frac{1}{m^2c^2r}\frac{dV}{dr}(S \cdot L) \tag{4.2.3a}$$

If we calculate $V_{\text{SO}}(r)$ with the proper Lorentz transformation for the field in a frame in which the electron is moving, then $V_{\text{SO}}(r)$ given by the above equation is reduced by a factor of 2. Thus, finally,

$$V_{\text{SO}}(r) = \frac{1}{2m^2c^2r}\frac{dV}{dr}(S \cdot L) \tag{4.2.3b}$$

Now,

$$J^2 = L^2 + S^2 + 2L \cdot S$$

Hence,

$$V_{\text{SO}}(r) = \frac{1}{4m^2c^2r}\frac{dV}{dr}(J^2 - L^2 - S^2) \tag{4.2.4}$$

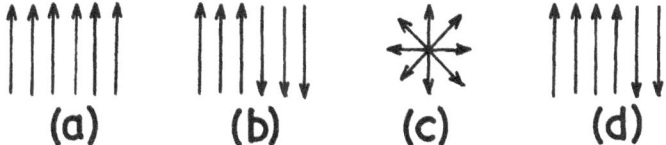

**FIGURE 4.1** Electron ensembles: (a) Polarized, (b) unpolarized, (c) unpolarized, and (d) partially polarized.

As $V_{SO}$ is usually small in comparison to $V(r)$, it is neglected in many investigations, but in this chapter its effect will be taken into consideration.

## 4.3   Ensemble of Polarized Electrons

Before considering the scattering of electrons by a potential $V(r) + V_{so}(r)$, a familiarization with the concept of the ensemble and the beam of polarized electrons is quite useful. If in an ensemble of $N$ electrons, the spin vectors of all the electrons point in the same direction (Fig. 4.1a), the ensemble is said to be fully polarized. In this case all the electrons are in the same spin state and the ensemble is said to be in a pure spin state. On the other hand, if the spins of half of the electrons point in one direction and the other half in the opposite direction (Fig. 4.1b), then the ensemble is completely unpolarized. An ensemble of electrons is also completely unpolarized if its spin vectors are distributed equally in all possible directions (Fig. 4.1c). Figure 4.1d depicts an ensemble of partially polarized electrons. The degree of polarization $P$ is defined by

$$P = \frac{N\!\uparrow - N\!\downarrow}{N\!\uparrow + N\!\downarrow} = \frac{N\!\uparrow - N\!\downarrow}{N} \qquad (4.3.1)$$

where $N\!\uparrow$ and $N\!\downarrow$ are the number of electrons having their spin up ($m_s = +\frac{1}{2}$) and down ($m_s = -\frac{1}{2}$), respectively. For $N\!\uparrow = N$ the ensemble is fully polarized ($P = 1$) and for $N\!\uparrow = N/2$ it is completely unpolarized ($P = 0$). For any other value of $N\!\uparrow$ the ensemble is partially polarized. For $N\!\uparrow < N/2$ the degree of polarization is negative. In conclusion, an ensemble (or a beam) of electrons is said to be polarized if its spins have a preferential orientation such that there exists a direction for which the two spin states ($\uparrow$ and $\downarrow$) with respect to quantization axis are not equally populated. It should be noted that due to the uncertainty principle, the vector $S$ is not stationary in the space. If we say that $S$ is in the $z$ direction, we mean that $S$ is somewhere on a cone in such a way that its component along the $z$-axis is $\hbar/2$ and $|S| = \sqrt{\frac{3}{4}}\,\hbar$ [see Eq. (4.3.9)].

The spin wave function $\chi$ of an electron in a pure spin state is represented by a vector in a two-dimensional Hilbert space. Taking $\alpha$ and $\beta$ to be two basis vectors we have

$$\chi = a\alpha + b\beta \tag{4.3.2}$$

The $\alpha$ and $\beta$ correspond to $m_s = +\frac{1}{2}$ and $-\frac{1}{2}$ states, respectively, and are represented by two component spinors:

$$\alpha = \begin{pmatrix} 1 \\ 0 \end{pmatrix} \quad \text{and} \quad \beta = \begin{pmatrix} 0 \\ 1 \end{pmatrix} \tag{4.3.3}$$

It is easy to verify that

$$\langle \alpha | \alpha \rangle = \langle \beta | \beta \rangle = 1 \quad \text{and} \quad \langle \alpha | \beta \rangle = 0 \tag{4.3.4}$$

Thus we get

$$\chi = a \begin{pmatrix} 1 \\ 0 \end{pmatrix} + b \begin{pmatrix} 0 \\ 1 \end{pmatrix} = \begin{pmatrix} a \\ b \end{pmatrix} \tag{4.3.5}$$

For a normalized $\chi$

$$\langle \chi | \chi \rangle = |a|^2 + |b|^2 = 1 \tag{4.3.6}$$

The spin angular momentum $S$ is defined by

$$S = \frac{1}{2}\hbar\sigma \tag{4.3.7}$$

where $\sigma$ is the Pauli spin operator. Its three components are

$$\sigma_x = \begin{pmatrix} 0 & 1 \\ 1 & 0 \end{pmatrix}, \quad \sigma_y = \begin{pmatrix} 0 & -i \\ i & 0 \end{pmatrix}, \quad \text{and} \quad \sigma_z = \begin{pmatrix} 1 & 0 \\ 0 & -1 \end{pmatrix} \tag{4.3.8}$$

With the above equation, it is easy to verify that

$$S_z \begin{pmatrix} \alpha \\ \beta \end{pmatrix} = \pm \frac{1}{2}\hbar \begin{pmatrix} \alpha \\ \beta \end{pmatrix} \quad \text{and} \quad S_z \begin{pmatrix} \alpha \\ \beta \end{pmatrix} = \frac{3}{4}\hbar^2 \begin{pmatrix} \alpha \\ \beta \end{pmatrix} \tag{4.3.9}$$

Furthermore,

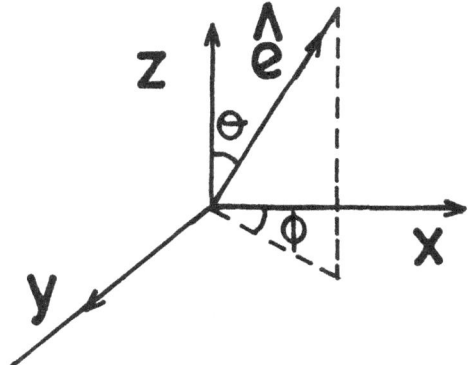

**FIGURE 4.2** The polar angles of the spin wave vectors $\chi$ are $(\theta, \phi)$.

$$S_z \chi = aS_z \begin{pmatrix} 1 \\ 0 \end{pmatrix} + bS_z \begin{pmatrix} 0 \\ 1 \end{pmatrix} = \tfrac{1}{2} \hbar \begin{pmatrix} a \\ -b \end{pmatrix}$$

Hence, $\chi$ is not an eigenfunction of $S_z$. However, if $S_\chi$ is the component of $S$ in the spin direction of $\chi$ then we should have

$$S_\chi \chi = \tfrac{1}{2} \hbar \chi \tag{4.3.10}$$

To verify the above equation let us take $(\theta, \phi)$ to be in the spin direction of $\chi$ and $\hat{e}$ as a unit vector along $(\theta, \phi)$; then with the help of (4.3.7) and (4.3.8),

$$S_\chi = S \cdot \hat{e}$$
$$= \tfrac{1}{2} \hbar \begin{pmatrix} e_z & e_x - ie_y \\ e_x + ie_y & -e_z \end{pmatrix} \tag{4.3.11}$$

Assuming (4.3.10) to be correct and using (4.3.11), we get

$$\begin{pmatrix} (e_z - 1)a & (e_x - ie_y)b \\ (e_x + ie_y)a & -(e_z + 1)b \end{pmatrix} = 0$$

For nontrivial values of $a$ and $b$ we have

$$\begin{vmatrix} e_z - 1 & e_x - ie_y \\ e_x + ie_y & -(e_z + 1) \end{vmatrix} = 0$$

which yields

$$1-(e_x^2 + e_y^2 + e_z^2) = 0$$

which is certainly correct, and thus, (4.3.10) is verified. Now from

$$(e_x - 1)a + (e_x - ie_y)b = 0$$

we have

$$\frac{b}{a} = \frac{1-e_z}{e_x - ie_y} \tag{4.3.12}$$

Furthermore $e_x = \sin\theta\cos\phi$, $e_y = \sin\theta\cos\phi$, and $e_z = \cos\theta$. Hence,

$$\frac{b}{a} = \frac{1-\cos\theta}{\sin\theta\, e^{-i\phi}} = \tan(\theta/2)e^{i\phi} \tag{4.3.13}$$

For a normalized $\chi$ we take

$$b = \sin(\theta/2)e^{i\phi} \qquad \text{and} \qquad a = \cos(\theta/2) \tag{4.3.14}$$

For a 100% polarized ensemble the polarization vector $P$ is the expectation value of the Pauli operator $\sigma$. Thus,

$$P = \langle \chi | \sigma | \chi \rangle \tag{4.3.15}$$

Using (4.3.5) for $\chi$ and (4.3.8) for $\sigma$, we obtain

$$P_x = (a^*\ b^*)\begin{pmatrix} b \\ a \end{pmatrix} = a^*b + b^*a = \sin\theta\,\cos\theta$$

$$P_y = (a^*\ b^*)\begin{pmatrix} -ib \\ ia \end{pmatrix} = i(ab^* - a^*b) = \sin\theta\,\sin\phi$$

$$P_z = (a^*\ b^*)\begin{pmatrix} a \\ -b \end{pmatrix} = |a|^2 - |b|^2 = \cos\theta \tag{4.3.16}$$

As expected $P^2 = 1$.

The density matrix $\rho$ is defined by

$$\rho = |\chi><\chi| = \begin{pmatrix} a \\ b \end{pmatrix}(a^* \; b^*)$$

or

$$\rho = \begin{pmatrix} aa^* & ab^* \\ ba^* & bb^* \end{pmatrix} \tag{4.3.17}$$

We note that

$$\mathrm{tr}\,\rho\sigma_z = \mathrm{tr}\begin{pmatrix} aa^* & ab^* \\ ba^* & bb^* \end{pmatrix}\begin{pmatrix} 1 & 0 \\ 0 & -1 \end{pmatrix} = P_z$$

Similar relations hold for the $x$ and $y$ components. Hence,

$$P = \mathrm{tr}(\rho\sigma) \tag{4.3.18}$$

Eqs. (4.3.16) and (4.3.17) give

$$\rho = \tfrac{1}{2}(1 + P \cdot \sigma) = \frac{1}{2}\begin{pmatrix} 1 + P_z & P_x - iP_y \\ P_x + iP_y & 1 - P_z \end{pmatrix} \tag{4.3.19}$$

Let us now consider an ensemble of electrons obtained by mixing a number of pure states, represented by $|\chi_i\rangle$. Suppose the $i^{\mathrm{th}}$ component of the mixture has $N_i$ electrons and its polarization vector is $P_i$. Since $|\chi_i\rangle$ is a pure state $|P_i| = 1$, but in the mixture $P_i$ and $P_j$ have different directions. The polarization vector of the mixture is

$$P = \frac{1}{N}\sum_i N_i P_i \tag{4.3.20}$$

where $N = \sum_i N_i$. Hence,

$$P^2 = P \cdot P = \frac{1}{N^2}\sum_i \sum_i N_i N_j P_i \cdot P_j$$

$$= \frac{1}{N^2}\left(\sum_i N_i^2 + \sum_i \sum_{i \neq 1} N_i N_j P_i \cdot P_j\right)$$

Since $P_i \cdot P_j$ is less than unity we get

$$P \cdot P < \frac{1}{N^2}\left(\sum_i N_i\right)^2 = 1 \tag{4.3.21}$$

i.e., the degree of polarization of the mixture is less than unity. Equation (4.3.19) is also valid for a mixture. This gives $\mathrm{tr}\rho = 1$, but

$$\mathrm{tr}\,\rho^2 = \tfrac{1}{4}\mathrm{tr}\begin{pmatrix} 1+P_z & P_x-iP_y \\ P_x+iP_y & 1-P_z \end{pmatrix}^2$$
$$= \tfrac{1}{2}(1+P^2) \tag{4.3.22}$$

is less than unity because $P$ is less than one. Thus we conclude that for a pure spin state $P = 1$, and $\mathrm{tr}\rho = \mathrm{tr}\rho^2 = 1$ but for a mixture $P < 1$ and $\mathrm{tr}\rho^2 < \mathrm{tr}\rho$.

For a 100% polarized beam in the $z$ direction (4.3.19) yields

$$\rho = \begin{pmatrix} 1 & 0 \\ 0 & 0 \end{pmatrix} \tag{4.3.23}$$

Similarly for a completely unpolarized beam ($P_x = P_y = P_z = 0$), we have

$$\rho = \frac{1}{2}\begin{pmatrix} 1 & 0 \\ 0 & 1 \end{pmatrix} \tag{4.3.24}$$

If a partially polarized beam is polarized in the $z$ direction with the degree of polarization $P$ then $P_z = P$ and $P_x = P_y = 0$. Hence, from (4.3.19)

$$\rho = \frac{1}{2}\begin{pmatrix} 1+P & 0 \\ 0 & 1-P \end{pmatrix}$$
$$= \frac{1}{2}(1-P)\begin{pmatrix} 1 & 0 \\ 0 & 1 \end{pmatrix} + P\begin{pmatrix} 1 & 0 \\ 0 & 0 \end{pmatrix} \tag{4.3.25}$$

According to the above equation, a partially polarized beam can be considered as being made up of a totally polarized beam and a completely unpolarized beam mixed in the ratio $P:(1-P)$. It is to be noted that the addition of density matrices is an incoherent addition. We may write (4.3.24) as

$$\rho = \frac{1}{2}\begin{pmatrix} 1 & 0 \\ 0 & 0 \end{pmatrix} + \frac{1}{2}\begin{pmatrix} 0 & 0 \\ 0 & 1 \end{pmatrix} \tag{4.3.26}$$

where the first term is the density matrix of a fully polarized ensemble in the $+z$ direction, whereas the second also represents a fully polarized ensemble but in the $-z$ direction. An incoherent addition of the two terms with the same weight factor $\frac{1}{2}$ results in a completely unpolarized ensemble. On the other hand, a coherent addition (addition of the amplitudes of the two oppositely polarized states) produces a new fully polarized state. For example, addition of two fully polarized states in the $+z$ and $-z$ directions, represented by

$$\begin{pmatrix} 1 \\ 0 \end{pmatrix} \quad \text{and} \quad \begin{pmatrix} 0 \\ 1 \end{pmatrix}$$

respectively, gives rise to

$$\frac{1}{\sqrt{2}}\begin{pmatrix} 1 \\ 0 \end{pmatrix} + \frac{1}{\sqrt{2}}\begin{pmatrix} 0 \\ 1 \end{pmatrix} = \frac{1}{\sqrt{2}}\begin{pmatrix} 1 \\ 1 \end{pmatrix} \tag{4.3.27}$$

A comparison of the above equation with (4.3.5) shows that the former equation represents a fully polarized state with $a = b = 1/\sqrt{2}$ or $\theta = \pi/2$ and $\phi = 0$.

## 4.4 Direct and Spin-Flip Scattering Amplitudes

Let us consider the scattering of the spin-up electrons by a central potential $V(r)$. Due to the spin–orbit interaction we have direct and spin-flip scatterings, represented by (4.1.1) and (4.1.2), respectively. The wave function of the system satisfies the following differential equation:

$$\left[ \nabla^2 + k^2 - U(r) - \frac{1}{4m^2c^2} \frac{1}{r} \frac{dU(r)}{dr} (J^2 - L^2 - S^2) \right] \psi(r,s) = 0 \tag{4.4.1}$$

where $U(r) = 2mV(r)/\hbar^2$. In the scattering $m_j = \frac{1}{2}$ is a constant of motion. Hence, for a given $l$, after the scattering, $m_s = \frac{1}{2}$ and $m_l = 0$ for direct scattering and $m_s = -\frac{1}{2}$ and $m_l = 1$ for spin-flip scattering. Furthermore, for each $l$, $j$ has two values given by $l \pm \frac{1}{2}$. We expand $\psi(r, s)$ in terms of a complete set represented by $y_{j,l,s,m_j}$. For $m_j = \frac{1}{2}$ and $s = \frac{1}{2}$ this function (Mott and Massey, 1965) is given by

$$y_{j,l,1/2,1/2} = C_{l,1/2}(j,\tfrac{1}{2},0,\tfrac{1}{2})Y_{l0}\alpha + C_{l,1/2}(j,\tfrac{1}{2},1,-\tfrac{1}{2})Y_{l1}\beta \tag{4.4.2}$$

where $C_{l,s}(j, m_j, m_\ell, m_s)$ are Clebsch–Gordon coefficients, given by

$$C_{l,1/2}(l + \tfrac{1}{2}, \tfrac{1}{2}, 0, \tfrac{1}{2}) = C_{l,1/2}(l - \tfrac{1}{2}, \tfrac{1}{2}, 1, -\tfrac{1}{2}) = [(l+1)/(2l+1)]^{1/2}$$

$$C_{l,1/2}(l + \tfrac{1}{2}, \tfrac{1}{2}, 1, -\tfrac{1}{2}) = -C_{l,1/2}(l - \tfrac{1}{2}, \tfrac{1}{2}, 0, \tfrac{1}{2}) = [l/(2l+1)]^{1/2} \qquad (4.4.3)$$

Hence,

$$y_{l+1/2,l,s,1/2} = [(l+1)/(2l+1)]^{1/2} Y_{l_0}\alpha + [l/(2l+1)]^{1/2} Y_{l_1}\beta \qquad (4.4.4a)$$

and

$$y_{l-1/2,l,s,1/2} = -[l/(2l+1)]^{1/2} Y_{l_0}\alpha + [(l+1)/(2l+1)]^{1/2} Y_{l_1}\beta \qquad (4.4.4b)$$

It is obvious that the $y_{l\pm1/2,l,s,1/2}$ are eigenfunctions $L^2$ and $S^2$. By using the relation

$$J^2 = L^2 + S^2 + \hbar(L_x\sigma_x + L_y\sigma_y + L_z\sigma_z) \qquad (4.4.5)$$

it can be shown that the above functions are eigenfunctions of $J^2$ as well with eigenvalues $j(j+1)\hbar^2$. Now the expansion of $\psi(r, s)$ in the complete set of $y_{j,\ell,s,1/2}$, similar to (3.9.14), is given by

$$\psi(r,s) = A\sum_i i^l (4\pi)^{1/2}(2l+1)^{1/2} \sum_{j=|l-1/2|}^{l+1/2} R_{j,l}(r)\exp(i\eta_{j,l})$$
$$\times C_{l,s}(j, \tfrac{1}{2}, 0, \tfrac{1}{2}) y_{j,l,s,1/2}(\hat{r}, s) \qquad (4.4.6)$$

The radial function $R_{j,\ell}(r)$ is the solution of the following one-dimensional differential equation:

$$\left[\frac{1}{r^2}\frac{d}{dr}\left(r^2\frac{d}{dr}\right) + k^2 - U(r) - \frac{l(l+1)}{r^2} - \frac{\hbar^2}{4m^2c^2}\lambda_{j,l}\frac{1}{r}\frac{dU(r)}{dr}\right]R_{j,l} = 0 \qquad (4.4.7)$$

where $\lambda_{j,l} = [j(j+1) - l(l+1) - s(s+1)]$, and $R_{j,l}(r)$ satisfies the following boundary conditions:

$$R_{j,l}(r) \underset{r\to0}{\to} r^l \qquad (4.4.8a)$$

and

$$R_{j,l}(r) \underset{r\to\infty}{\to} \frac{1}{kr}\sin(kr - l\pi/2 + \eta_{l,j}) \qquad (4.4.8b)$$

The phase shift $\eta_{j,l}$ depends upon both $j$ and $l$. Using (4.4.8b) in (4.4.6), we get

$$
\psi(r,s) \xrightarrow[r \to \infty]{} \frac{A\sqrt{\pi}}{ikr} \sum_l i^l (2l+1)^{1/2} \sum_{j=|l-1/2|}^{l+1/2} C_{l,s}(j,\tfrac{1}{2},0,\tfrac{1}{2}) y_{j,l,s,1/2} e^{i\eta_{j,l}}
$$
$$
\times \{\exp[i(kr - l\pi/2 + \eta_{j,l})] - \exp[-i(kr - l\pi/2 + \eta_{j,l})]\}
\tag{4.4.9}
$$

For electron–potential scattering, with $m_s = \tfrac{1}{2}$, Eq. (3.1.2) modifies to

$$
\psi^+(r,s) \xrightarrow[r \to \infty]{} A\left[ e^{ik\cdot r}\alpha + \frac{e^{ikr}}{r} f(\theta,\phi) \right]
\tag{4.4.10}
$$

where we have considered only outgoing scattered waves. The asymptotic expression for $A \exp(ik \cdot r)\alpha$ is easily obtained for (4.4.9) by taking $\eta_{j,l} = 0$. Use of this expression and (4.4.9) in (4.4.10) gives

$$
f(\theta,\phi) = \frac{\sqrt{\pi}}{ik} \sum_l i^l (2l+1)^{1/2} \sum_{j=|l-1/2|}^{l+1/2} \exp(-il\pi/2) C_{l,s}(j,\tfrac{1}{2},0,\tfrac{1}{2})
$$
$$
\times y_{j,l,1/2,1/2}[\exp(2i\eta_{l,j}-1)]
\tag{4.4.11}
$$

Now using (4.4.3) and (4.4.4a) in the above equation, we obtain

$$
f(\theta,\phi) = f(\theta)\alpha + g(\theta)e^{i\phi}\beta
\tag{4.4.12}
$$

where

$$
f(\theta) = \frac{1}{2ik} \sum_{l=0}^{\infty} \{(l+1)[\exp(2i\eta_{l+1/2,l})-1] + l[\exp(2i\eta_{l-1/2,l})-1]\} P_l(\cos\theta)
\tag{4.4.13}
$$

and

$$
g(\theta) = \frac{1}{2ik} \sum_{l=1}^{\infty} [\exp(2i\eta_{l-1/2,l}) - \exp(2i\eta_{l+1/2,l})] P_l^1(\cos\theta)
\tag{4.4.14}
$$

For the electron–atom collision $dV(r)/dr$ is quite large near the origin; hence the maximum contribution to $g(\theta)$ comes from the $l = 1$ partial wave (the $l = 0$ partial wave does not contribute to $g(\theta)$ because $l \cdot s = 0$ for this $l$). With the increase of $l$, $\eta_{l-1/2,l}$ approaches $\eta_{l+1/2,l}$. Hence, the contribution of the higher partial waves to $g(\theta)$ rapidly decreases. A relatively larger number of partial waves contribute to $f(\theta)$, so $f(\theta)$ as a function of $\theta$ exhibits more structure in comparison to that shown by $g(\theta)$. As the incident spin wave function is $\alpha$, then $f(\theta)$, being associated with $\alpha$, is known as the direct scattering amplitude. The association of $g(\theta)$ with the spin-down wave function $\beta$ shows that in the scattering some of the electrons flipped their spin from $\alpha$ to $\beta$. Hence, $g(\theta)$ is said to be the spin-flip scattering amplitude. Due to the spin–orbit interaction, the phase shifts $\eta_{l-1/2,l}$ and $\eta_{l+1/2,l}$ are different from each other, and $g(\theta)$ is nonzero. At large $r$, for the incident beam polarized upward $(+z)$, the scattered part of the wave function is a two-component spinor and is given by

$$\psi_{sc}(\uparrow) = A\begin{pmatrix} f(\theta) \\ g(\theta)e^{i\phi} \end{pmatrix}\frac{e^{ikr}}{r} \qquad (4.4.15)$$

Similarly for an incident electron beam polarized in $-z$ (spin-down) direction, the scattered wave at large $r$ is

$$\psi_{sc}(\downarrow) = A\begin{pmatrix} -g(\theta)e^{-i\phi} \\ f(\theta) \end{pmatrix}\frac{e^{ikr}}{r} \qquad (4.4.16)$$

The differential cross sections for the two polarization directions are given by

$$I(\uparrow,\downarrow) = [f^*(\theta) \pm g^*(\theta)\exp(\mp i\phi)]\begin{pmatrix} f(\theta) \\ \pm g(\theta)\exp(\pm i\phi) \end{pmatrix}$$

Thus

$$I\uparrow = I\downarrow = I(\theta) = |f(\theta)|^2 + |g(\theta)|^2 \qquad (4.4.17)$$

Further, due to cylindrical symmetry, both $I\uparrow$ and $I\downarrow$ are independent of $\phi$. A completely unpolarized beam is a mixture of two completely polarized beams (polarized in opposite directions) and the differential cross sections for each half is given by (4.4.17). The sum divided by 2 is again equal to $|f(\theta)|^2 + |g(\theta)|^2$. Thus (4.4.17) also gives the DCS for a completely unpolarized beam.

The scattered wave at any $r$ is given by

$$\psi_{SC}(r,s) = A(4\pi)^{1/2}\sum_l \sum_{j=|l-1/2|}^{l+1/2} i^l(2l+1)^{1/2} C_{ls}(j,\tfrac{1}{2},0,\tfrac{1}{2})y_{j,l,s,1/2}$$
$$\times [\exp(i\eta_{j,l})R_{j,l}(r) - j_l(kr)] \qquad (4.4.18)$$

To eliminate the first-order derivative from (4.4.7), we take $R_{j,l}(r) = f_{j,l}(r)/r$, and thus obtain the differential equation for $f_{j,l}$:

$$\left[\frac{d^2}{dr^2} + k^2 - U(r) - \frac{\hbar^2 \lambda_{j,l}}{4m^2c^2} \frac{1}{r} \frac{dU(r)}{dr} - \frac{l(l+1)}{r^2}\right] f_{j,l}(r) = 0 \qquad (4.4.19)$$

where the values of $\lambda_{j,l}$ are $l$ and $-(l+1)$, corresponding to $j = l \pm \frac{1}{2}$, respectively. As discussed in Chapter 3, Eq. (4.4.19) is solved numerically and the phase shifts $\eta_{l\pm1/2}$, are obtained from the asymptotic values of $f_{j,l}(r)$.

It may be noted that the spin of an electron is due to the relativistic effect. Hence, it is more appropriate to start from the Dirac scattering equation (Mott and Massey, 1965) rather than the Schrödinger scattering equation. However, in the energy range of our interest the results obtained from these two equations do not differ significantly.

### The Scattering Matrix and Left-Right Asymmetry

Let us assume that the incident electron beam is polarized in the $(\theta', \phi')$ direction; then the initial spinor is given by

$$\chi = \begin{pmatrix} a \\ b \end{pmatrix} = a\begin{pmatrix} 1 \\ 0 \end{pmatrix} + b\begin{pmatrix} 0 \\ 1 \end{pmatrix} \qquad (4.4.20)$$

where $a = \cos(\theta'/2)$ and $b = \sin(\theta'/2)e^{i\phi'}$. As discussed in the previous section, due to scattering, both the $\alpha$ and the $\beta$ components of $\chi$ change, and the scattered wave function is

$$\chi' = a\begin{pmatrix} f \\ ge^{i\phi} \end{pmatrix} + b\begin{pmatrix} -ge^{-i\phi} \\ f \end{pmatrix} \qquad (4.4.21a)$$

or

$$= (af - bge^{-i\phi})\alpha + (age^{i\phi} + bf)\beta \qquad (4.4.21b)$$

or

$$= \begin{pmatrix} f & -ge^{-i\phi} \\ ge^{i\phi} & f \end{pmatrix}\begin{pmatrix} a \\ b \end{pmatrix} \qquad (4.4.21c)$$

or

$$\chi' = S\chi \qquad (4.4.21d)$$

where the scattering matrix is defined by

$$S = \begin{pmatrix} f & -ge^{-i\phi} \\ ge^{i\phi} & f \end{pmatrix} \tag{4.4.22}$$

Similarly, the initial density matrix $\rho$ changes to

$$\rho' = \chi' \, \chi'^\dagger = S\chi\chi^\dagger S^\dagger = S\rho S^\dagger \tag{4.4.23}$$

and the DCS is given by

$$I(\theta,\phi) = \frac{\left|af - ge^{-i\phi}b\right|^2 + \left|bf + ge^{i\phi}a\right|^2}{\left|a\right|^2 + \left|b\right|^2} \tag{4.4.24}$$

or

$$I(\theta,\phi) = \frac{\mathrm{tr}\rho'}{\mathrm{tr}\rho} \tag{4.4.25}$$

Using (4.3.19) and (4.4.23) in the above equation, we get

$$I(\theta,\phi) = \tfrac{1}{2}\mathrm{tr}\{S(1+\boldsymbol{P}\cdot\boldsymbol{\sigma})S^\dagger\} \tag{4.4.26}$$

Since $\mathrm{tr}\rho = 1$ Eq. (4.4.26) yields

$$SS^\dagger = \begin{pmatrix} |f|^2 + |g|^2 & (fg^* - gf^*)e^{-i\phi} \\ (f^*g - fg^*)e^{i\phi} & |f|^2 + |g|^2 \end{pmatrix}$$

or

$$= I(\theta)\begin{pmatrix} 1 & -iS(\theta)e^{-i\phi} \\ iS(\theta)e^{i\phi} & 1 \end{pmatrix} \tag{4.4.27}$$

where $I(\theta)$ is given by (4.4.17) and the Sherman function $S(\theta)$ is defined by

$$S(\theta) = i\frac{(fg^* - f^*g)}{I(\theta)} \tag{4.4.28}$$

$$= -2\,Im(fg^*)/I(\theta) \tag{4.4.29}$$

Thus the Sherman function is a real quantity. Using (4.3.19) and (4.4.27) in (4.4.26) and noting that the trace is independent of the order of the matrices we obtain

$$I(\theta,\phi) = \tfrac{1}{2}\operatorname{tr}\left[ I(\theta)\begin{pmatrix} 1 & -iS(\theta)e^{-i\phi} \\ iS(\theta)e^{i\phi} & 1 \end{pmatrix}\begin{pmatrix} 1+P_z & P_x - iP_y \\ P_x + iP_y & 1-P_z \end{pmatrix}\right]$$

$$= \frac{I(\theta)}{2}\operatorname{tr}\begin{pmatrix} b_{11} & b_{12} \\ b_{21} & b_{22} \end{pmatrix} \tag{4.4.30}$$

where

$$\begin{aligned}
b_{11} &= 1 + P_z - iS(\theta)e^{-i\phi}(P_x + iP_y) \\
b_{12} &= P_x - iP_y - iS(\theta)e^{-i\phi}(1 - P_z) \\
b_{21} &= iS(\theta)e^{i\phi}(1 + P_z) + P_x + iP_y \\
b_{22} &= 1 - P_z + iS(\theta)e^{i\phi}(P_x - iP_y)
\end{aligned} \tag{4.4.31}$$

Let $P_t$ be the component of the polarization vector $P$ in the $x$–$y$ plane with $P_x$ and $P_y$ as its components. Then $P_x = P_t\cos\phi'$ and $P_y = P_t\sin\phi'$, where $\phi'$ is the angle between $P_t$ and $P_x$. Hence, from (4.4.30) and (4.4.31), we get

$$I(\theta,\phi) = I(\theta)[1 - S(\theta)P_t\sin(\phi - \phi')] \tag{4.4.32}$$

Assuming $P_t$ to be in the direction of the $x$-axis, we have

$$I(\theta,\phi) = I(\theta)[1 - S(\theta)P_t\sin\phi] \tag{4.4.33}$$

The asymmetry parameter is defined by

$$\begin{aligned}
A &= \frac{I(\theta,3\pi/2) - I(\theta,\pi/2)}{I(\theta,3\pi/2) + I(\theta,\pi/2)} \\
&= \frac{I_l(\theta) - I_r(\theta)}{I_l(\theta) + I_r(\theta)}
\end{aligned} \tag{4.4.34}$$

where $I_l(\theta)$ and $I_r(\theta)$ are the DCS for the scattering at an angle $\theta$ to the left and to the right, respectively. From (4.4.33) we get

$$A = S(\theta)P_t \tag{4.4.35}$$

Hence, the Sherman function is also known as the analyzing power or asymmetry function.

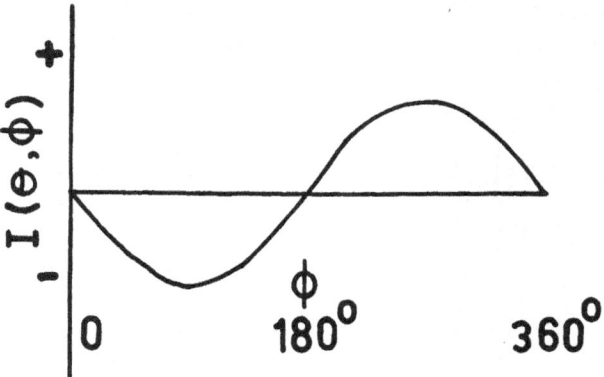

**FIGURE 4.3** The left–right asymmetry due to scattering of polarized electrons by a potential.

It is evident from (4.4.32) that for a given value of $\theta$,

$$I(\theta,\phi) \neq I(\theta,\pi+\phi)$$

This inequality is known as the left-right asymmetry. It arises only if the incident beam is polarized (even partially) and the polarization vector $P$ has a nonzero transverse component $P_t$. If $P$ is in the direction of $k_i$ then $P_t = 0$ and $I(\theta, \phi) = I(\theta)$, in agreement with (4.4.17). For the positive $S(\theta)$ a plot of $I(\theta, \phi)$ with $\phi$ is shown in Fig. 4.3.

Figure 4.3 shows that at $\phi = 0$, $\pi$, and $2\pi$, $I(\theta, \phi) = I(\theta)$. Its maximum and minimum values, at $\phi = 3\pi/2$ and $\pi/2$, are given by $I(\theta)[1 \pm S(\theta)P_t]$, respectively. Thus we see that the Sherman function plays an important role in determining the magnitude of the left-right asymmetry. To put (4.4.33) in a form that is independent of the choice of the coordinate system, we take a unit vector $\hat{n}$ perpendicular to the plane of scattering (a plane formed by the vectors $k_i$ and $k_j$). Let the polar angles of $\hat{n}$ be $\theta'$ and $\phi'$. Since $\hat{n} \cdot k_i = 0$ we have $\theta' = \pi/2$, and the polar coordinates of $\hat{n}$ are $(\cos\phi', \sin\phi', 0)$. It is also perpendicular to $k_j$. Hence,

$$\hat{n} \cdot k_j = (\cos\phi',\sin\phi',0) \cdot (\sin\theta\cos\phi,\sin\theta\sin\phi,\cos\theta) = 0 \qquad (4.4.36)$$

Thus we obtain $\phi' = \pi/2 + \phi$ and the polar coordinates of $\hat{n}$ are $(-\sin\phi, \cos\phi, 0)$. Since $P_t = P_x$ we have

$$P \cdot \hat{n} = (P_x, 0, P_z) \cdot (-\sin\phi, \cos\phi, 0) = -P_t \sin\phi \qquad (4.4.37)$$

Thus (4.4.33) reduces to

$$I(\theta,\phi) = I(\theta)[1 + S(\theta)\boldsymbol{P} \cdot \hat{\boldsymbol{n}}] \tag{4.4.38}$$

## 4.5   The Change in the Polarization Vector $\boldsymbol{P}$ Due to Scattering

Let an electron beam having a polarization vector $\boldsymbol{P}$ be scattered by a potential. The polarization vector $\boldsymbol{P'}$ for a unnormalized scattered beam, with the help of (4.3.18), is given by

$$\boldsymbol{P'} = \mathrm{tr}(\rho'\boldsymbol{\sigma})/\mathrm{tr}\,\rho' = \mathrm{tr}(S\rho S^\dagger \boldsymbol{\sigma})/\mathrm{tr}(S\rho S^\dagger) \tag{4.5.1}$$

Using (4.3.19), we get

$$\boldsymbol{P'} = \mathrm{tr}[S(1 + \boldsymbol{P} \cdot \boldsymbol{\sigma})S^\dagger \boldsymbol{\sigma}]/\mathrm{tr}\{S(1 + \boldsymbol{P} \cdot \boldsymbol{\sigma})S^\dagger]$$

$$= \frac{\mathrm{tr}\left[\begin{pmatrix} b_{11} & b_{12} \\ b_{21} & b_{22} \end{pmatrix}\begin{pmatrix} \hat{\boldsymbol{e}}_z & \hat{\boldsymbol{e}}_x - i\hat{\boldsymbol{e}}_y \\ \hat{\boldsymbol{e}}_x + i\hat{\boldsymbol{e}}_y & -\hat{\boldsymbol{e}}_z \end{pmatrix}\right]}{b_{11} + b_{22}} \tag{4.5.2}$$

where $b_{ij}$ are given by (4.4.31) and $\hat{\boldsymbol{e}}_x$ etc. are unit vectors. Hence,

$$\boldsymbol{P'} = \left[\frac{b_{11}\hat{\boldsymbol{e}}_z + b_{12}(\hat{\boldsymbol{e}}_x + i\hat{\boldsymbol{e}}_y) + b_{21}(\hat{\boldsymbol{e}}_x - i\hat{\boldsymbol{e}}_y) - b_{22}\hat{\boldsymbol{e}}_z}{b_{11} + b_{22}}\right] \tag{4.5.3}$$

The above equation simplifies to

$$\boldsymbol{P'} = \frac{[\boldsymbol{P} \cdot \hat{\boldsymbol{n}} + S(\theta)]\hat{\boldsymbol{n}} + T(\theta)[\boldsymbol{P} - (\boldsymbol{P} \cdot \hat{\boldsymbol{n}})\hat{\boldsymbol{n}}] + U(\theta)(\hat{\boldsymbol{n}} \times \boldsymbol{P})}{1 + \boldsymbol{P} \cdot \hat{\boldsymbol{n}}S(\theta)} \tag{4.5.4}$$

where $S(\theta)$ is the Sherman function and

$$T(\theta) = \frac{|f|^2 - |g|^2}{|f|^2 + |g|^2} \quad \text{and} \quad U(\theta) = \frac{fg^* + f^*g}{|f|^2 + |g|^2} \tag{4.5.5}$$

Hence, $T(\theta) \leq 1$ and $T^2 + U^2 + S^2 = 1$. For $g(\theta) = 0$ we have $S(\theta) = U(\theta) = 0$, $T(\theta) = 1$ and $\boldsymbol{P'} = \boldsymbol{P}$. Thus the change in the polarization vector is due to the spin-orbit interaction. If the initial beam is unpolarized, then $\boldsymbol{P} = 0$ and

$$\boldsymbol{P'} = S(\theta)\hat{\boldsymbol{n}} \tag{4.5.6}$$

Therefore an initially unpolarized beam gets partially polarized due to scattering, and the polarization vector $P'$ is perpendicular to the plane of scattering. The magnitude of $P'$ is equal to the Sherman function $S(\theta)$. Thus in the scattering of a polarized beam the Sherman function determines the extent of the left-right asymmetry, and in the scattering of an unpolarized beam, it gives the degree of polarization of the scattered beam. Hence,

$$S(\theta) = \frac{I\uparrow(\theta) - I\downarrow(\theta)}{I\uparrow(\theta) + I\downarrow(\theta)} = \frac{N\uparrow - N\downarrow}{N\uparrow + N\downarrow}$$

Thus the Sherman function is also equal to the difference in the fractions of the spin-up and spin-down electrons. At the critical angle $\theta_c$ (see Sec. 3.13) differential cross section $I_{un} = I\uparrow + I\downarrow$ has a deep and sharp minimum. Near this region, but on the opposite side of $\theta_c$, $I\uparrow$ and $I\downarrow$ also have deep and sharp minima. Thus the scattered electrons are highly upward polarized at the scattering angle where $I\downarrow$ is minimum. Similarly, at the scattering angle where $I\uparrow$ has its minimum, the scattered electrons are highly downward polarized. Thus $S(\theta)$ changes its sign in that region.

For a better physical understanding let us take

$$P = P_n + P_p \tag{4.5.7}$$

where $P_n$ is along $\hat{n}$ and $P_p$ is in the plane of scattering. Using (4.5.7) in (4.5.4) gives

$$P' = \frac{[P_n + S(\theta)]\hat{n} + T(\theta)P_p + U(\theta)(\hat{n} \times P_p)}{1 + P_n S(\theta)} \tag{4.5.8}$$

A comparison of (4.5.7) and (4.5.8) shows that $P$ has only two components $P_n$ and $P_p$ but $P'$ has three components. The component $P_n$ changes to

$$P'_n = \frac{P_n + S(\theta)}{1 + P_n S(\theta)} \hat{n} \tag{4.5.9}$$

The component $P_p$ reduces to

$$P'_p = \frac{T(\theta)P_p}{1 + P_n S(\theta)} \tag{4.5.10}$$

and a new component, perpendicular to the directions of $\hat{n}$ and $P_p$, is produced, whose magnitude is equal to $U(\theta)|P_p|/[1 + P_n S(\theta)] = P_S$, say, as shown in Fig. 4.4.

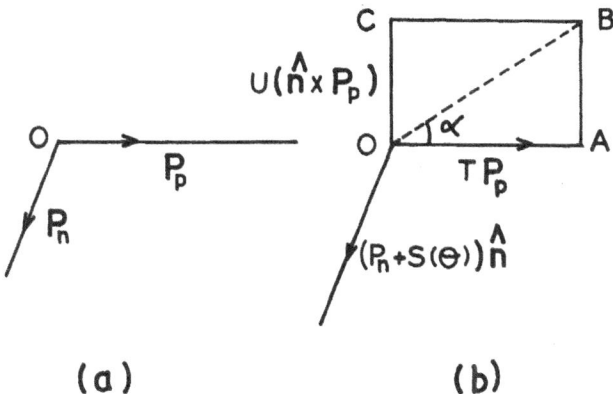

**(a)**            **(b)**

**FIGURE 4.4** Scattering of the partially polarized electrons by a potential: (a) $P_p$ and $P_n$ are the components of the initial polarization vector $P$ in the scattering plane OABC and perpendicular to the scattering plane, respectively. (b) The numerators of the three components of $P'$ are $TP_p$, $P_n + S(\theta)\,\hat{n}$, and $U(\hat{n} \times P_p)$. The denominator (not shown) of all the three terms is $(1 + P_n S)$.

We find that $P_p$ not only changes in magnitude but is also rotated by an angle $\alpha$ in the plane of the scattering such that $\tan \alpha = U(\theta)/T(\theta)$. Since $\hat{n}$, $P_p$ and $(\hat{n} \times P_p)$ are mutually perpendicular we have

$$|P'|^2 = \frac{[P_n + S(\theta)]^2 + [T^2(\theta) + U^2(\theta)]P_p^2}{[1 + P_n S(\theta)]^2} \tag{4.5.11}$$

If the initial beam is 100% polarized then $|P| = 1$ and $P_p^2 = 1 - P_n^2$. Then from (4.5.11)

$$P'^2 = \frac{P_n^2 + 2P_n S(\theta) + S^2(\theta) + (1 - P_n^2)[T^2(\theta) + U^2(\theta)]}{[1 + P_n S(\theta)]^2} \tag{4.5.12}$$

Using the relation $S^2 + T^2 + U^2 = 1$ we get

$$|P'|^2 = 1 \tag{4.5.13}$$

Hence, there is no change in the magnitude of the polarization vector but the directions of $P$ and $P'$ are different. However, for $P_p = 0$, the incident beam has only a transverse component, and (4.5.9) and (4.5.10) yield

$$P' = \frac{[P_n + S(\theta)]\hat{n}}{1 + P_n S(\theta)} \qquad (4.5.14)$$

In this case the directions of $P$ and $P'$ are the same but the magnitudes are different.

The spin-flip also takes place in electron–atom collisions. If we confine ourselves to spinless atoms and consider only elastic scattering, spin-flip is due solely to the spin–orbit interaction. In a perfect experiment one would like to determine module $|f|$ and $|g|$ and phases $\gamma_1$ and $\gamma_2$ of the scattering amplitudes $f$ and $g$. Thus we require a set of four observables, which is provided by $I(\theta)$, $S(\theta)$, $T(\theta)$, and $U(\theta)$. However, due to the relation $T^2 + U^2 + S^2 = 1$, the four observables are not independent of each other. Thus we can determine only three quantities namely $|f|$, $|g|$, and $\phi_{rel} = \gamma_1 - \gamma_2$. This is consistent with a concept of quantum mechanics, according to which an analysis of the scattered wave cannot determine the absolute values of the phases. From (4.4.28) and (4.5.5), $S(\theta)$ and $U(\theta)$ are proportional to $\sin\phi_{rel}$ and $\cos\phi_{rel}$, respectively, so for unambiguous values of $\phi_{rel}$ both $S(\theta)$ and $U(\theta)$ are required.

## 4.6 Measurement of the Sherman Function

To measure $S(\theta)$, Eqs. (4.4.38) and (4.5.1) are utilized and a double scattering experiment is performed. A monoenergetic beam of unpolarized electrons of energy $E$ is scattered by the material whose Sherman function is to be measured. The scattered beam traveling in the direction $(\theta_1, \phi_1)$ is allowed to be scattered for a second time by a second specimen identical to the first, and the differential cross section in the directions $(\theta_2, \phi_2)$ and $(\theta_2, \phi_2 + \pi)$ are measured. By the first scattering, the scattered beam gets partially polarized and the polarization vector is given by $P = S(\theta_1)\hat{n}_1$. The differential cross section for the scattered beam in the second scattering is given by

$$I(\theta_2, \phi_2) = I(\theta_2)[1 + S(\theta_2)S(\theta_1)\hat{n}_1 \cdot \hat{n}_2]$$

Now for $\phi_1 = 0$ we get

$$I(\theta_2, \phi_2) = I(\theta_2)[1 + S(\theta_2)S(\theta_1)\cos\phi_2]$$

Hence, for $\phi_2 = 0$ and $\phi_2 = \pi$ we have

$$\frac{I(\theta_2, 0) - I(\theta_2, \pi)}{I(\theta_2, 0) + I(\theta_2, \pi)} = S(\theta_2)S(\theta_1) \qquad (4.6.1)$$

For $\theta_2 = \theta_1 = \theta$, say, we obtain

$$\frac{I(\theta,0)-I(\theta,\pi)}{I(\theta,0)+I(\theta,\pi)} = S^2(\theta) \qquad (4.6.2)$$

Thus measuring $I(\theta, 0)$ and $I(\theta, \pi)$ of an electron beam, that is first converted into a partially polarized beam, we can determine the value of the Sherman function $S(\theta)$ of the chosen material. Usually to measure the left-right asymmetry [i.e., $I(\theta, \phi)$ and $I(\theta, \pi)$], a Mott detector is employed. Its Sherman function $S(120°)$ at $E = 120\,\text{keV}$ and $\theta = 120°$ has been determined with great accuracy. To use this detector, the beam obtained by the first scattering at $\theta_1$ is accelerated to $E = 120\,\text{keV}$ and allowed to fall on the Mott detector. Then $I(120°, 0°)$ and $I(120°, 180°)$ are measured. For these measurements (4.6.1) reduces to

$$\frac{I(120°,0°)-I(120°,180°)}{I(120°,0°)+I(120°,180°)} = S(\theta_1)\cdot S(120°) \qquad (4.6.3)$$

Since $S(120°)$ is already known $S(E, \theta_1)$ is evaluated. A change in $E$ and $\theta_1$, yields values of $S(E, \theta_1)$ at different values of $E$ and $\theta_1$. The Mott detector also measures the degree of polarization of a partially polarized beam with the help of (4.4.38). Since this equation contains only $P_l$, before scattering, the polarization vector $P$ is to be rotated in the direction of $\hat{n}$ (Kesseler, 1985). These detectors have been calibrated with an accuracy of 0.3% and are capable of detecting polarization as low as $10^{-3}$ (Mayer, 1995).

If $P$ of the incident electron beam is known, then (4.4.38) can be utilized to determine $S(\theta)$ of a given material. $I(\theta, \phi)$ is measured when $P$ and $\hat{n}$ are parallel to each other. Then the direction of $P$ (or $\hat{n}$) is reversed so that $P$ and $\hat{n}$ are antiparallel. The DCS is again measured. If we denote these cross sections by $I(\uparrow)$ and $I(\downarrow)$, from (4.4.38) it is easy to obtain

$$A = \frac{I(\uparrow)-I(\downarrow)}{I(\uparrow)+I(\uparrow)} = PS \qquad (4.6.4)$$

and since $P$ is already known, we now have the value of the Sherman function. Usually this method is more accurate than the double scattering method described earlier.

## Questions and Problems

4.1 Take electronic charge $e$ in the unit of $(\text{Joule}\cdot\text{meter})^{1/2}$ and calculate the value of the Bohr magneton in the unit that contains Joules and meters.

4.2 An electron having $l$ as its orbital angular momentum quantum number moves in a central potential and its potential energy is given by

$$V(r) = -e^2 \left( \frac{1}{r} + \frac{Z}{a_0} \right) \exp(-2Zr/a_0)$$

where $r$ is the distance of the electron from the center of the force, $e$ is electronic charge, and $Z$ and $a_0$ are constant. Show that the electric and magnetic fields acting on the electron are

$$\mathbf{E} = -e \frac{\mathbf{r}}{r} \exp\left( -\frac{2Zr}{a_0} \right) \left( \frac{1}{r^2} + \frac{2Z}{a_0 r} + \frac{2Z^2}{a_0^2} \right)$$

$$\mathbf{B} = \frac{e}{mc} \frac{1}{r} \exp\left( -\frac{2Zr}{a_0} \right) \left( \frac{1}{r^2} + \frac{2Z}{a_0 r} + \frac{2Z^2}{a_0^2} \right) \mathbf{L}$$

where $\mathbf{L}$ is the orbital angular momentum of the electron and $c$ is the velocity of light.

4.3  In the above problem obtain both values of the spin–orbit interaction energy $V_{so}$, corresponding to $j = l \pm \frac{1}{2}$, and show that for a $p$ electron ($l = 1$) one value is double the other but has the opposite sign.

4.4  In an ensemble of $6 \times 10^3$ electrons, 70% of them are polarized upward and rest are unpolarized. A measurement is made for the electrons polarized downward. What will be their number?

4.5  An electron beam is fully polarized in the $x$ direction. Show that the spin wave function of the electrons is equal to the linear superposition of the spin wave functions polarized in $+z$ and $-z$ directions having equal amplitude. Use the density matrix method to verify that $P_y = P_z = 0$ and $P_x = 1$.

4.6  An electron beam polarized in the $z$ direction is mixed incoherently with another beam polarized in the $y$ direction. If the intensity of the former beam is double that of the latter, obtain the density matrix $\rho$ of the mixed beam and show that

$$\mathrm{tr}\rho^2 < \mathrm{tr}\rho$$

Express $\rho$ in the digonalized form as well.

4.7 Show that $Y_{l0}\alpha$ is not an eigenfunction of the operator $J^2$ but that

$$\left(\frac{l+1}{2l+1}\right)^{1/2} Y_{l_0}\alpha + \left(\frac{l}{2l+1}\right)^{1/2} Y_{l_1}\beta$$

is an eigenfunction and that its eigenvalue is $j(j + 1)\hbar^2$.

4.8 Give a physical explanation for the reason that the $l = 0$ partial wave does not contribute to the spin-flip scattering amplitude $g$ and that the maximum contribution to $g$ usually comes from the $l = 1$ partial wave.

4.9 Verify Eq. (4.5.4).

4.10 A 60% polarized beam in the $y$ direction is traveling in the $z$ direction. It is scattered by a potential and the differential cross section $I(\theta, \phi)$ is measured in the $x$–$z$ direction. Now the polarization vector $P$ of the incident beam is reversed and the differential cross section is measured again in the same direction. If the latter cross section is 20% more than the former, calculate the value of the Sherman function.

# Collision between Two Particles

## 5.1   Introduction

In the first chapter we briefly considered a collision between two particles
$A$ and $B$ and obtained a relationship between the differential cross sections in the
center-of-mass frame and in the laboratory frame. However, to obtain the scattering amplitude we have to know the nature of the interaction between the two
particles. In this chapter we obtain the scattering amplitudes for collisions
between an incident particle $A$ and a target $B$ under the following different conditions: (1) $A$ and $B$ are distinguishable from each other. (2) $A$ and $B$ are identical but follow classical mechanics; hence, they can be distinguished by their
trajectories. (3) $A$ and $B$ are identical and are either bosons (follow Bose–
Einstien statistics) or fermions (follow Fermi–Dirac statistics). For the bosons,
the total wave function (including spins) of the system is symmetric, i.e., the
wave function is unchanged on the exchange of $A$ and $B$. On the other hand, for
the fermions, the total wave function is antisymmetric and changes its sign on
the exchange of $A$ and $B$.

## 5.2   Reduction of the Two-Particle Problem

A collision between $A$ and $B$ is obviously a two-particle problem. However,
if the interaction between $A$ and $B$ depends only upon their relative coordinates
then, like the problem of the hydrogen atom, the present problem can also be
decomposed into two one-body problems.

Under steady state conditions, $\psi(r_A, r_B)$, the space part of the wave
function of the system, satisfies the following time-independent Schrödinger
equation:

$$\left[-\frac{\hbar^2}{2m_A}\nabla_A^2 - \frac{\hbar^2}{2m_B}\nabla_B^2 + V(r_A - r_B)\right]\psi(r_A, r_B) = E_T\psi(r_A, r_B) \qquad (5.2.1)$$

where $r_A$ and $r_B$ are the coordinates of $A$ and $B$, respectively, the potential energy has been assumed to depend only upon the relative coordinate $r_A - r_B$, and $E_T$ is the total energy of the system. To decompose (5.2.1) into two one-body Schrödinger equations, we denote the coordinates of the center of mass and the relative coordinates by $R$ and $r$, respectively. Then

$$R = \frac{m_A r_A + m_B r_B}{m_A + m_B} \qquad \text{and} \qquad r = r_A - r_B \qquad (5.2.2)$$

The use of (5.2.2) in (5.2.1), just as was done for the hydrogen atom problem, yields two differential equations. We take

$$\psi(R, r) = \phi(R)\psi(r) \qquad (5.2.3)$$

and put it into (5.2.1) along with (5.2.2). Thus we get

$$-\frac{\hbar^2}{2M}\nabla_R^2\phi(R) = (E_T - E)\phi(R) \qquad (5.2.4)$$

and

$$-\frac{\hbar^2}{2\mu}\nabla^2\psi(r) + V(r)\psi(r) = E\psi(r) \qquad (5.2.5)$$

where $M = m_A + m_B$ and $\mu = m_A m_B/M$ is the reduced mass of the system. It is evident that (5.2.4) describes the motion of a single free particle of mass $M$. Hence, its solution is a plane wave given by

$$\phi(R) = -\frac{1}{(2\pi)^{3/2}}e^{iK \cdot R} \qquad (5.2.6)$$

Equation (5.2.5) describes the motion of a fictitious particle of mass $\mu$ in the center-of-mass frame. Its potential energy is $V(r)$. Thus to study the collision between two particles, we have to solve the one-body equation given by (5.2.5).

Let $A$ and $B$ be elementary particles (we consider a particle to be elementary whose structure in a given physical situation can be ignored) and consider their collisions under the three different conditions mentioned in Sec. 5.1.

## 5.3    Collision between Two Distinguishable Particles

If $V(r)$ is central and falls faster than $1/r^2$ at large values of $r$, then the scattering amplitude $f(\theta_C,\phi_C)$ and the differential cross section $I(\theta_C,\phi_C)$ for the scattering of the particle $A$ can be calculated by using the different integral and differential approaches discussed in Chapter 3. To compare the theory with experiment, (1.6.10) can be utilized to obtain $I(\theta_L,\phi_L)$, the differential cross section in the laboratory frame, from the calculated $I(\theta_C,\phi_C)$.

However, for charged elementary particles having charges $z_1e$ and $z_2e$, respectively, $V(r) = z_1z_2e^2/r$. This interaction is Coulombic, and even at large $r$ $V(r)$ does not fall faster than $1/r^2$. In this case the solution of (5.2.5) is a Coulomb wave. At large values of $r$ the Coulomb wave function is given by

$$\psi_C \underset{|r-z|\to\infty}{\to} A\exp\{i[kz+\gamma\ln(kr)(1-\cos\theta)]\}\{1+\gamma^2/[ikr(1-\cos\theta)]\}$$

$$+ A\frac{f_c(\theta)}{r}\exp\{i[kr-\gamma\ln(2kr)]\}\{1+(1+i\gamma)^2/[ikr(1-\cos\theta)]\} \qquad (5.3.1)$$

where $\gamma = \mu z_1 z_2 e^2/\hbar k$, and

$$f_C(\theta) = -\gamma\exp(2i\sigma_0)\frac{\exp\{-i\gamma\ln[\sin^2(\theta/2)]\}}{2k\sin^2(\theta/2)} \qquad (5.3.2)$$

and $\sigma_0 = \arg\Gamma(1 + i\gamma)$. It should be noted that the above relation does not hold for $\theta = 0$, in which case $|r - z|$ cannot tend to infinity. It is also evident that (5.3.1) does not reduce to (3.1.2). However, as the differences are mainly in the phases, $f_C(\theta)$ is still defined as the scattering amplitude, and the differential cross section in the CM frame is

$$I(\theta_C) = |f_C|^2 = [z_1z_2e^2/4E\sin^2(\theta_C/2)]^2 \qquad (5.3.3)$$

The above equation is identical to the formula derived by Rutherford with the help of classical mechanics, which bears his name. Since $I(\theta_C)$ diverges at $\theta_C = 0$, the integrated cross section is infinite.

## 5.4    Collision between Two Identical Classical Particles

In the CM frame, $A$ and $B$ will always move in opposite directions. According to (5.2.2), an exchange of $A$ and $B$ will change $r$ to $-r$. Let us place

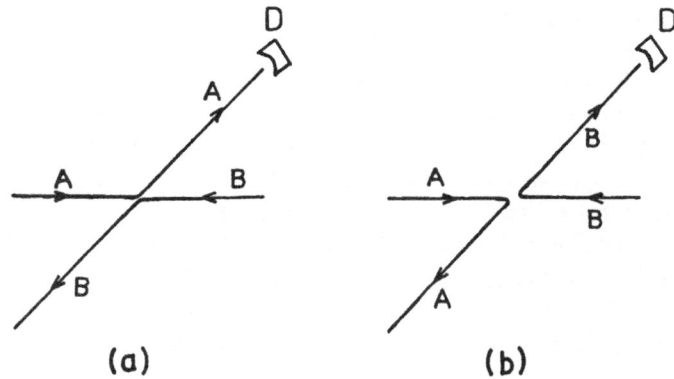

**FIGURE 5.1** Collision of two identical particles A and B in the center-of-mass frame: (a) shows direct scattering, and (b) represents exchange scattering.

a detector $D$ in the CM frame to detect those particles that are scattered in the direction $(\theta_C, \phi_C)$ with respect to the initial direction of $A$, as shown by Fig. 5.1(a).

Now, $B$ will recoil in the opposite direction, i.e., it will move in the direction $(\pi - \theta_C, \pi + \phi_C)$. Hence, the detector will detect particle $B$ when $A$ is scattered in the direction $(\pi - \theta_C, \pi + \phi_C)$. Since $A$ and $B$ are indistinguishable, the detector detects $A$-like particles for two sets of scattering angles for $A$; namely $(\theta_C, \phi_C)$ and $(\pi - \theta_C, \pi + \phi_C)$. However, as $A$ and $B$ are classical particles, they can be distinguished by their trajectories in spite of being identical. Hence, the effective differential cross section will be the sum of the differential cross sections obtained by the direct and exchange processes, represented by Figs. 5.1(a) and (b), respectively. Thus

$$I(\theta_C, \phi_C) = I_d(\theta_C, \phi_C) + I_{ex}(\pi - \theta_C, \pi + \phi_C) \qquad (5.4.1)$$

The first term in the above equation is due to the direct scattering of $A$, whereas the second corresponds to the situation in which $B$ is detected by the detector, i.e., $A$ is exchanged by $B$ and $A$ itself is scattered in the direction $(\pi - \theta_C, \pi + \phi_C)$.

## 5.5   Collision between Two Identical Bosons

As mentioned in Sec. 5.1, the total wave function of a system consisting of bosons is symmetric and so does not change when two of its bosons are exchanged. Let us consider a collision between two spinless identical bosons. Now $m_A = m_B$ and $\psi(r)$ of (5.2.5) denotes the wave function of a particle of mass

$\mu$ in the CM frame. Exchange of two bosons means $r \to -r$. Hence, if $v(r)$ is the wave function of the system, we must have $v(r) = v(-r)$. To achieve this, we take

$$v(r) = \psi(r) + \psi(-r) \qquad (5.5.1)$$

where $\psi(r)$ is the solution of (5.2.5).

Let us assume that the interaction between the two bosons depends only upon $r$, which falls faster than $1/r^2$ at large $r$; then from (3.1.2) and (5.5.1),

$$v(r) \underset{r \to \infty}{\to} A\left\{ e^{ik \cdot r} + e^{-ik \cdot r} + \frac{e^{ikr}}{r} [f(\theta_C, \phi_C) + f(\pi - \theta_C, \pi + \phi_C)] \right\} \qquad (5.5.2)$$

Hence, the differential cross section for the scattering of two identical bosons is given by

$$I(\theta_C, \phi_C) = |f(\theta_C, \phi_C) + f(\pi - \theta_C, \pi + \phi_C)|^2 \qquad (5.5.3)$$

Again, $f(\theta_C, \phi_C)$ is the scattering amplitude for the process when the incident boson is detected by the detector and the other is scattered by $(\pi - \theta_C, \pi + \phi_C)$; while $f(\pi - \theta_C, \pi + \phi_C)$ is the scattering amplitude for the process when the incident boson is scattered by $(\pi - \theta_C, \pi + \phi_C)$ and the target boson enters the detector. Using the notation of (5.4.1), we get

$$I(\theta_C, \phi_C) = I_d(\theta_C, \phi_C) + I_{ex}(\pi - \theta_C, \pi + \phi_C)$$
$$+ 2\text{Re}[f(\theta_C, \phi_C)f^*(\pi - \theta_C, \pi + \phi_C)] \qquad (5.5.4)$$

For central potentials, $I(\theta_C, \phi_C)$ is independent of $\phi_C$ and (5.5.4) reduces to

$$I(\theta_C) = I_d(\theta_C) + I_{ex}(\pi - \theta_C) + 2\text{Re}[f(\theta_C)f^*(\pi - \theta_C)] \qquad (5.5.5)$$

Quite often $f(\pi - \theta_C)$ is denoted by $g(\theta_C)$ and so we also have

$$I(\theta_C) = |f(\theta_C)|^2 + |g\theta_C|^2 + 2\text{Re}[f(\theta_C)g^*(\theta_C)] \qquad (5.5.6)$$

A comparison of (5.4.1) with (5.5.5) shows that the latter equation has an extra term, which arises from the interference of the direct and exchange scattering amplitudes. This term is due to the coherent addition of the direct and exchange scattering amplitudes, whereas in (5.4.1) the amplitudes are added in an incoherent manner. It should be noted that the normalization of $v(r)$ is taken in such a way that if the interference term is neglected (5.5.6) reduces to the classical equation (5.4.1).

For a Coulomb potential equal to $z_1 z_2 e^2/r$, $\psi(r)$ is a Coulomb wave (given by 5.3.1) and $f_c(\theta_c)$ is given by (5.3.2). Hence, from (5.5.6) we obtain

$$I(\theta_c) = \left(z_1 z_2 e^2/4E\right)^2 \left| \frac{\exp\{-2i\gamma \ln[\sin(\theta_c/2)]\}}{\sin^2(\theta_c/2)} + \frac{\exp\{-2i\gamma \ln[\cos(\theta_c/2)]\}}{\cos^2(\theta_c/2)} \right|^2$$

(5.5.7)

or

$$I(\theta_c) = \left(z_1 z_2 e^2/4E\right)^2 \left[ \frac{1}{\sin^4(\theta_c/2)} + \frac{1}{\cos^4(\theta_c/2)} + \frac{2\cos\{2\gamma \ln[\tan(\theta_c/2)]\}}{\sin^2(\theta_c/2)\cos^2(\theta_c/2)} \right]$$

(5.5.8)

The above equation represents the Mott scattering formula for the Coulomb scattering between two identical bosons.

## 5.6   Collision between Two Eelectrons

The collision between two electrons is once again equivalent to the scattering of a particle of mass $\mu$ by a central potential in the center-of-mass frame. Since electrons are fermions with $s = \frac{1}{2}$, the spins $S$ of the particle of mass $\mu$ are 1 and 0, with $M_s = 1$, 0, $-1$ and 0, respectively. The corresponding four spin wave functions are given by:

| $S$ | $M_S$ | Wave function |
|-----|-------|---------------|
| 1 | 1 | $\alpha(A)\alpha(B)$ |
| 1 | 0 | $\frac{1}{\sqrt{2}}[\alpha(A)\beta(B) + \alpha(B)\beta(A)]$ |
| 1 | $-1$ | $\beta(A)\beta(B)$ |
| 0 | 0 | $\frac{1}{\sqrt{2}}[\alpha(A)\beta(B) - \alpha(B)\beta(A)]$    (5.6.1) |

It is easy to verify that all of the above four wave functions are orthonormal and are eigenfunctions of the operators $S^2$ and $S_z$, with eigenvalues $S(S + 1)\hbar^2$ and $M_s\hbar$, respectively.

The first three spin wave functions given by (5.6.1) are symmetric with respect to the exchange of $A$ and $B$. The corresponding triplet space wave function is given by

$$v_t(r) = \psi(r) - \psi(-r) \tag{5.6.2}$$

and the singlet wave function corresponding to $S = 0$ and $Ms = 0$ is given by

$$v_s(r) = \psi(r) + \psi(-r) \tag{5.6.3}$$

For the central potentials, differential cross sections are independent of $\phi$. Hence, as before, the triplet and singlet differential cross sections are given by

$$I_{t,s}(\theta_C) = |f(\theta_C) \mp f(\pi - \theta_C)|^2 \tag{5.6.4}$$

For completely unpolarized beams of electrons the weight factors of the triplet and singlet states are $\frac{3}{4}$ and $\frac{1}{4}$, respectively. Hence,

$$I(\theta_C) = \tfrac{3}{4} I_t(\theta_C) + \tfrac{1}{4} I_s(\theta_C) \tag{5.6.5}$$

or

$$= \tfrac{3}{4}|f(\theta_C) - g(\theta_C)|^2 + \tfrac{1}{4}|f(\theta_C) + g(\theta_C)|^2 \tag{5.6.6}$$

The interaction between the two electrons is Coulombic; hence, as before, $f(\theta_C)$ is given by (5.3.2) with $\gamma = e^2\mu/\hbar^2 k$. Further, $g(\theta_C)$ is given by the same equation but with $\theta_C$ being replaced by $\pi - \theta_C$. Thus we get

$$I_{t,s}(\theta_C) = (e^2/4E)^2 \left[ \frac{1}{\sin^4(\theta_C/2)} + \frac{1}{\cos^4(\theta_C/2)} \mp \frac{2\cos\{2\gamma \ln[\tan(\theta_C/2)]\}}{\sin^2(\theta_C/2)\cos^2(\theta_C/2)} \right] \tag{5.6.7}$$

It is interesting to note that $I(\theta_C)$ as given by (5.5.8) and $I_{t,s}(\theta_C)$ as given by (5.6.7) are symmetric about $\theta_C = \pi/2$. In all three cases the effect of the interference term is most pronounced at $\theta_C = \pi/2$ and becomes more and more noticeable as the value of $\gamma$ increases. Since the scattering angle $\theta_L$ in the laboratory frame is half of $\theta_C$ for $m_A = m_B$, the symmetry in the laboratory frame is at $\theta_L = \pi/4$.

Finally it should be noted that to obtain the total scattering cross section $\sigma$, the DCS given by (5.4.1), (5.5.6), and (5.6.4) is to be integrated over $\theta_C$ from 0 to $\pi$. However, the limits of integration should be from 0 to $\pi/2$ to avoid double counting as they cover the counting of both the projectiles and target particles. Since the DCS diverges at $\theta_C = 0$, $\sigma$ is infinite in all three cases.

## Questions and Problems

[Note: Neglect relativistic effects in all the following problems.]

5.1 $\alpha$-particles having 2 MeV energy are scattered elastically by stationary lead nuclei. If the angle of scattering in the CM frame is 30°, what is the differential cross section in the L frame?

5.2 A proton of 10 keV energy collides elastically with a stationary $\alpha$-particle in the laboratory. The angle of scattering in the CM frame is 60°. Calculate the real and imaginary parts of the scattering amplitude and the differential cross section in the CM frame.

5.3 An $\alpha$-particle of energy 4 keV collides with a stationary $\alpha$-particle and is scattered by 22.5° in the L frame. Calculate the magnitudes of the direct and exchange scatterings in the CM frame and the differential cross section in the L frame.

5.4 Electrons of 5 keV energy collide with stationary electrons in the L frame. For the scattering angle of 60° in the L frame, calculate the differential cross sections in the same frame under following conditions: (a) The projectile and target electrons have same spins. (b) The spins of the projectile electrons are opposite to those of the target electrons. (c) The spins of the projectile and the target electrons are random.

5.5 Consider the elastic scattering between two electrons in the CM frame. Take the initial wave function of the system as

$$\psi_i(r_1, r_2) = \tfrac{1}{\sqrt{2}}\left[\varphi_{k1}(r_1)\varphi_{k2}(r^2) + \varphi_{k1}(r_2)\varphi_{k2}(r_1)\right]$$

where the $\varphi_{ki}(r_i)$ are plane waves. The final wave function is also given by the above equation, with $k_1$ and $k_2$ being replaced by $k_1'$ and $k_2'$, respectively. Obtain the direct and exchange scattering amplitudes in the CM frame and show

$$g(\theta_C) = f_d(\pi - \theta_C)$$

where $\theta_C$ is the angle of scattering in the CM frame.

# Collision of Photons with Atoms

## 6.1 Introduction

A knowledge of the cross sections for photon-induced processes is of importance in a number of fields, including dosimetry, radiation therapy, and health physics; space physics and chemistry; laser physics; environmental protection; fusion; plasmas; radiation-induced decomposition; and electron and X-ray microscopy (Brion, 1985). Photoionization cross sections control the temperature of the solar corona and are needed to determine the rate of ionization in the ionosphere. The existence of the ionic layers in our upper atmosphere is partially due to the interaction of photons with the atmospheric gases. Hence, it is appropriate to devote a chapter to the collision of photons with atomic systems. We shall see later that such a study is also helpful in the discussion of the collision of electrons with atomic systems.

A collision between a photon and an atom can be elastic as well as inelastic. Rayleigh and Thomson scattering are examples of elastic collisions, where the incident and the scattered photons have the same energy. Excitation and ionization of atoms by photons, e.g., Raman scattering are examples of the inelastic collisions. A molecule may also dissociate due to photon impact.

## 6.2 Photons and Electromagnetic Waves

According to the quantum theory of fields, every field is associated with a particle of finite mass and spin. Following the same general features, the quantum mechanical excitation of electromagnetic waves of angular frequency $\omega$ gives rise to photons of energy $E = \hbar\omega$ and momentum $|p| = \hbar\omega/c$ or their integral multiples. The spin of a particle can be defined as the angular momentum it possesses in its rest frame. However, for a photon, the relativistic relation

$$E^2 = m^2 c^4 + p^2 c^2 \tag{6.2.1}$$

and the relations between $E$, $|p|$, and $\omega$, given above, show that $m$, the rest mass of a photon, is zero. As a matter of fact there is no frame in which a photon is at rest. In all frames, in vacuum, a photon moves with the velocity of light $c$. Thus, the spin of a photon needs a different definition, and we shall come back to it.

To develop a quantum mechanical theory for collisions between photons and atomic systems, Maxwell's equations have to be quantized, but, we shall continue to use them. In most situations they are not a bad approximation because even for weak electromagnetic (EM) fields of wavelength $\lambda$ the number of photons in a volume $\lambda^3$ is very large. Hence, the number can be treated as a continuous variable. Under such a condition a semiclassical theory, in which the EM field is described by Maxwell's equations and the atomic system is treated quantum mechanically, should be adequate. In such a theory the EM field disturbs the atomic system but it is assumed that the latter, even by its emission or absorption of a photon, does not disturb the field. Obviously, this assumption is valid when there are a large number of photons in the field, so the stimulated emission and absorption of photons can be successfully described by semiclassical theory. However, spontaneous emission takes place even in vacuum (no photons in the field). In such a situation it is incorrect to neglect the disturbance of the field by the atomic system. Nevertheless, it is possible to obtain an expression for the transition rate for spontaneous emission using the transition rates for absorption and stimulated emission and Planck's formula for blackbody radiation without resorting to a full quantum treatment.

## 6.3   The Electromagnetic Field in Free Space

Maxwell's equations for the EM field are given in terms of the electric field $E$ and the magnetic field $B$. The vectors $E$ and $B$ are perpendicular to each other and also to the momentum vector $k$. With proper gauge transformations, $E$ and $B$, in free space, are expressed in terms of a vector potential $A$ by the following equations (Schiff, 1968):

$$B = \nabla \times A \qquad E = -\frac{1}{c}\frac{\partial A}{\partial t} \tag{6.3.1}$$

The vector potential $A$ satisfies

$$\nabla^2 A - \frac{\partial^2 A / \partial t^2}{c^2} = 0 \tag{6.3.2}$$

and obeys the transversality condition

$$\nabla \cdot A = 0 \tag{6.3.3}$$

For a linearly polarized monochromatic plane wave, the solution of (6.3.2) is given by

$$A = A_0 \varepsilon e^{i(k \cdot r - \omega t)} + A_0^* \varepsilon e^{-i(k \cdot r - \omega t)} \tag{6.3.4}$$

where $\varepsilon$ is the polarization unit vector and is perpendicular to the direction of propagation. Since $k = -i\nabla$, the solution (6.3.4) automatically satisfies (6.3.3). For such a wave, the energy flux (intensity) is given by

$$F = \frac{\omega^2 |A_0|^2}{2\pi c} \tag{6.3.5}$$

## 6.4 Excitation and De-Excitation of Atoms Due to the Electromagnetic Field

Let us consider a one-electron atom. In a stationary state it satisfies the following time-independent Schrödinger equation:

$$H_0 v_0 = \left[ \frac{p^2}{2m} + V(r) \right] v_0 = \varepsilon_0 v_0 \tag{6.4.1}$$

where $v_0$ and $\varepsilon_0$ are the eigenfunction and eigenenergy of the Hamiltonian $H_0$, respectively. At time $t = 0$, we put this atom under the EM field represented by (6.3.4). This field causes the Hamiltonian to change to

$$H = \frac{1}{2m}(p - eA/c)^2 + V(r)$$

$$= H_0 - \frac{e}{mc} A \cdot p + \frac{e^2}{2mc^2} |A|^2 + V(r) \tag{6.4.2}$$

where we have used the fact that the operator $p$ acts on every object on its right and from (6.3.3) $(p \cdot A) = 0$. Thus the Hamiltonian, given by (6.4.2), contains new time-dependent terms, which perturb the atom. This perturbation gives rise to the possibility that the atom makes a transition from its initial state $v_0$ to a new state, say, $v_q$. Up to first order the perturbation is

$$H' = -\frac{e}{mc} A \cdot p \tag{6.4.3}$$

According to first-order time-dependent perturbation theory, the transition probability amplitude $a_q(t)$ for finding the atom in the stationary state $\nu_q$ after a time $t$ is given by (Schiff, 1968)

$$a_q(t) = \frac{1}{i\hbar} \int_0^t \langle q|H'(t')|0\rangle > e^{i\omega_{q0}t'} dt' \tag{6.4.4}$$

where $\omega_{q0} = (\varepsilon_q - \varepsilon_0)/\hbar$. Using (6.4.3) and (6.4.4) and integrating over $t'$ we get

$$a_q(t) = \frac{1}{\hbar}\left( \langle q|Y|0\rangle \frac{1 - e^{i(\omega_{q0}-\omega)t}}{\omega_{q0} - \omega} + \langle q|Y^\dagger|0\rangle \frac{1 - e^{-i(\omega_{q0}+\omega)t}}{\omega_{q0} + \omega} \right) \tag{6.4.5}$$

with

$$Y = \frac{e}{mc} A_0 e^{ik\cdot r} \varepsilon \cdot p \tag{6.4.6}$$

The above equation shows that the probability of finding the atom in the state $|q\rangle$ is appreciable only when the denominator in one of the two terms of (6.4.5) is practically equal to zero. The first term dominates when $\varepsilon_q - \varepsilon_0 \approx \hbar\omega$ and represents the absorption of a quantum $\hbar\omega$ by an atom from the field by which the atom makes a transition from the lower state $|0\rangle$ to an excited state $|q\rangle$. Similarly, the second term is of importance when $\varepsilon_q - \varepsilon_0 \approx -\hbar\omega$ and it represents a process in which the atom makes a transition from an excited state $|0\rangle$ to a lower state $|q\rangle$ by emitting a photon, due to the presence of other similar photons (stimulated emission). There is no interference between the two terms, and the two processes (absorption and stimulated emission) can be treated independently.

Let us concentrate on the excitation of an atom due to absorption of a photon from the field. Equation (6.4.5) shows that the excitation probability is given by

$$|a_q^{ex}(t)|^2 = \frac{4}{\hbar^2}|Y_{q0}|^2 \frac{\sin^2[(\omega_{q0} - \omega)t/2]}{(\omega_{q0} - \omega)^2} \tag{6.4.7}$$

where $Y_{q0} = \langle q|Y|0\rangle$. Now if we plot $\sin^2[(\omega_{q0} - \omega)t/2]/[(\omega_{q0} - \omega)t/2]^2$ as a function of $(\omega_{q0} - \omega)$ we get the curve shown in the Fig. 6.1. The maximum value of $\sin^2[(\omega_{q0} - \omega)t/2]/[(\omega_{q0} - \omega)]^2$ is $t^2/4$ at $\omega_{q0} = \omega$ and it goes to zero at $\omega_{q0} - \omega =$

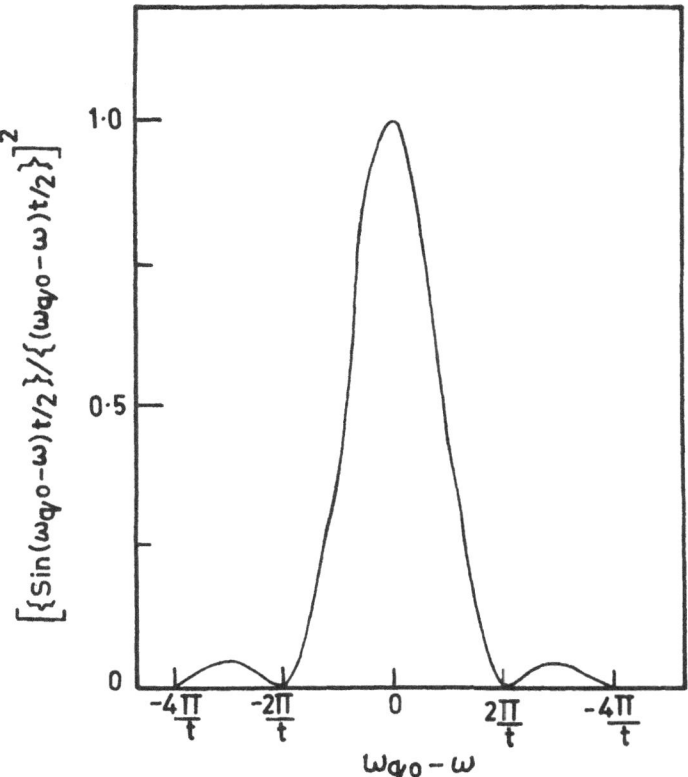

**FIGURE 6.1** Variation of $[\sin(\omega_{q0} - \omega)t/2]/[(\omega_{q0} - \omega)t/2]^2$ with $\omega_{q0} - \omega$.

$\pm 2\pi/t$. Thus the area of the main loop is proportional to $t$, and so the transition probability when $\omega_{q0} - \omega$ ranges from $-2\pi/t$ to $+2\pi/t$ is also proportional to $t$. Hence, the transition probability per unit time $W_{q0}^{ex}$ will be independent of time. Let us assume that $t$ is quite large; then from the definition of the Dirac delta function

$$\lim_{t \to \infty} \frac{\sin^2[(\omega_{q0} - \omega)t/2]}{(\omega_{q0} - \omega)^2} = \frac{\pi t}{2} \delta(\omega_{q0} - \omega) \qquad (6.4.8)$$

Hence,

$$W_{q0}^{ex} = \frac{1}{t}|a_{q0}^{ex}|^2 = \frac{2\pi}{\hbar^2}|Y_{q0}|^2 \delta(\omega_{q0} - \omega) \qquad (6.4.9)$$

In this equation either $\omega_{q0}$ or $\omega$ can be treated as a variable. We shall consider both cases. Let us first take $\omega_{q0}$ as a variable by assuming that the final state is made up of a group of states having eigenenergies very close to $\varepsilon_q$ and that these eigenenergies vary in a continuous manner. In other words, the final state $|q\rangle$ is assumed to have a width. This assumption is quite appropriate because of the uncertainty principle and the broadening of the spectral lines due to temperature, pressure, and collisions. Let $\rho(\varepsilon_q)d\varepsilon_q$ represent the number of the final states with $\rho(\varepsilon_q)$ as the number of states per unit energy range. Then, for the whole group,

$$W_{(q)0}^{\text{ex}} = \int W_{(q)0}^{\text{ex}} \rho(\varepsilon_q)d\varepsilon_q$$

$$= \frac{2\pi}{\hbar} \int |Y_{q0}|^2 \rho(\varepsilon_q)\delta(\omega_{q0} - \omega)d(\omega_{q0}) \tag{6.4.10}$$

Neglecting the variation of $Y_{q0}$ with $\varepsilon_q$ within the group, we obtain

$$W_{(q)0}^{\text{ex}} = \frac{2\pi}{\hbar} |Y_{q0}|^2 \rho(\varepsilon_q) \tag{6.4.11}$$

The above relation has been found so useful that it is known as the Fermi's Golden Rule 2.

Let us now consider the second case, in which $\omega$ is treated as a variable. We assume that the EM wave is not strictly monochromatic but has a width $d\omega$. Then, from (6.3.5),

$$|A_0|^2 = \frac{2\pi c}{\omega^2} I(\omega)d\omega \tag{6.4.12}$$

where $I(\omega)$ is the intensity of the wave per unit frequency range. Using (6.4.12) in (6.4.9), we get

$$dW_{q0}^{\text{ex}} = \frac{4\pi^2 c}{\hbar^2 \omega^2} |X_{q0}|^2 I(\omega)\delta(\omega_{q0} - \omega)d\omega \tag{6.4.13}$$

where $X = Y/A_0$. Integration over $\omega$ yields

$$W_{q0}^{\text{ex}} = \frac{4\pi^2 c}{\hbar^2 \omega_{q0}^2} |X_{q0}|^2 I(\omega_{q0}) \tag{6.4.14}$$

The transition probability per unit time per atom divided by the incident photon flux gives the photo cross section per photon. Hence, the excitation (absorption) cross section is given by

$$\sigma_{q0}^{ex} = \frac{W_{q0}^{ex} \hbar \omega}{F} \qquad (6.4.15)$$

where $F$ is the energy flux of the EM field and the energy of each photon is $\hbar \omega$. Using (6.3.5) for $F$, the relationship $Y = XA_0$ and (6.4.11) for $W_{q0}^{ex}$, we obtain

$$\sigma_{q0}^{ex} = \frac{4\pi^2 c}{\omega_{q0}} |X_{q0}|^2 \rho(\varepsilon_q) \qquad (6.4.16)$$

## 6.5 The Electric Dipole Approximation

Since the interaction between the EM field and the atom takes place over the area of the atom, the interaction length is of the order of the radius of the atom. Furthermore, for excitation or de-excitation to take place, the energy of the photons should be equal to the energy spacing between the two corresponding atomic levels. Hence, between the $n$ and $m$ levels of a hydrogenic atom we have

$$\hbar \omega = \frac{Z^2 e^2}{2a_0} \left( \frac{1}{n^2} - \frac{1}{m^2} \right)$$

Thus $\hbar \omega \approx Z^2 e^2 / a_0$ and $k \approx Z^2 e^2 / \hbar c a_0$. The radius of the atom can be approximated by $a_0 / Z$; hence $|k \cdot r| \approx Z\alpha$, where the fine-structure constant $\alpha = (e^2 / \hbar c)$ is equal to $\frac{1}{137}$. Now

$$e^{ik \cdot r} = 1 + ik \cdot r + \frac{(ik \cdot r)^2}{2!} + \cdots \qquad (6.5.1)$$

Since $|k \cdot r|$ is quite small for light atoms (small $Z$), for most cases we may replace $e^{ik \cdot r}$ occurring in $X_{q0}$ by its first term (dipole term), i.e., by unity. Hence in the dipole approximation, (6.4.14) reduces to

$$W_{q0}^{ex} = \frac{4\pi^2 e^2}{(\hbar \omega_{q0} m)^2 c} |\varepsilon \cdot \langle q|p|0 \rangle|^2 I(\omega_{q0}) \qquad (6.5.2)$$

The matrix element $(p)_{q0}$ involves a momentum (or velocity) operator and, is thus known as the velocity form of the matrix element. However, it is more convenient to use the length form of the matrix element given by $(r)_{q0}$. For the atomic system whose eigenfunctions are known exactly(such as a hydrogenic atom), an exact conversion of $(p)_{q0}$ into $(r)_{q0}$ is possible. To achieve this let us consider

$$[r, H_0] = [r, p^2/2m + V(r)] = \frac{1}{2m}[r, p^2] = i\hbar \frac{p}{m}$$

Hence

$$\langle q|[r, H_0]|0\rangle = \frac{i\hbar}{m}\langle q|p|0\rangle \tag{6.5.3}$$

Furthermore,

$$\langle q|rH_0 - H_0 r|0\rangle = (\varepsilon_0 - \varepsilon_q)\langle q|r|0\rangle \tag{6.5.4}$$

Equating (6.5.3) to (6.5.4), we get

$$\langle q|p|0\rangle = im\omega_{q0}\langle q|r|0\rangle \tag{6.5.5}$$

Hence, in the dipole approximation

$$W_{q0}^{ex} = \frac{4\pi^2 e^2}{\hbar^2 c}|\varepsilon\cdot\langle q|r|0\rangle|^2 I(\omega_{q0}) \tag{6.5.6}$$

For a given polarization direction the vector $\langle q|r|0\rangle$ makes all possible angles with the polarization vector $\varepsilon$. Hence, taking all the directions into consideration, we have to obtain an average value of $W_{q0}^{ex}$. Let $\mu$ be the cosine of the angle between $\varepsilon$ and $\langle q|r|0\rangle$. Now, the average value of $\mu^2$ is $\frac{1}{3}$. Finally,

$$W_{q0}^{ex} = \frac{4\pi^2 e^2}{3\hbar^2 c}I(\omega_{q0})|\langle q|r|0\rangle|^2 \tag{6.5.7}$$

$$= \frac{4\pi^2\alpha}{\hbar}I(\omega_{q0})|x_{q0}|^2 \tag{6.5.8}$$

because $|r_{q0}|^2 = 3|x_{q0}|^2$. Similarly, it is easy to show that the excitation (absorption) cross section in the dipole approximation is given by

$$\sigma_{(q)0}^{ex} = 4\pi^2\hbar\alpha\omega_{q0}|x_{q0}|^2 \rho(\varepsilon_q) \tag{6.5.9}$$

## 6.6   The Einstein B and A Coefficients

The Einstein $B$ coefficient is obtained by dividing the transition rate given by (6.5.7) by the energy density of the radiation per unit angular frequency $u(\omega)$ $= I(\omega)/c$. Hence, $B_{q0}$ for a transition from $|0\rangle$ to $|q\rangle$ is given by

$$B_{q0} = \frac{4\pi^2 e^2}{3\hbar^2} |\langle q|r|0\rangle|^2 \qquad (6.6.1)$$

It is evident that $B_{0q}$, the Einstein $B$ coefficient for stimulated emission, is equal to that for absorption. However, if $|0\rangle$ and $|q\rangle$ states are $g_0$- and $g_q$-fold degenerate, then $g_0 B_{q0} = g_q B_{0q}$.

To obtain an expression for the Einstein $A$ coefficient (transition rate for spontaneous emission) we take an ensemble of atoms in statistical equilibrium at a temperature $T$ and consider transitions between two quantum states $|0\rangle$ and $|q\rangle$. Let $N_0$ and $N_q$ be the number of atoms in the $|0\rangle$ and $|q\rangle$ states, respectively. Equating the number of atoms going from $|0\rangle$ to $|q\rangle$ per unit time by absorbing radiation to the number of atoms making the transition from $|q\rangle$ to $|0\rangle$ per unit time by stimulated and spontaneous radiation we get

$$N_0 B_{q0} u(\omega) = N_q [B_{0q} u(\omega) + A_{0q}] \qquad (6.6.2)$$

Since $N_q/N_0 = g_q/g_0 \exp(-\hbar\omega_{q0}/kT)$, where $k$ is Bolzmann's constant and

$$B_{q0}/B_{0q} = g_q/g_0$$

We get from (6.6.2)

$$A_{0q} = \frac{g_0}{g_q}(e^{\hbar\omega/kT} - 1)u(\omega)B_{q0} \qquad (6.6.3)$$

Further, according to the Planck's blackbody formula, the energy density of radiation per unit angular frequency at thermal equilibrium is given by

$$u(\omega) = \frac{\hbar\omega^3}{\pi^2 c^3 [\exp(\hbar\omega/kT) - 1]} \qquad (6.6.4)$$

Hence, putting (6.6.4) into (6.6.3) at $\omega = \omega_{q0}$, we obtain

$$A_{0q} = \frac{\hbar\omega_{q0}^3}{\pi^2 c^3} B_{q0} \frac{g_0}{g_q} = \frac{\hbar\omega_{q0}^3}{\pi^2 c^3} B_{0q} \qquad (6.6.5)$$

It is interesting to see that the ratio $A/B$ is independent of $e$, $m$, and the matrix element $r_{q0}$. In order that the stimulated emissions may dominate over the spontaneous emissions, $\omega_{q0}$ should be small. Based on this observation the first amplifier *MASER* (microwave amplification by stimulated emission of radiation) used $\omega_{q0}$ in the microwave region. Now with improved techniques we have X-ray lasers and even free-electron lasers. Using (6.6.1) in (6.6.5), we finally obtain

$$A_{0q} = \frac{4e^2\omega_{q0}^3}{3\hbar c^3}|\langle q|r|0\rangle|^2 \frac{g_0}{g_q} \tag{6.6.6}$$

## 6.7   Dipole Selection Rules

It is evident from (6.5.7) and (6.6.6.) that in the dipole approximation an atom will radiate or absorb EM waves only if the matrix element $\langle q|r|0\rangle$ is nonzero. The operator

$$r = (\hat{i}\sin\theta\cos\varphi + \hat{j}\sin\theta\cos\varphi + \hat{k}\cos\theta)r$$

is an odd-parity operator with $l = 1$ and $m_l = 0, \pm 1$. Therefore, for a nonzero value of the matrix element $\langle\psi_{n'l'm'l'}|r|\psi_{nlml}\rangle$ we must have

(1)   $\Delta l = \pm 1$

and

(2)   $\Delta m_l = 0, \pm 1$                                                                    (6.7.1)

which are the dipole selection rules. Rule 1 shows that the initial and final atomic states have to be of opposite parity. This is known as the Laporte rule. Since the operator $r$ has no effect on the spin wave functions, the spin quantum numbers $s$ and $m_s$ remain conserved in the dipole transition. Hence, for the quantum numbers $j$ and $m_j$ we have the following selection rules:

(3)   $\Delta j = 0, \pm 1$

(4)   $\Delta m_j = 0, \pm 1$

For a multielectron atom, (an atom with a number of charged particles, the interactions among which cannot be neglected) the selection rules are based on the quantum numbers $J$, $L$, and $S$ of the whole atom. Quantum electrodynamics shows that the Laporte rule and rule 1 are still valid. In addition, we have

(5)   $\Delta J = 0, \pm 1$ but not from 0 to 0

(6)   $\Delta M = 0, \pm 1$

(7)   $\Delta S = 0$

(8)   $\Delta L = 0, \pm 1$ but not from 0 to 0                                                (6.7.2)

All those transitions that obey the dipole selection rules are said to be allowed transitions, while the rest of them are known as forbidden transitions. The matrix element $\langle q|e^{ik\cdot r}\varepsilon\cdot p|0\rangle$ for forbidden transitions may have nonzero values when higher terms of the expansion of $e^{ik\cdot r}$, given by (6.5.1), are taken into account. For example, a transition from $3^2{}_{D_{3/2}}$ to $1^2{}_{S_{1/2}}$ is a forbidden transition but the matrix element $\langle 3^2{}_{D_{3/2}}|ik\cdot r|1^2{}_{S_{1/2}}\rangle$ is nonzero. Transitions that take place due to the operator $(ik\cdot r)$ are known as electric quadrupole transitions. Similarly, we have transitions due to the higher electric poles as well as to magnetic interactions. They are all examples of forbidden transitions and their intensities are much lower than those of electric dipole transitions.

For those excited states that decay by electric dipole transition, the inverse of the Einstein $A$ coefficient gives their lifetime $\tau$. The excited states, which do not decay by a dipole transition, have a much longer lifetime and are known as metastable states.

A transition from a state $s$ ($l = 0$) to another $s$ state cannot take place even when the full operator $e^{ik\cdot r}\varepsilon\cdot p$ is taken into account. To verify the above statement let us take $\varepsilon$ in the $x$ direction; then the above operator reduces to $e^{i(k_y y+k_z z)}p_x$. The term $e^{i(k_y y+k_z z)}$ is of even parity whereas $p_x$ is of odd parity with respect to a reflection about the $x$-axis. Hence, the integrand is of odd parity and the integration yields zero. Such transitions are said to be strictly forbidden. For example, a transition from $2^2{}_{S_{1/2}}$ to $1^2{}_{S_{1/2}}$ is a strictly forbidden transition. Consequently, the metastable state $2^2{}_{S_{1/2}}$ has a very long lifetime compared to the lifetime of the $2^2{}_{P_{1/2}}$ state. The metastable state $2^2{}_{S_{1/2}}$ decays to the ground state by the second-order perturbation term $|A|^2 e^2/(2mc^2)$, occurring in (6.4.2). Such a transition emits two photons. The angular frequencies $\omega_1$ and $\omega_2$ of the emitted photons satisfy the relation

$$h(\omega_1 + \omega_2) = \varepsilon(2^2{}_{S_{1/2}}) - \varepsilon(1^2{}_{S_{1/2}})$$

Thus a large number of combinations of $\omega_1$ and $\omega_2$ are possible.

## 6.8   Spin and Spin States of Photons

We have noted that the angular momentum of a particle in its rest frame is its spin angular momentum. However, the above definition fails for the photon because it has no rest frame. Hence, we adopt alternative procedures.

According to the dipole transition selection rules, the emission or absorption of a photon by an atomic system changes the orbital angular momentum of the system by one unit. To conserve the angular momentum, the same degree of change must occur in the radiation field. In the dipole approximation, the variation in the EM field vector potential $A$ with direction over the atomic size has been neglected by assuming it to be spherically symmetric. This field remains spherically symmetric even after the emission (or absorption) of a photon. Hence, the orbital angular momentum released by the atomic system becomes the internal (spin) angular momentum of the emitted photon. Thus the spin angular momentum of a photon is $\hbar$. When a photon is absorbed its spin angular momentum of one unit increases the orbital angular momentum of the atomic electron that makes the transition from one atomic orbital to another.

We also note that $\boldsymbol{\varepsilon}$ of (6.3.4) transforms like a vector. Hence, following the general theory of angular momenta we associate one unit of the angular momentum with it. Let us take $\boldsymbol{k}$ along the $z$-axis. Since $\boldsymbol{\varepsilon}$ is perpendicular to the direction of propagation $\boldsymbol{k}$, we can have two linearly polarized waves having polarization vectors $\varepsilon_x$ and $\varepsilon_y$, which are perpendicular to each other as well as to $\boldsymbol{k}$. Linear combinations of $\varepsilon_x$ and $\varepsilon_y$ give rise to two linearly independent circularly polarized waves. The circularly polarized vectors are given by

$$\varepsilon_{\pm} = \mp \frac{1}{\sqrt{2}}(\varepsilon_x \pm i\varepsilon_y) \tag{6.8.1}$$

The change in $\varepsilon$ due to infinitesimal rotation $\delta\varphi$ about the $\boldsymbol{k}$-axis is given by

$$\delta\varepsilon_{\pm} = \mp i\delta\varphi\varepsilon_{\pm} \tag{6.8.2}$$

The above equation shows that the components of the spin associated with $\boldsymbol{\varepsilon}$ along the $\boldsymbol{k}$-axis are $m = \pm 1$. Hence, we again find that the spin angular momentum of a photon is $\hbar$. However, its $z$ component $m_s$ has only two values given by $\pm\hbar$. The third component $m_s = 0$ does not exist because $\boldsymbol{\varepsilon} \cdot \boldsymbol{k} = 0$ and the rest mass of the photons is zero. Thus the quantum mechanical excitation of the radiation field are photons having zero rest mass, spin $s = 1$, and $m_s = \pm 1$. The quantum number $m_s$ represents helicity of the state of the photons. A photon beam of a definite helicity corresponds to circularly polarized light. We shall refer to the light of positive helicity ($m_s = +1$) as right-handed circularly polarized light. A left-handed circularly polarized light has negative helicity ($m_s = -1$).

To discuss the spin states of the photons let us once again consider a monochromatic plane EM wave, polarized in the $x$ direction. According to (6.3.4), it is given by

$$A = A_0 a\varepsilon_x e^{i(k\cdot r - \omega t)} \tag{6.8.3}$$

We combine the above wave with another monochromatic plane wave having the same frequency and wave vector with amplitude $A_0 b$ but polarized in the $y$ direction and differing from (6.8.3) by a definite phase $\delta$. The resultant wave is also polarized and its unit polarization vector is given by

$$\varepsilon = a\varepsilon_x + b\varepsilon_y e^{i\delta} \qquad (6.8.4)$$

with

$$|a|^2 + |b|^2 = 1 \qquad (6.8.5)$$

If we take $a = \cos\alpha$ and $b = \sin\alpha$, then (6.8.5) is automatically satisfied and (6.8.4) reduces to

$$\varepsilon(\alpha) = \cos\alpha\varepsilon_x + \sin\alpha\varepsilon_y e^{i\delta} \qquad (6.8.6)$$

which is a vector in a two-dimensional vector space with $\varepsilon_x$ and $\varepsilon_y$ as two basis vectors. Hence, we represent a photon state having a polarization vector $\varepsilon(\alpha)$ by

$$|\varepsilon(\alpha)\rangle = \cos\alpha|\varepsilon_x\rangle + \sin\alpha e^{i\delta}|\varepsilon_y\rangle \qquad (6.8.7)$$

Since $|\varepsilon(\alpha)\rangle$, $|\varepsilon_x\rangle$, and $|\varepsilon_y\rangle$ are state vectors in a two-dimensional vector space, they can be represented by two-component spinors, involving $a$ and $b$. To obtain the values of $a$ and $b$ let us use (6.8.1), according to which

$$|\varepsilon_x\rangle = -\frac{1}{\sqrt{2}}(|\varepsilon_+\rangle - |\varepsilon_-\rangle) \qquad (6.8.8)$$

and

$$|\varepsilon_y\rangle = \frac{i}{\sqrt{2}}(|\varepsilon_+\rangle + |\varepsilon_-\rangle) \qquad (6.8.9)$$

where the photon states $|\varepsilon_+\rangle$ and $|\varepsilon_-\rangle$ correspond to the helicity $+1$ and $-1$, respectively. Hence, like the spin-up and spin-down electron states (see 4.3.3), we have

$$|\varepsilon_+\rangle = \begin{pmatrix} 1 \\ 0 \end{pmatrix} \quad \text{and} \quad |\varepsilon_-\rangle = \begin{pmatrix} 0 \\ 1 \end{pmatrix} \qquad (6.8.10)$$

Thus we get

$$|\varepsilon_x\rangle = \frac{1}{\sqrt{2}}\begin{pmatrix} -1 \\ 1 \end{pmatrix} \quad \text{and} \quad |\varepsilon_y\rangle = \frac{i}{\sqrt{2}}\begin{pmatrix} 1 \\ 1 \end{pmatrix} \qquad (6.8.11)$$

Putting the above equation into (6.8.7), we get

$$|\varepsilon_\alpha\rangle = \frac{1}{\sqrt{2}} \begin{pmatrix} -\cos\alpha + i\sin\alpha\, e^{i\delta} \\ \cos\alpha + i\sin\alpha\, e^{i\delta} \end{pmatrix} \equiv \begin{pmatrix} a_\alpha \\ b_\alpha \end{pmatrix} \tag{6.8.12}$$

Similar to an electron beam, a beam of photons having all the photons in the same state of polarization is fully polarized and its polarization properties can be described by a single polarization vector, say, $\varepsilon(\alpha)$. The state (6.8.12) is said to be a pure state. If we mix two or more fully polarized beams that do not have definite phase relationships, a mixture is obtained. To describe the polarization properties of a mixed beam we consider its density matrix operator, defined by

$$\rho = \sum_j I_j |\varepsilon_j\rangle\langle\varepsilon_j| \tag{6.8.13}$$

where $I_j$ is the intensity of the $j^{th}$ pure component of the mixed beam and

$$|\varepsilon_j\rangle = \begin{pmatrix} a_j \\ b_j \end{pmatrix} \tag{6.8.14}$$

With the help of the above equation, we get from (6.8.13)

$$\rho = \begin{pmatrix} \rho_{11} & \rho_{12} \\ \rho_{21} & \rho_{22} \end{pmatrix} = \sum_j I_j \begin{pmatrix} |a_j|^2 & a_j b_j^* \\ a_j^* b_j & |b_j|^2 \end{pmatrix} \tag{6.8.15}$$

Since $|\varepsilon_j\rangle$ is a normalized ket

$$\mathrm{tr}\,\rho = \sum_j I_j = I \tag{6.8.16}$$

The above normalization of the mixed photon beam is different from the normalization of a mixed electron beam, where we have $\mathrm{tr}\,\rho = 1$. For the mixed photon beam

$$\mathrm{tr}\,\rho^2 < (\mathrm{tr}\,\rho)^2 \tag{6.8.17}$$

Only for the pure photon beams is $\mathrm{tr}\,\rho^2 = I^2$.

Since $a_i$ and $b_j$ are complex, in general, it takes four independent parameters to completely determine the polarization state of a mixed beam: $I$, $\eta_1$, $\eta_2$, and $\eta_3$. The $\eta_i$ are known as the Stokes parameters and are defined as follows:

$$\eta_1 = \frac{I(45°) - I(135°)}{I} \qquad (6.8.18a)$$

$$\eta_2 = \frac{I_+ - I_-}{I} \qquad (6.8.18b)$$

and

$$\eta_3 = \frac{I(0°) - I(90°)}{I} \qquad (6.8.18c)$$

where $I(\alpha)$ is the intensity of the transmitted light when a mixed light beam of intensity $I$ moving along the $z$-axis is passed through a Nicol prism whose axis of complete transmission makes an angle $\alpha$ with the $x$-axis. $I_+$ ($I_-$) is the intensity of the transmitted light when the mixed beam is passed through a filter which fully transmits photons of helicity $+1$ ($-1$). The intensity $I(\alpha)$ for the mixed beam is given by

$$I(\alpha) = \langle \varepsilon(\alpha)|\rho|\varepsilon(\alpha)\rangle \qquad (6.8.19)$$

Using (6.8.10) and (6.8.15), we get from the above equation

$$I_+ = \rho_{11} \qquad \text{and} \qquad I_- = \rho_{22} \qquad (6.8.20)$$

Similarly, using (6.8.12) and (6.8.15) in (6.8.19) gives

$$I(\alpha) = \rho_{11}|a_\alpha|^2 + \rho_{12}a_\alpha^* b_\alpha + \rho_{21}a_\alpha b_\alpha^* + \rho_{22}|b_\alpha|^2 \qquad (6.8.21)$$

For $\delta = 0$ and $\alpha = 0$ we have $a_0 = -1/\sqrt{2}$ and $b_0 = 1/\sqrt{2}$. Hence,

$$I(0) = \tfrac{1}{2}(\rho_{11} - \rho_{12} - \rho_{21} + \rho_{22}) \qquad (6.8.22)$$

Similarly,

$$I(45°) = \tfrac{1}{2}(\rho_{11} - i\rho_{12} + i\rho_{21} + i\rho_{22}) \qquad (6.8.23)$$

$$I(90°) = \tfrac{1}{2}(\rho_{11} + \rho_{12} + \rho_{21} + \rho_{22}) \qquad (6.8.24)$$

and

$$I(135°) = \tfrac{1}{2}(\rho_{11} + i\rho_{12} - i\rho_{21} + \rho_{22})$$ (6.8.25)

With the help of above equations, we obtain

$$\eta_1 = \frac{i}{I}(\rho_{21} - \rho_{12})$$ (6.8.26)

$$\eta_2 = \frac{1}{I}(\rho_{11} - \rho_{22})$$ (6.8.27)

and

$$\eta_3 = -\frac{1}{I}(\rho_{12} + \rho_{21})$$ (6.8.28)

where

$$I = \operatorname{tr}\rho = \rho_{11} + \rho_{22}$$ (6.8.29)

Further,

$$\rho_{11} = \frac{I}{2}(1 + \eta_2)$$ (6.8.30)

$$\rho_{12} = \frac{I}{2}(-\eta_3 + i\eta_1)$$ (6.8.31)

$$\rho_{21} = \frac{I}{2}(-\eta_3 - i\eta_1)$$ (6.8.32)

$$\rho_{22} = \frac{I}{2}(1 - \eta_2)$$ (6.8.33)

Thus,

$$\rho = \frac{I}{2}\begin{pmatrix} 1 + \eta_2 & i\eta_1 - \eta_3 \\ -i\eta_1 - \eta_3 & 1 - \eta_2 \end{pmatrix}$$ (6.8.34)

and

$$\text{tr}\rho^2 = \frac{I^2}{2}(1 + \eta_1^2 + \eta_2^2 + \eta_3^2) \qquad (6.8.35)$$

Now for a mixed beam $\text{tr}\,\rho^2 < I^2$. Hence,

$$\eta_1^2 + \eta_2^2 + \eta_3^2 < 1 \qquad (6.8.36)$$

Sometimes the Stokes parameters are also represented by $P_1$, $P_2$, and $P_3$ with the relations $P_1 = \eta_3$, $P_2 = \eta_1$, and $P_3 = \eta_2$ (Kesseler, 1991).

The degree of polarization of the beam $P$ satisfies

$$P^2 = P_1^2 + P_2^2 + P_3^2 \qquad (6.8.37)$$

and, as expected, for a partially polarized (mixed) beam $P < 1$. Using (6.8.12) and the expressions for $\rho_{ij}$ in (6.8.21), we get

$$I(\alpha, \delta) = \frac{I}{2}(1 + \eta_2 \sin 2\alpha \sin \delta + \eta_3 \cos 2\alpha + \eta_1 \sin 2\alpha \cos \delta) \quad (6.8.38)$$

## 6.9   Optical Oscillator Strength

The optical oscillator strength $f^o$ for a transition from $|0\rangle$ to $|q\rangle$ atomic states due to EM waves, in the length form, is defined by

$$f_{q0}^{oL} = \frac{\varepsilon_{q0}}{R}\frac{1}{a_0^2}|\langle q|x|0\rangle|^2 \qquad (6.9.1)$$

where $\varepsilon_{q0}$ is the excitation energy and $R$ is the Rydberg energy; $f_{q0}^{oL}$ is a dimensionless quantity and is related to the strength of the transition from the $|0\rangle$ state to the $|q\rangle$ state. It is positive for excitation and negative for the de-excitation. Due to the length operator $x$, Eq. (6.9.1) gives the length form of the optical oscillator strength. It can be converted into the velocity form by using (6.5.5) in (6.9.1). This gives

$$f_{q0}^{oV} = \frac{2}{m\varepsilon_{q0}}|\langle q|p_x|0\rangle|^2 \qquad (6.9.2)$$

To obtain $f_{q0}^{oL}$ in the acceleration form we consider

$$\langle q|[H_0, p_x]|0\rangle = \langle q|[p^2/2m + V(r), p_x]|0\rangle$$
$$= \langle q|V(r)p_x - p_x V(r)|0\rangle$$

Since

$$p_x V(r)\psi = \psi p_x V(r) + V(r) p_x \psi$$

we get

$$\langle q|[H_0, p_x]|0\rangle = -\langle q|(p_x V)|0\rangle = i\hbar\langle q|dV/dx|0\rangle$$

We also have

$$\langle q|[H_0, p_x]|0\rangle = \varepsilon_{q0}\langle q|p_x|0\rangle$$

Hence, in the acceleration form

$$f_{q0}^{oA} = \frac{4Ra_0^2}{\varepsilon_{q0}^3}|\langle q|dV/dx|0\rangle|^2 \tag{6.9.3}$$

For the exact atomic wave function we have

$$f_{q0}^{oL} = f_{q0}^{oV} = f_{q0}^{oA} \tag{6.9.4}$$

However, it is only for the hydrogen atom that exact atomic wave functions are known. With approximate wave functions the calculated values of the $f_{q0}^o$ in the three different forms are found to be different. The magnitude of the differences among the three values indicates the inaccuracy of the employed approximate wave functions: the smaller the differences, the better the wave functions. It is evident from (6.9.1), that $f_{q0}^o$ is nonzero only for optically allowed transitions. For the excited $2p$ state of a hydrogen-like atom having $m = 0, \pm1$, it is easy to show that

$$f_{2p,1s}^o = \frac{2^{15}}{3^{10}}\frac{\varepsilon_{2p,1s}}{R}\frac{1}{Z^2} \tag{6.9.5}$$

The optical oscillator strength for a transition from the $1s$ to the $np$ state is given by

$$f_{np,1s}^o = \frac{2^8}{3}\frac{\varepsilon_{np,1s}}{R}\frac{1}{Z^2}\frac{n^7(n^2-1)(n-1)^{2n-6}}{(n+1)^{2n+6}} \tag{6.9.6}$$

Let us sum $f_{q0}^o$ over all the final states $|q\rangle$. To obtain the value of $S_q f_{q0}^o$ for a one-electron atom we note that

$$[x, [x, H_0]] = x^2 H_0 - 2x H_0 x + H_0 x^2 \tag{6.9.7}$$

Hence,

$$\begin{aligned}\langle 0|[x, [x, H_0]]|0\rangle &= 2\varepsilon_0 \langle 0|x^2|0\rangle - 2\langle 0|n H_0 x|0\rangle \\ &= 2\varepsilon_0 \underset{q}{S} \langle 0|x|q\rangle\langle q|x|0\rangle - 2\underset{q}{S}\langle 0|x|q\rangle\langle q|H_0 x|0\rangle\end{aligned} \tag{6.9.8}$$

where the $|q\rangle$ form a complete set and we have used $\underset{q}{S}|q\rangle\langle q| = 1$. Hence,

$$\langle 0|[x, [x, H_0]]|0\rangle = 2\underset{q}{S}(\varepsilon_0 - \varepsilon_q)|\langle 0|x|q\rangle|^2 \tag{6.9.9}$$

We also have $\langle 0|[x, [x, H_0]]|0\rangle = \langle 0|[x, i\hbar\, p_x/m]|0\rangle$

$$= -\frac{\hbar}{m} \tag{6.9.10}$$

Equating (6.9.9) to (6.9.10), we get

$$\underset{q}{S}\,\omega_{q0}|\langle 0|x|q\rangle|^2 = \frac{\hbar}{2m} \tag{6.9.11}$$

or

$$\underset{q}{S} f_{q0}^o = 1 \tag{6.9.12}$$

Similarly for an $N$-electron atom we get

$$\underset{q}{S} f_{q0}^o = N \tag{6.9.13}$$

This important equation is known as the Thomas–Reiche–Kuhn sum rule.

## 6.10 Photoionization of Atoms

So far we have discussed the excitation and de-excitation of an atom due to photon impact from one discrete state $|0\rangle$ to another discrete state $|q\rangle$. Since the initial and final states are bound states these collisions give rise to bound–bound transitions. In this section we extend our study to those collisions in which the final state is a continuum state and consider ionization of an atom by photon impact. We represent the collision by

$$hv + A \rightarrow A^+ + e \qquad (6.10.1)$$

and the energy of the ejected electron is

$$\varepsilon_k = \frac{\hbar^2 k^2}{2m} = W - I \qquad (6.10.2)$$

where $W = hv$ and $I$ is the ionization potential of the atom. For reaction (6.10.1) the photon can have all values of the energy $W$ greater than $I$.

For a given ejected energy $\varepsilon_k$, the ionized electron can come out in all possible directions. Hence, when (6.5.9) is extended to ionization, the cross section for a given $\varepsilon_k$ becomes a differential with respect to the direction. Thus

$$d\sigma_{ph}(W) = 4\pi^2 \hbar \alpha \omega_{k0} |x_{k0}|^2 \rho(\varepsilon_k) \qquad (6.10.3)$$

where $\sigma_{ph}(W)$ is the photoionization cross section of the atom for the photon energy $W$. To obtain an expression for the density of states $\rho(\varepsilon_k)$, we consider the ejected electron in a box of length $L$:

$$k_x = \left(\frac{2\pi}{L}\right) n_x$$

where $n_x$ is an integer. The number of states when $k_x$ varies between $k_x$ and $k_x + dk_x$ is given by

$$dn_x = \left(\frac{L}{2\pi}\right) dk_x$$

Hence, in three dimensions, the number of states between $\varepsilon_k$ and $\varepsilon_k + d\varepsilon_k$ is equal to $(L/2\pi)^3 dk_x dk_y dk_z$. Thus

$$\rho(\varepsilon_k) d\varepsilon_k = \left(\frac{L}{2\pi}\right)^3 k^2 dk \sin\theta d\theta d\varphi$$

where $(\theta, \varphi)$ is the direction of the vector $\mathbf{k}$ with respect to some chosen axis. Now,

$$d\varepsilon_k = \frac{\hbar^2 k}{m} dk$$

Hence,

$$\rho(\varepsilon_k) = \left(\frac{L}{2\pi}\right)^3 \frac{mk}{\hbar^2} \sin\theta \, d\theta \, d\varphi \qquad (6.10.4)$$

We employ the Dirac delta function normalization for plane waves as well as for other continuum waves, such as Coulomb waves. For a box normalized plane wave the normalization constant $A$ is $L^{-3/2}$. This changes to $(2\pi)^{-3/2}$ for a delta function normalized plane wave. Hence, $\rho(\varepsilon_k)$ with the above normalization is obtained by replacing $L$ by $2\pi$ in (6.10.4). With this replacement we get from (6.10.3)

$$d\sigma_{\rm ph}(\omega) = \frac{4\pi^2}{\hbar} \alpha\omega k m |x_{k0}|^2 \, d\Omega_k \qquad (6.10.5)$$

To evaluate the above matrix element for a plane wave it is more convenient to take the matrix element in the velocity form. Hence, the use of (6.5.5) gives

$$d\sigma_{\rm ph}(W) = \frac{4\pi^2 \alpha k}{m\hbar\omega} |\langle k|p_x|0\rangle|^2 \, d\Omega_k$$

$$= \frac{4\pi^2 \alpha \hbar k k_x^2}{m\omega} |\langle k|0\rangle|^2 \, d\Omega_k \qquad (6.10.6)$$

Taking

$$v_0(r) = (Z^3/\pi a_0^3)^{1/2} e^{-Zr/a_0} \qquad (6.10.7)$$

we get

$$\langle k|0\rangle = \frac{(2Za_0)^{3/2} Z}{\pi(Z^2 + k^2 a_0^2)^2} \qquad (6.10.8)$$

Hence,

$$\sigma_{\rm ph}(\omega) = \frac{32\alpha\hbar k}{m\omega} \frac{Z^5 a_0^3}{(Z^2 + k^2 a_0^2)^4} \int k_x^2 \, d\Omega_k \qquad (6.10.9)$$

Noting that

$$k_x = k\sin\theta\cos\varphi$$

and

$$\int \sin^2\theta \cos^2\varphi \sin\theta \, d\theta \, d\varphi = 4\pi/3$$

we finally obtain

$$\sigma_{\mathrm{ph}}(\omega) = \frac{128\pi}{3} \frac{\alpha\hbar k^3}{m\omega} \frac{Z^5 a_0^3}{(Z^2 + k^2 a_0^2)^4} \qquad (6.10.10)$$

It is evident from (6.10.1) that the ejected electron moves in the field of the ion $A^+$. Hence, a better wave function for the ejected electron is a Coulomb wave instead of a plane wave. A partial wave expansion of the Coulomb wave is given by (Joachain, 1987)

$$v_k(r) = \sum_{l,m} R_l(kr) Y_{lm}^*(\hat{r}) Y_{lm}(\hat{k}) \qquad (6.10.11)$$

with the radial wave function

$$R_l(kr) = \sqrt{2/\pi} \, i^l \exp(-\pi\gamma/2) |\Gamma(l+1+i\gamma)| e^{ikr} (kr)^l \, 2^l \exp(i\sigma_l)/(2l+1)!$$
$$\times {}_1F_1(l+1+i\gamma, 2l+2, -2ikr) \qquad (6.10.12)$$

where $\gamma = Z/ka_0$ and the Coulomb phase shift $\sigma_l = \arg\Gamma(l+1+i\gamma)$. $\Gamma(x)$ and ${}_1F_1$ are the gamma function and the hypergeometric series, respectively. Let us first consider the angular integration over the direction of $r$. Now

$$x = (2\pi/3)^{1/2} r[Y_{1,-1}(\hat{r}) - Y_{11}(\hat{r})]$$

and

$$\sum_{l,m} \int Y_{lm}^*(\hat{r}) Y_{lm}(\hat{k})[Y_{1,-1}(\hat{r}) - Y_{11}(\hat{r})] d\hat{r} = Y_{1,-1}(\hat{k}) - Y_{11}(\hat{k})$$

Hence, in (6.10.11), only $l = 1$ and $m_l = \pm 1$ terms are to be considered. It is easy to see, with the help of the above equation, that

$$\int |\langle k|x|0\rangle|^2 \, d\Omega_k = 2|A_{0k}|^2 \qquad (6.10.13)$$

where

$$A_{0k} = \sqrt{2\pi/3} \int R_1(kr) r v_0(r) r^2 dr \tag{6.10.14}$$

From (6.10.5) after integration over $\Omega_k$, we get

$$\sigma_{ph}(W) = \frac{8\pi^2 e^2 \omega km}{\hbar^2 c} |A_{0k}|^2 \tag{6.10.15}$$

Evaluation of the radical integral $A_{0k}$ finally gives (Saksena, 1994)

$$\sigma_{ph}(W) = \frac{2^{10}}{3} \pi^2 \frac{\omega}{c} \frac{Z^6 a_0^3}{(Z^2 + k^2 a_0^2)^5}$$
$$\exp(-4/\alpha \tan^{-1} \alpha)[1 - \exp(-2\pi/\alpha)]^{-1} \tag{6.10.16}$$

where $\alpha = 1/\gamma = k a_0/Z$ (not to be confused with the fine-structure constant).

Just as we extended (6.5.9) to get the differential photoionization cross section, we can extend (6.9.1) to obtain the differential oscillator strength for the bound-to-continuum transitions. Replacing excitation energy $\varepsilon_{q0}$ in (6.9.1) by the photon energy $W = \hbar\omega$ and taking $dk = k^2 dk d\Omega_k$, we obtain the differential oscillator strength per unit energy range

$$\frac{df}{dW} = \frac{\hbar\omega}{R} \frac{k^2}{a_0^2} \left( \frac{dk}{dW} \right) \int |x_{k0}|^2 d\Omega_k \tag{6.10.17}$$

To simplify the notation we drop the superscript $o$ on $f$.

Now,

$$k\,dk = (m/\hbar^2)dW$$

Hence,

$$\frac{df}{dW} = \frac{km\omega}{R\hbar a_0^2} \int |x_{k0}|^2 d\Omega_k \tag{6.10.18}$$

A comparison of (6.10.5) and (6.10.18) gives

$$\frac{df}{dW} = \frac{\hbar c}{4\pi^2 R a_0^2 e^2} \sigma_{ph}(W) \tag{6.10.19}$$

If we express $\sigma_{ph}(W)$ in megabarns ($10^{-18}\,cm^2$) and $df/dW$ in $eV^{-1}$ then

$$\frac{df}{dW}(eV^{-1}) = \frac{\sigma_{ph}(Mb)}{109.75} \tag{6.10.20}$$

Putting (6.10.16) into (6.10.19), we get

$$\frac{df}{dW} = \frac{2^7}{3}\frac{W}{R^2}\frac{Z^6}{(Z^2+k^2a_0^2)^5}\exp\left(-\frac{4}{\alpha}\tan^{-1}\alpha\right)\left[1-\exp\left(-\frac{2\pi}{\alpha}\right)\right]^{-1} \tag{6.10.21}$$

As the ionization potential energy for $I$ of the ground state of the hydrogenic atom is $Z^2R$ and $W = I(1 + \alpha^2)$, the above equation changes to

$$I\frac{df}{dW} = \frac{2^7}{3}\frac{1}{(1+a^2)^4}\exp\left(-\frac{4}{\alpha}\tan^{-1}\alpha\right)\left[1-\exp\left(-\frac{2\pi}{\alpha}\right)\right]^{-1} \tag{6.10.22}$$

## 6.11  K-Shell Photoionization of Atoms

So far we have considered the photoionization of the ground state of a one-electron atom. It can be extended to $K$-shell photoionization of multielectron atoms, which contain two $K$-shell electrons along with other electrons in higher shells. Since these $K$-electrons are tightly bound, we may represent each of them by

$$v_0(r) = \sqrt{Z_s^3/\pi a_0^3}\,e^{-Z_s r/a_0} \tag{6.11.1}$$

where due to inner screening $Z_s < Z$. We may take $Z_s = Z - s$ with the screening parameter $s = 0.3$, as derived by Slater (1930). Further, the matrix element $\langle k|x|0\rangle$ receives a large contribution from the small values of $x$, where the replacement of $Z$ by $Z_s$ in the expression of a Coulomb wave function is also justified. However, due to external screening, the experimental ionization potential $I_K$ of the atom is less than $I_s = Z_s^2R$. Thus the ratio $p = I_K/I_s$ is less than unity and its value increases with an increase in $Z$. Since there is no screening in a hydrogen atom, $p = 1$. Now we modify the definition of $\alpha$ and take $\alpha = \delta a_0/Z_s$, where $\delta^2 a_0^2R$ is the apparent energy of the ejected electron. Hence,

$$W = I_s + \delta^2 a_0^2 R = I_s(1+\alpha^2)$$

For $W$ less than $I_s$, $\alpha^2$ as well as $\delta^2$ are negative. At the threshold of photoionization $W = I_K$, So $\alpha^2$ is negative for $I_K \leq W \leq I_s$. Incorporating the above changes and noting that the $K$-shell contains two electrons, the $K$-shell photoionization cross section, with the help of (6.10.19) and (6.10.21), is given by

$$I_s\sigma_K = \frac{2^9}{3}\frac{\pi^2\hbar e^2}{mc}\frac{1}{(1+\alpha^2)^4}F(\alpha^2) \qquad (6.11.2)$$

where

$$F(\alpha^2) = \exp\left(-\frac{4}{\alpha}\tan^{-1}\alpha\right)\left[1-\exp\left(-\frac{2\pi}{\alpha}\right)\right]^{-1} \quad \text{for} \quad \alpha^2 \geq 0 \quad (6.11.3)$$

$$= \exp\left\{-\frac{2}{(-\alpha^2)^{1/2}}\ln\left[\frac{1+(-\alpha^2)^{1/2}}{1-(-\alpha^2)^{1/2}}\right]\right\} \quad \text{for} \quad \alpha^2 < 0 \quad (6.11.4)$$

In the derivation of (6.11.4) the relation

$$\ln(x+iy) = \tfrac{1}{2}\ln(x^2+y^2) + i\tan^{-1}(y/x) \qquad (6.11.5)$$

has been utilized and the value of the normalization constant $[1 - \exp(-2\pi/\alpha)]$ is taken to be unity.

Equation (6.11.2) shows that $I_s\sigma_k$, as a function of $\alpha^2$ is the same for all the atoms. It represents a universal curve for $K$-shell photoionization (Khare et al., 1992). Figure 6.2 compares $I_s df/dW = df/d\alpha^2$, obtained with the help of (6.10.19) and (6.11.2), with the tabulated values of Veigele (1973) for a number of atoms. Good agreement between the universal curve and the data of Veigele is noted for low values of $\alpha^2$. However, at higher values of $\alpha^2$, particularly for the heavier atoms, the curve lies below the data. This indicates the need for a better wave function to represent the ejected electron and the inclusion of the relativistic effect.

## 6.12 The Fano Effect

So far we have considered photoionization of unpolarized atoms by unpolarized light. Ejected electrons are also unpolarized. However, if we take polarized alkali metals in their ground state with $M_s = \tfrac{1}{2}$ then photoionization by unpolarized light produces polarized electrons with $M_s = \tfrac{1}{2}$. Fano (1969) predicted that electrons produced by photoionization of unpolarized atoms by circularly polarized light would also be partially polarized. Since it is much easier to produce polarized light than it is to produce polarized atoms, the

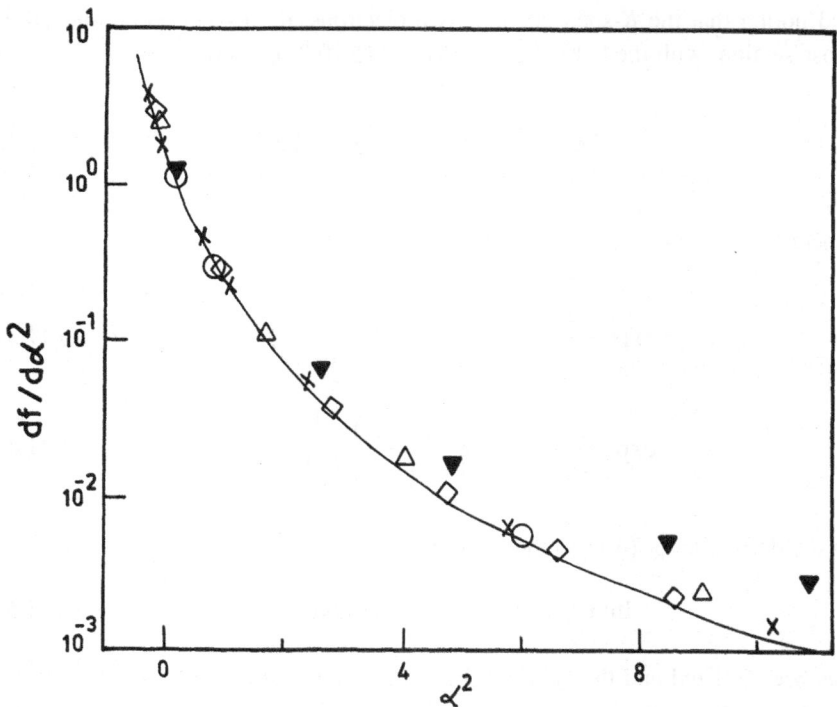

**FIGURE 6.2** Curve showing the variation of $df/d\alpha^2$ with $\alpha^2$ in the hydrogenic approximation. The values of $df/d\alpha^2$, obtained from the table of Veigele (1973) for different atoms, are as follows: $X$, carbon; $\bigcirc$, argon; $\Diamond$, nickle; $\triangle$, silver; $\blacktriangledown$, gold. Reproduced from "A scaling relation for $K$-shell photoionization cross sections of atoms" S. P. Khare, V. Saksena, and S. P. Ojha, *J. Phys. B* **25**, 2001, 1992, with permission from IOP, Publishing Ltd., UK.

Fano's prediction was found to be interesting. The very next year Kessler and Lorenz (1970) produced polarized electrons experimentally by photoionizing ground state alkali metals by circularly polarized light and named the phenomenon the Fano effect. In this section we discuss the theory of the Fano effect briefly.

Let us consider the photoionization of alkali metals from their ground state ($^2S_{1/2}$). Since the atoms are unpolarized, half of them have $M_j = +\frac{1}{2}$ and for the other half, $M_j = -\frac{1}{2}$. The dipole operator for the photoionization is $\hat{\boldsymbol{\varepsilon}} \cdot \boldsymbol{r}$, where $\hat{\boldsymbol{\varepsilon}}$ is a unit polarization vector for linearly polarized light. For right-handed circularly polarized light, the operator changes to $(\boldsymbol{\varepsilon}^{(1)} + i\boldsymbol{\varepsilon}^{(2)}) \cdot \boldsymbol{r}$, where $\boldsymbol{\varepsilon}^{(1)}$ and $\boldsymbol{\varepsilon}^{(2)}$ are two linearly independent unit polarization vectors. If $z$ is the direction of propagation of the light, the dipole operator for right-handed circularly polarized light is $(x + iy) = -\sqrt{(8\pi/3)} Y_{11} r$. Hence, during the ionization, the photon having $s = j = m_j = 1$ gives up one unit of angular momentum to the ejected photoelectron. Initially the alkali atoms are in the state $^2S_{1/2,\pm1/2}$. Therefore the produced photoelectrons, due to the photoionization of the alkali atoms by a right-handed circularly polarized light, are in states $^2P_{3/2,3/2}$, $^2P_{3/2,1/2}$, and $^2P_{1/2,1/2}$. For photoelectrons with $m_j' = \frac{3}{2}$ the collision can be represented by

$$\text{Na}(^2S_{1/2,1/2}, J = M_j = M_s = \tfrac{1}{2}) + h\nu(j = m_j = 1)$$
$$\rightarrow \text{Na}^+(^1S_0) + e(^2P_{3/2,3/2}, j' = m_j' = \tfrac{3}{2}, m_l' = 1, m_s' = \tfrac{1}{2}) \qquad (6.12.1)$$

Thus the spin-up sodium atoms on being ionized produce only spin-up photoelectrons. Hence, (6.12.1) represents direct collisions. However, the photoionization of the spin-down sodium atoms ($M_s = -\frac{1}{2}$) produces photoelectrons having $m_j' = \frac{1}{2}$. For such collisions the following four combinations are possible:

$$
\left.
\begin{array}{cccc}
j' & m_l' & m_s' & \\
\frac{3}{2} & 1 & -\frac{1}{2} & (a) \\
\frac{1}{2} & 1 & -\frac{1}{2} & (b) \\
\frac{3}{2} & 0 & \frac{1}{2} & (c) \\
\frac{1}{2} & 0 & \frac{1}{2} & (d)
\end{array}
\right\} \qquad (6.12.2)
$$

It is evident that in (a) and (b) $m_s'$ of the photoelectrons is same as that of the initial sodium atom, i.e., $-\frac{1}{2}$. Hence, they also represent direct collisions. However, for (c) and (d) the spin of the photoelectrons is $+\frac{1}{2}$, different from the spin of the sodium atom, which was $-\frac{1}{2}$. Thus (c) and (d) represent spin-flip collisions. Like (6.12.1), for reaction (d) we have

$$\text{Na}(^1S_{1/2,-1/2}, J = \tfrac{1}{2}, M_j = M_s = -\tfrac{1}{2}) + h\nu(j = m_j = 1)$$
$$\rightarrow \text{Na}^+(^1S_0) + e(^2P_{1/2,1/2}, j' = m_j' = m_s' = \tfrac{1}{2}) \qquad (6.12.3)$$

Because of the spin-flip collisions, represented by (c) and (d), the produced photoelectrons are partially polarized. A similar effect takes place if we consider left-handed circularly polarized light.

Let us now proceed to calculate the degree of polarization of the photoelectrons. For (6.12.1), the wave function of the initial system is given by $R(r)Y_{00}\alpha$ and that of the ejected electron is $R_3(r)Y_{11}\alpha$, where $\alpha$ represents a spin-up electron and $R_3(r)$ is the radical function with $j' = \frac{3}{2}$. Hence, the transition matrix element is

$$a_1 = \langle R_3(r)Y_{11}\alpha | -\sqrt{8\pi/3}\, r Y_{11} | R(r)Y_{00}\alpha \rangle$$

or

$$a_1 = -\sqrt{\tfrac{2}{3}} A_3 \qquad (6.12.4)$$

where

$$A_3 = \langle R_3(r)|r|R(r)\rangle \qquad (6.12.5)$$

For the ejected electron having $j' = \tfrac{3}{2}$ and $m_j' = \tfrac{1}{2}$, the wave function is obtained by taking a linear combination of (a) and (c) of (6.12.2). The matrix element for such a transition is

$$a_2 = \langle R_3(r)(C_a Y_{11}\beta + C_c Y_{10}\alpha)|-\sqrt{8\pi/3}\,rY_{11}|R(r)Y_{00}\beta\rangle$$
$$= -\sqrt{\tfrac{2}{3}}C_a A_3 = -\frac{\sqrt{2}}{3} A_3 \qquad (6.12.6)$$

where the $C_i$ are the Clebsch–Gordon coefficients, and their values are obtained from (4.4.3) by taking $l = 1$. Thus $C_a = -C_d = 1/\sqrt{3}$ and $C_b = C_c = \sqrt{\tfrac{2}{3}}$. $\beta$ represents a spin-down electron. Similarly for (b) and (d), $j' = \tfrac{1}{2}$, $m_j' = \tfrac{1}{2}$ and

$$a_3 = \langle R_1(r)(C_b Y_{11}\beta + C_d Y_{10}\alpha)|-\sqrt{8\pi/3}\,rY_{11}|R(r)Y_{00}\beta\rangle$$
$$= -\sqrt{\tfrac{2}{3}}C_b A_1 = -\frac{2}{3} A_1 \qquad (6.12.7)$$

where $R_1(r)$ is the radial wave function of the electron having $j' = \tfrac{1}{2}$ and

$$A_1 = \langle R_1(r)|r|R(r)\rangle \qquad (6.12.8)$$

The electrons produced by the four reactions of (6.12.2) cannot be distinguished from one another (all of them have $m_j' = m_l' + m_s' = \tfrac{1}{2}$). Hence, the wave function of the system is obtained by taking a linear superposition of all four combinations:

$$\chi_{1/2} = [a_2 R_3(r)C_c Y_{10} + a_3 R_1 C_d Y_{10}]\alpha$$
$$+ [a_2 R_3(r)C_a Y_{11} + a_3 R_1 C_b Y_{11}]\beta \qquad (6.12.9)$$

The number of photoelectrons with $m_s' = \tfrac{1}{2}$ and $m_j' = \tfrac{1}{2}$ is proportional to square of

$$\langle [a_2 R_3(r)C_c + a_3 R_1(r)C_d]\alpha Y_{10}|[a_2 R_3(r)C_c + a_3 R_1(r)C_d]Y_{10}\alpha\rangle$$

The difference between $R_3$ and $R_1$ is due to the spin–orbit interaction and is small; hence, we take $\langle R_3/R_1 \rangle = 1$ in the above relation and obtain

$$N_{1/2}(\uparrow) \propto |(a_2 C_c + a_3 C_d)|^2 \qquad (6.12.10)$$

where $N_{1/2}(\uparrow)$ is the number of spin-up electrons. Similarly $N_{1/2}(\downarrow)$, the number of spin-down photoelectrons (with $m'_s = -\frac{1}{2}$, $m'_j = \frac{1}{2}$), is

$$N_{1/2}(\downarrow) \propto |(a_2 C_a + a_3 C_b)|^2 \qquad (6.12.11)$$

All the electrons produced in the collisions represented by (6.12.1) are spin-up electrons but with $m'_j = \frac{3}{2}$. Hence,

$$N_{3/2}(\uparrow) \propto |a_1|^2 \qquad (6.12.12)$$

Therefore the degree of polarization $P$ of the ejected photoelectron is

$$P = \frac{N_{1/2}(\uparrow) + N_{3/2}(\uparrow) - N_{1/2}(\downarrow)}{N_{1/2}(\uparrow) + N_{3/2}(\uparrow) + N_{1/2}(\downarrow)} \qquad (6.12.13)$$

or

$$P = \frac{(a_2 C_c + a_3 C_d)^2 + a_1^2 - (a_2 C_a + a_3 C_b)^2}{(a_2 C_b + a_3 C_d)^2 + a_1^2 + (a_2 C_a + a_3 C_b)^2} \qquad (6.12.14)$$

Putting the values of $a_i$ and $C_i$ into (6.12.14), we obtain

$$P = \frac{9A_3^2 + 2(A_1 - A_3)^2 - (A_3 + 2A_1)^2}{9A_3^2 + 2(A_1 - A_3)^2 + (A_3 + 2A_1)^2} \qquad (6.12.15)$$

As expected, the numerator of Eq. (6.12.15) vanishes for $A_1 = A_3$, showing no spin flip in the photoionization. However, for $A_3 = -2A_1$, the degree of polarization $P$ is unity, i.e., the ejected electrons are 100% polarized. The values of $A_1$ and $A_3$ depend upon the wavelength of the circularly polarized light. For the sodium atom $\lambda = 2900 \times 10^{-10}$ m produces a very high degree of polarization (Kessler, 1985). We note that the degree of polarization is nonzero only because $A_1 \neq A_3$. These matrix elements are different only because of the spin–orbit interaction occurring in the photoelectrons. We have already seen in Sec. 4.5 that polarization of an unpolarized beam of electrons that is the result of its being scattered by a potential is also due to the interaction between the spin and orbital angular momenta of the incident electron. Hence, in both these phenomena the spin–orbit interaction is responsible for the spin flip.

The Fano effect has been utilized in the construction of sources of polarized electrons. GaAs is the most widely used source and the polarization vector of its photoelectrons coincides with the axis of the circularly polarized visible light beam. The degree of the polarization $P$ of the photoelectrons is about 30 to 40% and the photocurrent is of the order of $\mu A$ (Kesseler, 1985, 1991). Recently, strained GaAs crystals have given $P$ as high as 0.9 (Maruyama et al., 1992).

## Questions and Problems

6.1 Calculate the lifetime of the $2P$ excited state of the hydrogen atom.

6.2 The resonance angular frequency $\omega_{2P-2S}$ for the lithium atom is $2.81 \times 10^{15}$ Hz. In the thermal equilibrium at $T = 4000°K$ the number of atoms in $2P$ and $2S$ states is $N_1$ and $N_2$, respectively. Obtain the value of the ratio $N_1/N_2$ and the value of the energy density of radiation per unit angular frequency $u(\omega_{2P-2S})$.

6.3 Gaseous lithium atoms are in thermal equilibrium with its surrounding EM waves. Find the temperature at which the probabilities of induced and spontaneous radiations are equal for the $2P - 2S$ transition. Take $\omega_{2P-2S} = 2.81 \times 10^{15}$ Hz.

6.4 Using (6.8.10) as the basis vectors, express the photon state $|\varepsilon_\beta\rangle$ as a two-component spinor. The polarization vector of the state $|\varepsilon_\beta\rangle$ makes an angle $\beta$ with the $x$-axis. Also obtain the density matrix $\rho_\beta$. Take $\beta = 0°, 45°, 90°$ and $135°$.

6.5 Consider a linearly polarized light beam of intensity $I$ propagating along the $z$ direction. Its photons are represented by (6.8.12) with fixed values of $\alpha$ and $\delta$. Show that the Stokes parameters are given by

$$\eta_1 = \sin 2\alpha \cos \delta, \quad \eta_2 = \sin 2\alpha \cos \delta, \quad \text{and} \quad \eta_3 = \cos 2\alpha$$

Show also that

$$\rho = \frac{I}{2} \begin{pmatrix} 1 + \sin 2\alpha \sin \delta & i \sin 2\alpha \cos \delta - \cos 2\alpha \\ -i \sin 2\alpha \cos \delta - \cos 2\alpha & 1 - \sin 2\alpha \sin \delta \end{pmatrix}$$

and

$$\text{tr} \rho^2 = I^2$$

6.6 In the interaction of EM waves with an atomic system, the quadrupole operator is $(i\mathbf{k}\cdot\mathbf{r})(\boldsymbol{\varepsilon}\cdot\mathbf{p})$. Take $\mathbf{k}$ and the unit vector $\boldsymbol{\varepsilon}$ parallel to the $z$- and $x$-axis, respectively, and show that

$$\langle q|(i\mathbf{k}\cdot\mathbf{r})(\boldsymbol{\varepsilon}\cdot\mathbf{p})|s\rangle = \frac{ik}{2}\langle q|L_y|s\rangle - \frac{km}{2\hbar}(\varepsilon_q - \varepsilon_s)\langle q|xz|s\rangle$$

where $L_y$ is the $y$ component of the orbital angular momentum, $m$ is the mass of the electron, and $|q\rangle$ and $|s\rangle$ are the eigenkets of the unperturbed atomic Hamiltonian with eigenvalues $\varepsilon_q$ and $\varepsilon_s$, respectively. Show also that for the second term to be nonzero the two atomic states must differ by two units of orbital angular momentum.

6.7 Show that the optical oscillator strength for $3p - 1s$ excitation of the hydrogen atom is 0.0791 and that this is about 0.19 times the oscillator strength for $2p - 1s$ excitation.

6.8 Verify Eq. (6.10.16).

6.9 Using the theory given in Sec. 6.11 obtain the value of the continuum optical oscillator strength $df/dW$ at the threshold of the $K$-shell ionization of a silver atom. Take $Z = 47$ and $I_K = 25.52\,\text{keV}$.

6.10 Using the theory discussed in Sec. 6.12 show that the photoelectrons produced by the photoionization of unpolarized ground state alkali atoms by a linearly polarized light are unpolarized.

# Collision of Electrons with Atoms:
# The Integral Approach

## 7.1 Introduction

In the Chapter 3 the collision of a free particle with a potential was discussed. In this chapter we take an electron as a projectile and replace the potential by an atom. The projectile is still a structureless particle but the atom, as a target, is a composite particle having a nucleus and a number of electrons. Thus the electron–atom collision is a many-body problem. Even the electron–hydrogen-atom collision is a three-body problem. Due to the many-body nature of the electron–atom system, an exact evaluation of the electron–atom collision cross section is not yet possible. In this chapter we shall consider a number of approximate methods based on the integral approach. To start with, both the electron and the atom are considered to be spinless particles. The exchange scattering due to the spin of the particles is discussed later on. Since atoms have a structure, the collisions can be elastic as well as inelastic.

## 7.2 The Basic Equations

Let us consider the collision of an electron with a neutral atom, having $Z$ electrons, in a frame of reference in which the nucleus is at rest. We assume that the center of mass of the system coincides with the nucleus. Thus we neglect the small difference between the rest mass of the electron $m$ and its reduced mass $\mu$. In the steady state the above system is described by the following time-independent Schrödinger equation:

$$H\psi_{ki,i}(\boldsymbol{r}, \boldsymbol{X}) = E_{ki,i}\psi_{ki,i}(\boldsymbol{r}, \boldsymbol{X}) \qquad (7.2.1)$$

where $r$ is the coordinate of the incident electron and $X$ represents the collective coordinates of all the atomic electrons. $\psi_{k_i,i}$ is the eigenfunction of the Hamiltonian $H$ with $E_{k_i,i}$ as its eigenenergy. We further take

$$H = H_e + H_A + V \qquad (7.2.2)$$

with

$$H_e\phi_{kn}(r) = \frac{\hbar^2 k_n^2}{2m}\phi_{kn}(r) \qquad (7.2.3)$$

$$H_A v_n(X) = \varepsilon_n v_n(X) \qquad (7.2.4)$$

$$E_{kn,n} = \varepsilon_n + \frac{\hbar^2 k_n^2}{2m} \qquad (7.2.5)$$

and

$$V = -\frac{Ze^2}{r} + \sum_{l=1}^{Z}\frac{e^2}{|r - r_l|} \qquad (7.2.6)$$

In the above equations $\phi_{kn}(r)$ and $v_n(X)$ represent the free electron and the atom and are the eigenfunctions of the Hamiltonians $H_e$ and $H_A$, respectively. The corresponding eigenenergies are $\hbar^2 k_n^2/2m$ and $\varepsilon_n$, and $V$ is the interaction energy.

The conversion of the Schrödinger equation (7.2.1) into the Lippmann–Schwinger integral equation for the outgoing scattered wave gives [similar to (3.3.2)]

$$\psi_{k_i,i}^+(r, X) = \phi_{ki}(r)v_i(X)$$
$$+ \int G_0^+(r, X; r', X') U(r', X')\psi_{k_i,i}^+(r', X')dr'dX' \qquad (7.2.7)$$

where $G_0^+$ is the free-particle Green's function for the noninteracting projectile and the target. Analogous to (3.3.7), $G_0^+$ is given by

$$G_0^+ = \lim_{\varepsilon \to 0} S\sum_n \int \frac{|k_q, n\rangle\langle k_q, n|}{k_n^2 - k_q^2 + i\varepsilon}dk_q \qquad (7.2.8)$$

where $|k_q\rangle$ and $|n\rangle$ are the intermediate states of the projectile and the target, respectively, and $k_n^2 - k_i^2 = (\varepsilon_i - \varepsilon_n)2m/\hbar^2$. As before, the reduced interaction energy $U$ is $2mV/\hbar^2$. A generalization of (3.3.12) for an electron–atom collision gives

$$f_{ji}(k_j, k_i) = -2\pi^2\langle\phi_{kj}, v_j|U|\psi_{k_i,i}^+\rangle \qquad (7.2.9)$$

Equation (3.3.13) modifies to

$$f_{ji}(\mathbf{k}_j, \mathbf{k}_i) = -\frac{4\pi^2 m}{\hbar^2} T_{ji}(\mathbf{k}_j, \mathbf{k}_i) \tag{7.2.10}$$

with the transition matrix element equal to

$$T_{ji}(\mathbf{k}_j, \mathbf{k}_i) = \langle \phi_{kj}, v_j | T | \phi_{ki}, v_i \rangle \tag{7.2.11}$$

and the $T$ operator being defined by

$$T | \phi_{ki}, v_i \rangle = V | \psi_{ki,i}^+ \rangle \tag{7.2.12}$$

The Born series of $\psi_{ki,i}^+$, like that of $\psi_{ki}^+$ [see (3.4.7)], is given by

$$\psi_{ki,i}^+ = \sum_{n=1}^{\infty} (G_0^+ U)^{n-1} \phi_{ki}(\mathbf{r}) v_i(\mathbf{X}) \tag{7.2.13}$$

Putting (7.2.13) into (7.2.9) yields

$$f_{ji}(\mathbf{k}_j, \mathbf{k}_i) = -2\pi^2 \langle \phi_{kj}, v_j | UP | \phi_{ki}, v_i \rangle \tag{7.2.14}$$

where

$$P = \sum_{n=1}^{\infty} (G_0^+ U)^{n-1}$$

We may interpret the above equation by visualizing that the initial object, represented by $\phi_{ki}(\mathbf{r}) v_i(\mathbf{X})$, is converted into a new object by the operator $UP$. This new object is a vector in Hilbert space whose basis vectors are $\phi_{kl}(\mathbf{r}) v_l(\mathbf{X})$. The integral (7.2.14) gives the projection of the new object on the basis vector $\phi_{kj}(\mathbf{r}) v_j(\mathbf{X})$. Hence, the integral is proportional to the transition probability amplitude for the transition of the object from its state $|\mathbf{k}_i, i\rangle$ to a new state $|\mathbf{k}_j, j\rangle$. As pointed out in the Chapter 3, the original object is changed into a new object due to multiple interactions of the object with the interaction potential (energy) $V$.

Similar to the Born series for potential scattering we also have the Born series for electron–atom scattering, given by (3.4.10); but now the $n$th Born term is

$$\bar{f}_{ji}^{Bn}(\mathbf{k}_j, \mathbf{k}_i) = -2\pi^2 \langle \phi_{kj}, v_j | U(G_0^+ U)^{n-1} | \phi_{ki}, v_i \rangle \tag{7.2.15}$$

As before, the $n$th Born scattering amplitude is given by (3.4.12). Taking $n = 1$ in (7.2.15) we get the first Born amplitude:

$$f_{ji}^{B1}(\mathbf{k}_j, \mathbf{k}_i) = -2\pi^2 \langle \phi_{kj}, v_j | U | \phi_{ki}, v_i \rangle \qquad (7.2.16)$$

Putting (7.2.7) into (7.2.9) we get

$$f_{ji}(\mathbf{k}_j, \mathbf{k}_i) = -2\pi^2 \langle \phi_{kj}, v_j | U | \phi_{ki}, v_i \rangle - 2\pi^2 \langle \phi_{kj}, v_j | U G_0^+ U | \psi_{kij}^+ \rangle \qquad (7.2.17)$$

Now using (7.2.8) in (7.2.17), we get

$$f_{ji}(\mathbf{k}_j, \mathbf{k}_i) = f_{ji}^{B1}(\mathbf{k}_j, \mathbf{k}_i) - \frac{1}{2\pi^2} S \lim_n \lim_{\varepsilon \to 0} \int f_{jn}^{B1}(\mathbf{k}_j, \mathbf{k}_q)$$

$$\times \frac{1}{k_n^2 - k_q^2 + i\varepsilon} f_{ni}(\mathbf{k}_q, \mathbf{k}_i) d\mathbf{k}_q \qquad (7.2.18)$$

The above equation is known as the Fredholm integral equation. We shall consider its application to the electron–atom collision later on.

The differential cross section for the transition of an atom from the $|i\rangle$ to the $|j\rangle$ state as a result of its collision with an electron is given by

$$I_{ji}(\mathbf{k}_j, \mathbf{k}_i) d\Omega = k_j / k_i |f_{ji}|^2 d\Omega \qquad (7.2.19)$$

The above equation is slightly different from (3.2.11) because in electron–atom collisions there is also inelastic scattering. In such a collision the flux of the scattered particles going to the detector with momentum $\hbar k_j$ is proportional to $k_j$, whereas the incident flux is proportional to $k_i$. The change in the momentum vector $\mathbf{K}$ due to scattering is given by (3.5.2), but

$$K = \left( k_i^2 + k_j^2 - 2k_i k_j \cos \theta \right)^{1/2} \qquad (7.2.20)$$

Thus

$$K \, dK = k_i k_j \sin \theta \, d\theta \qquad (7.2.21a)$$

and

$$d\Omega = K \, dK \, d\phi / k_i k_j \qquad (7.2.21b)$$

where $\theta$ and $\phi$ are the scattering angles. For a system having cylindrical symmetry, the scattering amplitude does not depend upon $\phi$. Then

$$\sigma_{ji} = \frac{2\pi}{k_i^2} \int_{K_{min}}^{K_{max}} |f_{ji}|^2 K \, dK \qquad (7.2.22)$$

where

$$K_{max,min}^2 = \frac{1}{Ra_0^2} \left(2E - \varepsilon_{ji} \pm 2\sqrt{E(E - \varepsilon_{ji})}\right) \qquad (7.2.23)$$

and $\varepsilon_{ji}$ is the excitation energy. For $E \gg \varepsilon_{ji}$,

$$K_{max}^2 = \frac{4E}{Ra_0^2} = 4k_i^2 \qquad (7.2.24a)$$

and

$$K_{min}^2 = \varepsilon_{ji}^2 / (4Ra_0^2 E) \qquad (7.2.24b)$$

For elastic scattering $\varepsilon_{ji} = 0$; hence, in this case $K_{min}$ is zero and $K_{max}$ is $2k_i$.

## 7.3   The First Born Approximation

It is evident from (7.2.16) that in the first Born approximation (FBA) effects due to distortions in the wave functions of the projectile and the target are completely neglected. Hence, as noted in Sec.3.5, the FBA is a weak potential approximation.

Let us now proceed to obtain $f_{ji}$ in the FBA. Putting plane waves for $\langle r|k_i\rangle$ and $\langle r|k_j\rangle$ into (7.2.16), we obtain

$$f_{ji}^{B1}(k_j, k_i) = -\frac{1}{4\pi} \int e^{iK \cdot r} v_j^*(X) U(r, X) v_i(X) dr dX \qquad (7.3.1)$$

For a neutral atom having $Z$ electrons

$$U(r, X) = \frac{2}{a_0} \left(-\frac{Z}{r} + \sum_{l=1}^{Z} \frac{1}{|r - r_l|}\right) \qquad (7.3.2)$$

where $r_l$ represents the coordinates of the $l$th atomic electron. Using the Bethe integral

$$\int e^{iK \cdot r} \frac{1}{|r - r_l|} dr = \frac{4\pi}{K^2} e^{iK \cdot \eta} \qquad (7.3.3)$$

in (7.3.1) along with (7.3.2) and noting that $v_j$ is orthogonal to $v_i$, we get

$$f_{ji}^{B1}(K) = \frac{2}{K^2 a_0}[Z\delta_{ji} - F_{ji}] \qquad (7.3.4)$$

where the form factor $F_{ji}$ is defined by

$$F_{ji}(K) = \left\langle v_j(X) \left| \sum_{l=1}^{Z} \exp(iK \cdot r_l) \right| v_i(X) \right\rangle \qquad (7.3.5)$$

The form factor arises due to the scattering of the incident electron by the atomic electrons. On the other hand, the first term $Z\delta_{ji}$ in (7.3.4) is due to the projectile–nucleus interaction. For elastic scattering

$$f_{ii}^{B1}(K) = \frac{2}{K^2 a_0}(Z - F_{ii}) \qquad (7.3.6)$$

and the form factor $F_{ii}$ depends upon the charge density $\rho = |v_i(X)|^2$. From (7.3.1) we also have

$$f_{ii}^{B1}(K) = -\frac{1}{4\pi} \int e^{iK \cdot r} U_{SF}(r) dr \qquad (7.3.7)$$

where

$$U_{SF}(r) = \langle v_i | U | v_i \rangle \qquad (7.3.8)$$

is the average value of $U$ for a fixed value of $r$. In the evaluation of $U_{SF}$ the projectile is frozen or made static at $r$. Hence, $U_{SF}$ is known as the reduced static potential of the atom and the electron–atom collision is reduced to the scattering of the electrons by a static potential.

For inelastic collisions, (7.3.4) reduces to

$$f_{ji}^{B1}(K) = -\frac{2}{K^2 a_0} F_{ji} \qquad (7.3.9)$$

The above equation shows that in the FBA the nucleus does not play any role in the electron impact excitation of the atom from $|i\rangle$ to $|j\rangle$. Use of (7.3.4) and (7.3.5) in (7.2.22) gives the integrated cross section $\sigma_{ji}$ in the FBA. This is independent of the sign of the charge of the incident particle. Since positrons and electrons have the same mass, spin, and magnitude of charge, the collision cross sections due to electrons and positrons are identical in the FBA. Thus the FBA cannot distinguish between matter–matter (atom–electron) and the matter–antimatter (atom–positron) interactions.

From (7.2.22) and (7.3.9) the integrated inelastic cross sections summed over all the excited states is given by

$$\underset{j\neq i}{S}\,\sigma_{ji}^{B1} = \frac{8\pi}{k_i^2 a_0^2}\,\underset{j\neq i}{S}\int \frac{|F_{ji}|^2}{K^4}K\,dK$$

Neglecting the dependence of $K_{max,min}^2$ on $j$ we take

$$\underset{j\neq i}{S}\,\sigma_{ji}^{B1} = \frac{4\pi}{k_i^2 a_0^2}\int_{\ln(K^2 a_0^2)_{min}}^{\ln(K^2 a_0^2)_{max}}\underset{j\neq i}{S}\frac{|F_{ji}|^2}{K^2}d[\ln(K^2 a_0^2)] \qquad (7.3.10)$$

To evaluate $\underset{j}{S}|F_{ji}|$ we note that for a one-electron atom

$$|j\rangle\langle j|e^{iK\cdot X}|i\rangle = |j\rangle F_{ji} \qquad (7.3.11)$$

and $\underset{j}{S}|j\rangle\langle j| = 1.$ Hence,

$$e^{iK\cdot X}|i\rangle = \underset{j}{S}|j\rangle F_{ji} \qquad (7.3.12)$$

Converting (7.3.12) into a bra equation, we get

$$\langle i|e^{-iK\cdot X} = \underset{j}{S}F_{ji}^*\langle j| \qquad (7.3.13)$$

Hence,

$$\underset{j}{S}|F_{ji}|^2 = \langle i|i\rangle = 1$$

So that,

$$\underset{j\neq i}{S}|F_{ji}|^2 = 1 - |F_{ii}|^2 \qquad (7.3.14)$$

and

$$\underset{j \neq i}{S}\,\sigma_{ji}^{B1} = \frac{4\pi}{k_i^2} \int_{\ln\left(K^2 a_0^2\right)\min}^{\ln\left(K^2 a_0^2\right)\max} \left(1 - |F_{ii}|^2\right) d\left[\ln(K^2 a_0^2)\right]/(K a_0)^2 \tag{7.3.15}$$

Thus with the help of the charge density of the $|i\rangle$ state the total inelastic cross section in the FBA can be evaluated.

Bethe (1930), extended the concept of the optical oscillator strength to electron–atom collisions and defined the generalized oscillator strength (GOS), for $j \neq i$, as

$$f_{ji}^G(K) = \frac{\varepsilon_{ex}}{R}\,\frac{1}{K^2 a_0^2}\,|F_{ji}(K)|^2 \tag{7.3.16}$$

This is also a dimensionless quantity and for optically allowed transitions $f_{ji}^G(K = 0)$ is equal to the optical oscillator strength $f^o$. As expected, for optically forbidden transitions $f_{ji}^G(K = 0)$ is zero. The term $f_{ji}^G(K)$ also obeys the Thomas–Reiche–Kuhn sum rule at each $K$, i.e., for an atom having $N$ electrons,

$$\underset{j}{S}\,f_{ji}^G(K) = N \tag{7.3.17}$$

Let us prove the above relation for a one-electron atom. We take $K$ along the $x$-axis and obtain

$$[e^{iKx}, H] = -\frac{\hbar K}{2m}(e^{iKx} p + p e^{iKx}) \tag{7.3.18}$$

Further,

$$\langle j|p e^{iKx}|i\rangle = \hbar \langle j|K e^{iKx}|i\rangle + \langle j|e^{iKx} p|i\rangle \tag{7.3.19}$$

and

$$\langle j|[e^{iKx}, H]|i\rangle = (\varepsilon_i - \varepsilon_j)\langle j|e^{iKx}|i\rangle \tag{7.3.20}$$

Hence, with the help of (7.3.18) to (7.3.20), we get

$$(\varepsilon_i - \varepsilon_j)\langle j|e^{iKx}|i\rangle = -\frac{\hbar K}{2m}(2\langle j|e^{iKx} p|i\rangle + \hbar K \langle j|e^{iKx}|i\rangle) \tag{7.3.21}$$

Now we multiply the above equation by $\langle i|e^{-iKx}|j\rangle$ from the left, sum over all the values of $j$, and use $\underset{j}{S}|j\rangle\langle j| = 1$ to obtain

$$\underset{j}{S}(\varepsilon_i - \varepsilon_j)|\langle j|e^{iKx}|i\rangle|^2 = -\frac{\hbar K}{2m}(2\langle i|p|i\rangle + \hbar K\langle i|i\rangle) \qquad (7.3.22)$$

As $p$ is an odd-parity operator, $\langle i|p|i\rangle = 0$. Furthermore, $\langle i|i\rangle = 1$, so that

$$\frac{2m}{\hbar^2 K^2}\underset{j}{S}(\varepsilon_j - \varepsilon_i)|F_{ji}(K)|^2 = 1 \qquad (7.3.23)$$

Similarly, we get (7.3.17) for an $N$-electron atom.

## 7.3.1  Elastic Scattering in the FBA

Let us investigate elastic collisions of electrons with some simple atoms such as hydrogen and helium in the FBA. We take the atoms in their ground states. For the hydrogen atom, from (7.3.6),

$$f_{el}^{B1}(\boldsymbol{k}_j, \boldsymbol{k}_i) = \frac{2}{K^2 a_0}[1 - \langle v_i(X)|e^{i\boldsymbol{K}\cdot\boldsymbol{X}}|v_i(X)\rangle] \qquad (7.3.24)$$

Keeping our future convenience in mind, we represent the hydrogen atom by the hydrogen-like wave function given by (6.11.1). This yields

$$\langle i|e^{i\boldsymbol{K}\cdot\boldsymbol{X}}|i\rangle = \frac{16Z_s^4}{(4Z_s^2 + K^2 a_0^2)^2} \qquad (7.3.25)$$

Hence, in the FBA

$$I_{el}^{B1}(Ka_0) = |f_{el}^{B1}(Ka_0)|^2 = \frac{4a_0^2(8Z_s^2 + K^2 a_0^2)^2}{(4Z_s^2 + K^2 a_0^2)^4} \qquad (7.3.26)$$

Thus for $Z_s = 1$ the differential cross section in the forward direction ($K = 0$) is $a_0^2$ and it falls monotonically with the increase in $Ka_0$. At large values of $Ka_0$ the differential cross section varies as $K^{-4}$. Using (7.3.25) in (7.2.22) and carrying out the required integration over $K$, we get $\sigma_{el}$, the total elastic cross section:

$$\sigma_{el} = \frac{7(k_i a_0)^4 + 18Z_s^2(k_i a_0)^2 + 12Z_s^4}{3Z_s^2[Z_s^2 + (k_i a_0)^2]^3}\pi a_0^2 \qquad (7.3.27)$$

Hence, for large $E$, $\sigma_{el} \approx 7\pi/3(k_i Z_s)^2$ and it falls as $E^{-1}$.

For the helium atom, $Z = 2$ and it has two electrons. We represent its ground state ($1^1S_0$) by

$$v(r_1, r_2) = v_1(r_1)v_2(r_2) \tag{7.3.28}$$

Each $v(r)$ is given by (6.11.1). The value of $Z_s$ as obtained from the variational principle is $\frac{27}{16}$ (Schiff, 1968). This value is between 1 and 2 and indicates that the nucleus is partially shielded from each electron by the other electron. For (7.3.28), the form factor is given by

$$F_{ii} = 2\langle v(r)|e^{iK \cdot r}|v(r)\rangle \tag{7.3.29}$$

and is double (7.3.25). The differential cross section for the helium atom with (6.11.1) and (7.3.29) is

$$I_{el}^{B1}(Ka_0) = \frac{16a_0^2(8Z_s^2 + K^2a_0^2)^2}{(4Z_s^2 + K^2a_0^2)^4} \tag{7.3.30}$$

In the forward direction,

$$I_{el}^{B1}(0) = \frac{4}{Z_s^4}a_0^2 = 0.4933a_0^2 \tag{7.3.31}$$

For the next atom in the periodic table $Z = 3$, its ground state electronic configuration is $1s^22s$, and the ground state term is $^2S_{1/2}$. Hence, its Hartree wave function is given by

$$v(r_1, r_2, r_3) = v_1(r_1)v_2(r_2)v_3(r_3) \tag{7.3.32}$$

where both $v_1$ and $v_2$ are $1s$ orbitals and $v_3$ is a $2s$ orbital. Suitable values of the exponential parameters $Z'$ and $Z''$ are obtained by the Hartree self-consistent field method (Weissbluth, 1978). In this case

$$F_{ii} = 2\langle v_1(r)|e^{iK \cdot r}|v_1(r)\rangle + \langle v_3(r)|e^{iK \cdot r}|v_3(r)\rangle \tag{7.3.33}$$

In a similar manner, the elastic collision of electrons with a multielectron atom can be investigated in the FBA.

Figures 7.1(a) to (d) show the differential cross sections for the elastic scattering of electrons by the ground states of H, He, Ne, and Ar in the FBA as obtained by Shobha (1972). The wave functions taken by her were the same as employed by Khare and Moiseiwitsch (1965) for the He atom and by Khare and

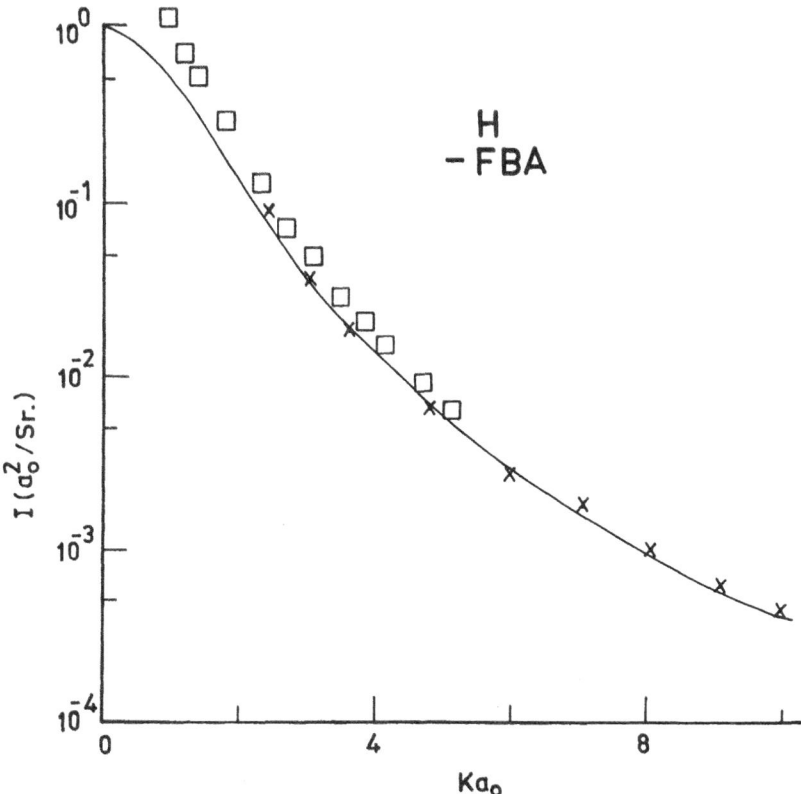

**FIGURE 7.1(a)** Curve showing the variation of the differential cross section $I$(in $a_0^2/Sr$) with $Ka_0$ for elastic scattering of electrons by hydrogen atoms in the first Born approximation. □ and $X$ represent the experimental data of Williams (1975) for $E$ equal to 200 and 680 eV, respectively.

Shobha (1974) for Ne and Ar atoms. Experimental data for one or two investigations are also shown for comparison. For the lightest atom, namely the H atom, the agreement between theory and experiment is satisfactory even at 200 eV, although the theory has a tendency to underestimate the cross sections at smaller values of $Ka_0$. At 650 eV, the agreement between theory and experiment is quite good over the whole range of $Ka_0$.

We also observe a similar trend for the He atom. At 500 eV the agreement between theory and experiment is better than at 200 eV. This shows that the experimental cross sections, plotted as a function of $Ka_0$, are not independent of $E$, as demanded by the FBA. Although there is qualitative agreement between theory and experiment, the underestimation of the cross sections at 200 eV by the FBA is clear, particularly at small values of $Ka_0$.

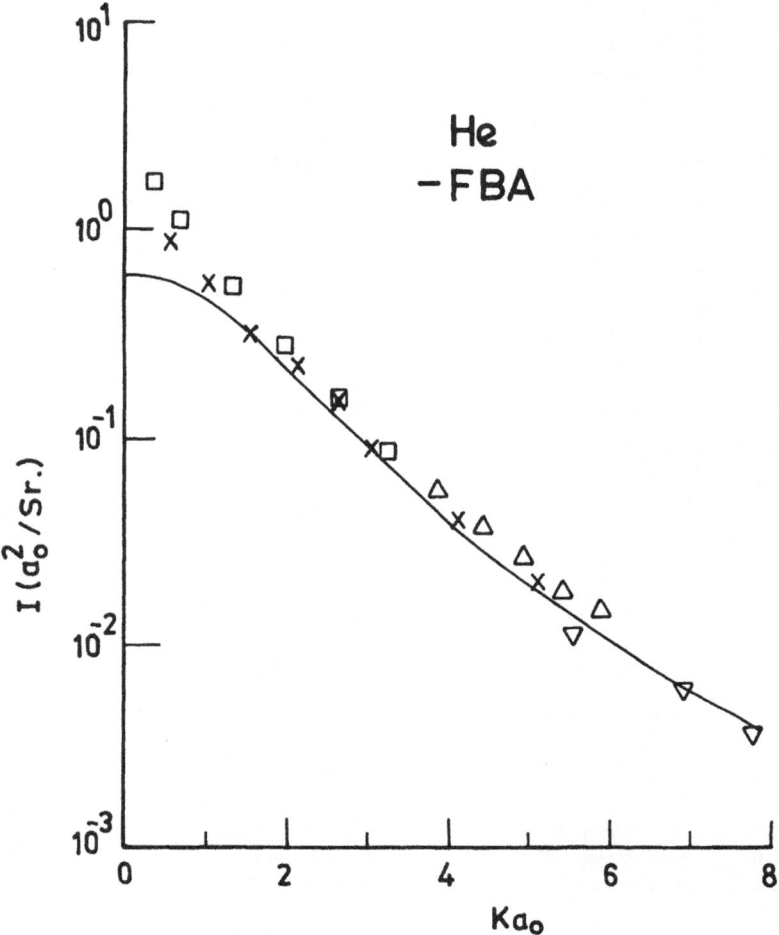

**FIGURE 7.1(b)** Same as Fig. 7.1(a) but for the helium atom. $\square$ and $X$ represent the experimental data of Jansen et al. (1976) at 200 and 500 eV, respectively. $\triangle$ and $\nabla$ represent the data of Sethuraman et al. (1974) for the same energies.

Figures 7.1(c) and (d) show that for heavier atoms the FBA overestimates the cross sections over most of the range of $Ka_0$ by an appreciable amount. There is not even qualitative agreement between the FBA cross sections and the experimental data. This shows that the effects due to the distortions in the wave functions of the projectile and the target by the atomic field and the Coulomb field of the incident electron (neglected by the FBA) are quite large. These distortions should be included in the theory to obtain better agreement with the experimental data. We shall consider these distortions in Sec. 7.7 and in Chapter 8.

As expected, the accuracy of the FBA increases with $E$. Hence, at high $E$ a comparison between the FBA cross sections, obtained with the Hartree–Fock (Weissbluth, 1978) target wave functions (which do not include correlation) and the experimental data yields information about the correlation between the atomic electrons.

### 7.3.2    Inelastic Scattering in the FBA

For the excitation of the hydrogen atom from the $1s$ to the $2s$ state we employ hydrogen-like wave functions and readily find that

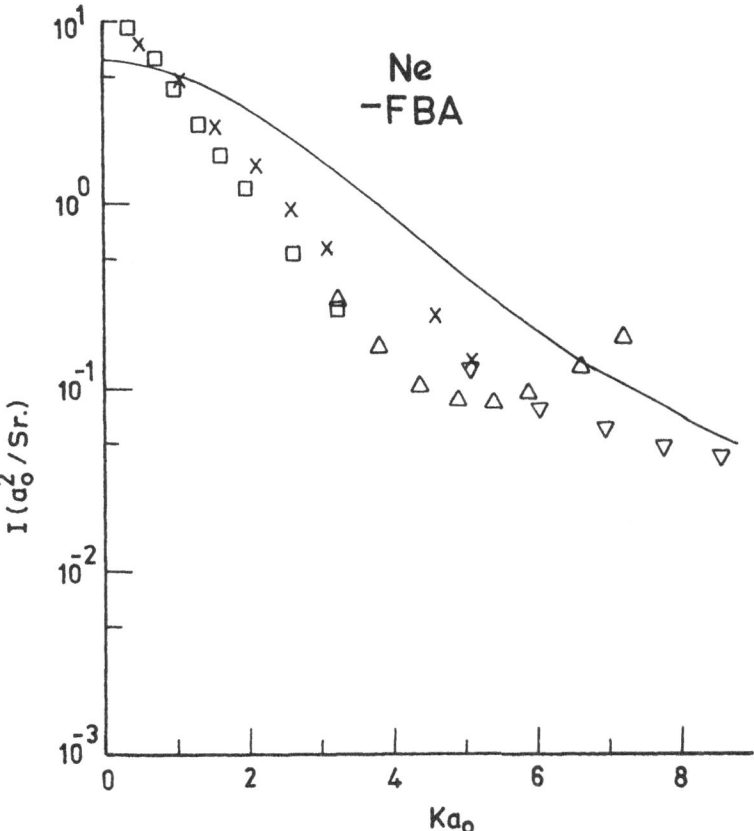

**FIGURE 7.1(c)** Same as Fig. 7.1(a) but for the neon atom. $\square$ and $X$ represent the experimental data of Jansen et al. (1976) at 200 and 500 eV, respectively. $\triangle$ and $\nabla$ represent the data of Gupta and Rees (1975a) for the same energies.

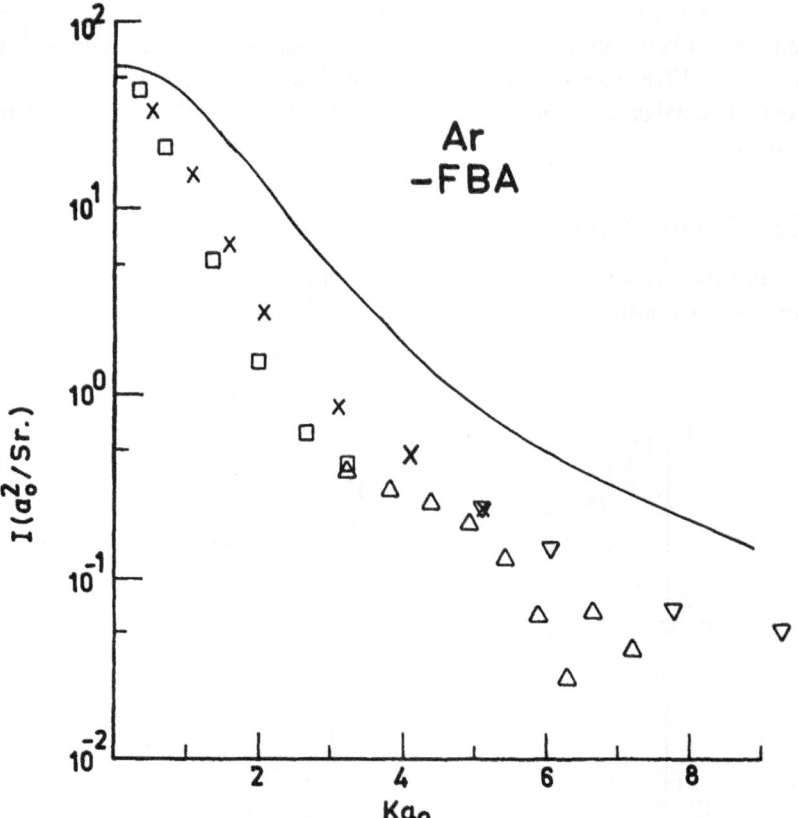

**FIGURE 7.1(d)** Same as Fig. 7.1(a) but for the argon atom. $\square$ and $X$ represent the experimental data of Jansen et al. (1976) at 100 and 500 eV, respectively. $\triangle$ represents the data of Gupta and Rees (1975b) for 100 eV and $\nabla$ that of Dubois and Rudd (1975) for 500 eV.

$$F_{2s,1s}(Ka_0) = \frac{2^8 \sqrt{2}\, Z_s^4 (Ka_0)^2}{\left(9Z_s^2 + 4K^2 a_0^2\right)^3} \tag{7.3.34}$$

Using (7.3.34) in (7.3.4), with the help of (7.2.19) and (7.3.9), we obtain the differential cross section as

$$I_{2s,1s}(Ka_0) = \frac{k_{2s}}{k_{1s}} \frac{2^{19} Z_s^8}{\left(9Z_s^2 + 4K^2 a_0^2\right)^6} a_0^2 \tag{7.3.35}$$

For $K = 0$ we have

$$I_{2s,1s}(K=0) = \frac{k_{2s}}{k_{1s}} \frac{2^{19}}{3^{12}} \frac{a_0^2}{Z_s^4} \tag{7.3.36}$$

At large $K$ the differential cross section falls as $K^{-12}$ in contrast with the $1s$–$1s$ elastic scattering, where the fall is proportional to $K^{-4}$. The generalized oscillator strength for the above transition is given by

$$f_{2s,1s}^G(Ka_0) = \frac{3 \, 2^{15} Z_s^{10} K^2 a_0^2}{(9Z_s^2 + 4K^2 a_0^2)^6} \tag{7.3.37}$$

As expected $f_{2s,1s}^G(K) \to 0$ at $K = 0$ because a $1s \to 2s$ transition is optically forbidden.

Let us now consider the excitation of a hydrogen atom from the $1s$ to $2p$ states. As the final state is now a $p$ state ($l = 1$), we have $m_l = 0, \pm1$ and there are three final states. However, we choose $K$ to be along the $z$-axis. Then the form factor $F_{2p,1s}$ is finite only for $m_l = 0$, and with a hydrogen-like wave function it is given by

$$F_{2p,1s} = i3 \times 2^7 \sqrt{2} \, \frac{Z_s^5 Ka_0}{(9Z_s^2 + 4K^2 a_0^2)^3} \tag{7.3.38}$$

The use of the above equation yields the differential cross section

$$I_{2p,1s} = \frac{k_{2p}}{k_{1s}} \frac{2^{17} 3^2 Z_s^{10}}{K^2 (9Z_s^2 + 4K^2 a_0^2)^6} \tag{7.3.39}$$

and the generalized oscillator strength is

$$f_{2p,1s}^G = \frac{2^{13} 3^3 Z_s^{12}}{(9Z_s^2 + 4K^2 a_0^2)^6} \tag{7.3.40}$$

Equation (7.3.39) shows that $I_{2p,1s} \to \infty$ at $K = 0$. This is a general feature of all optically allowed transitions. However, for inelastic collisions $K$ is never zero. Even in the forward direction it is finite. At large $K$ the differential cross section falls as $K^{-14}$. This fall is faster than that for elastic scattering and the $1s$–$2s$ excitation. Figure 7.2 shows the variation of the energy-independent quantities

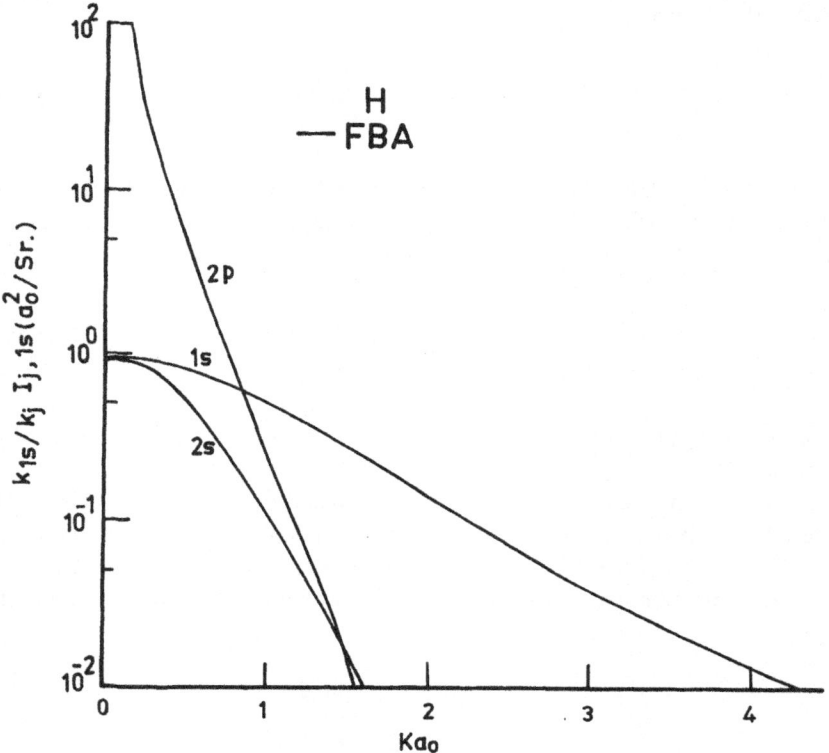

**FIGURE 7.2** Variation of $(k_{1s}/k_j)I_{j,1s}$(in $a_0^2/Sr$), the differential cross section multiplied by the weight factor $k_{1s}/k_j$, with $Ka_0$ for the ground state of the hydrogen atom due to electron impact in the FBA. The final states represented by $j$ are the $1s$, $2s$, and $2p$ states of the atom.

$(k_{1s}/k_j)I_{j,1s}$ (for $j = 1s$, $2s$, and $2p$) with $Ka_0$ in the FBA. From the figure we not that for small values of $Ka_0$, the reduced differential cross section $(k_{1s}/k_j)I_{j,1s}$ is largest for the $2p$ excitation and those for the elastic scattering and the $2s$ excitation are almost the same. All three curves fall with $Ka_0$, but the rate of fall is fastest for the $2p$ state and slowest for elastic scattering. As a result, for large values of $Ka_0$, elastic scattering dominates.

In Fig. 7.3 the variations of the GOS with $\ln Q$ (where $Q = Ra_0^2K^2$ is the recoil energy) are shown for the $1s$–$2s$ and $1s$–$2p$ excitations. The figure shows that $f_{2p,1s}^G$ is much larger than $f_{2s,1s}^G$ for all values of $Q$. At $Q = 0$, $f_{2p,1s}^G$ is finite (equal to the optical oscillator strength) and falls monotonically with the increase in $Q$. On the other hand, $f_{2s,1s}^G = 0$ at $Q = 0$. It increases with the increase in $Q$, reaches a peak, and then falls with further increase in $Q$.

The generalized oscillator strength for $|1s\rangle \rightarrow |nlm_l\rangle$ summed over all the allowed values of $m_l$ and $l$ for a hydrogenic atom is given by (Wadehra and Khare, 1993)

$$f_{n,1s}^{G} = \sum_{lm_l} f_{nlm_l,1s}^{G} = \frac{2^8 \varepsilon_{ex} n^7}{RZ_s^2} \left[ \tfrac{1}{3}(n^2 - 1) + \frac{n^2 Q}{Z_s^2 R} \right]$$

$$\frac{\left[(n-1)^2 + n^2 Q / Z_s^2 R\right]^{n-3}}{\left[(n+1)^2 + n^2 Q / Z_s^2 R\right]^{n+3}} \qquad (7.3.41)$$

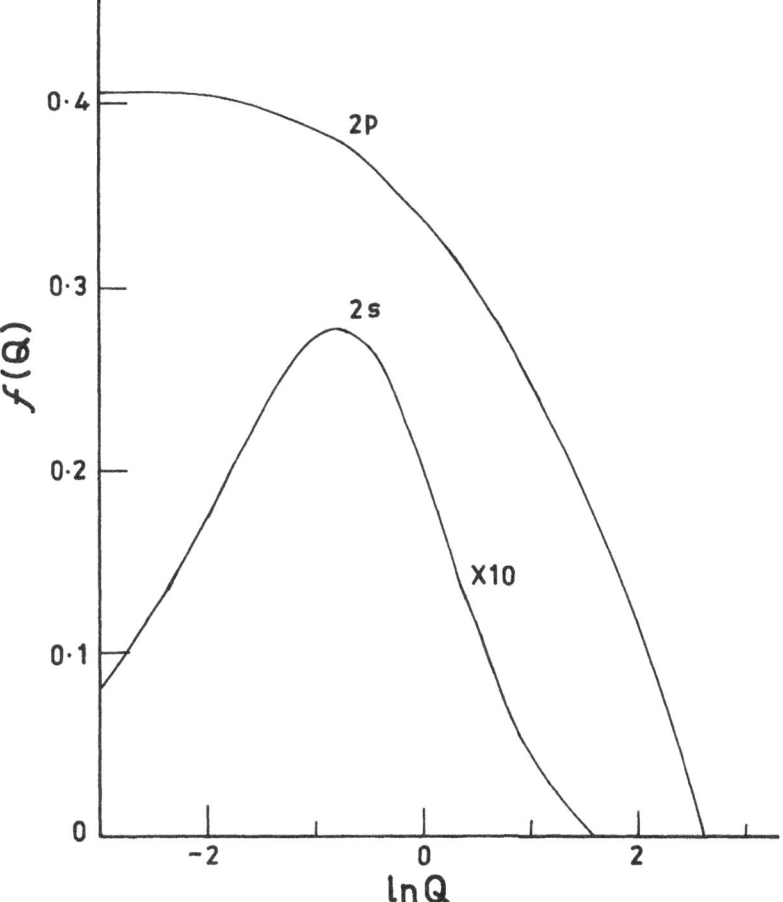

**FIGURE 7.3** Variation of the generalized oscillator strength for the 2s and 2p states for the hydrogen atom with $Q$ $(= Ra_0^2 K^2)$.

Hence, for $n = 2$,

$$f_{2p,1s}^G + f_{2s,1s}^G = 2^{15} \frac{\varepsilon_{ex}}{R} \frac{Z_s^8}{(9Z_s^2 + 4Q/R)^5} \tag{7.3.42}$$

As expected the above equation agrees with the sum of (7.3.37) and (7.3.40).

In terms of the GOS and the recoil energy $Q$ the integrated inelastic cross section in the FBA is given by

$$\sigma_{ji} = \frac{4\pi Ra_0^2}{E} \frac{R}{\varepsilon_{ex}} \int_{\ln Q_-}^{\ln Q_+} f_{ji}^G(Q) d(\ln Q) \tag{7.3.43}$$

where $Q_{+,-}$ are $Ra_0^2$ times $K_{max,min}^2$, which are given by (7.2.23). Hence,

$$\sigma_{2s,1s} = \frac{2^{17} RZ_s^8 \pi a_0^2}{5E} \left[ (9Z_s^2 + 4Q_-/R)^{-5} - (9Z_s^2 + 4Q_+/R)^{-5} \right] \tag{7.3.44}$$

At large $E$ the minimum value of $Q$ is close to zero and $Q_+$ is quite large. Hence, asymptotically,

$$\sigma_{2s,1s} = \frac{128}{5} \left( \frac{2}{3} \right)^{10} \frac{R}{E} \frac{1}{Z_s^2} \pi a_0^2 \tag{7.3.45}$$

Thus like $\sigma_{el}$, the excitation cross section $\sigma_{2s,1s}$ also falls as $E^{-1}$ at large $E$.

Let us consider optically allowed transitions. Figure 7.3 shows that for such transitions a large contribution to $\sigma_{ji}$ comes from small values of $Q$. In that region the $f_{ji}^G(Q)$ are nearly equal to the $f_{ji}^o$. We replace $f_{ji}^G(Q)$ by $f_{ji}^o$ in (7.3.43) and choose the upper limit of integration to be $\overline{Q}$ in such a way that the resultant cross section

$$\sigma_{ji} = \frac{4\pi a_0^2 R^2}{E\varepsilon_{ex}} f_{ji}^o \ln(\overline{Q}/Q_-) \tag{7.3.46}$$

is the same as that given by (7.3.43), i.e.,

$$\overline{Q} = Q_- \exp\left[ \frac{1}{f_{ji}^o} \int_{\ln Q_-}^{\ln Q_+} f_{ji}^G(Q) d(\ln Q) \right]$$

It is evident from the above equation that an evaluation of $\overline{Q}$ requires the distribution of $f_{ji}^G(Q)$ as a function of $Q$. At large $E$, the minimum value of $Q$ is approximately equal to $\varepsilon_{ex}^2/4E$. Hence, asymptotically,

$$\sigma_{ji} = \frac{4\pi a_0^2 R^2}{E\varepsilon_{ex}} f_{ji}^o \ln(c_{ex}E) \qquad (7.3.47)$$

where $c_{ex}(= 4\overline{Q}/\varepsilon_{ex}^2)$ is known as the Bethe collision parameter. The cross section given by (7.3.47) is referred to as the Bethe or the Bethe–Born cross section. It gives reasonable values at large $E$. A plot of $\sigma_{ji}E$ vs. ln $E$, known as the Bethe plot, is a straight line. Using experimental values of $\sigma_{ji}$ at large $E$, we can determine the value of the collision parameter $c_{ex}$ and $f_{ji}^o$ from the Bethe plot.

Equation (7.3.47) shows that for optically allowed transitions, $\sigma_{ji}$ falls as $E^{-1}$ ln $E$ at large $E$. On the other hand, $\sigma_{el}$ and $\sigma_{ji}$ for optically forbidden transitions fall as $E^{-1}$. Hence, at large $E$, the optically allowed transitions dominate. It is easy to see that in the Bethe–Born approximation

$$\sigma_{2p,1s} = \frac{128}{Z_s^2}\left(\frac{2}{3}\right)^{10} \frac{R}{E} \ln(c_{ex}E)\pi a_0^2 \qquad (7.3.48)$$

Figure 7.4 shows the variation of the total FBA cross section $\sigma_{ji}$ for the excitation of the ground state hydrogen atom due to electron impact, with $E$ in the intermediate energy range. The final states are $1s$, $2s$, and $2p$. The figure shows that for all values of $E$ the optically allowed excitation cross section $\sigma_{2p,1s}$ is the largest and the optically forbidden cross section $\sigma_{2s,1s}$ is the smallest. The elastic cross section $\sigma_{1s,1s}$ lies in between.

## 7.4  Effect of Electron Spin on Collisions

So far in this chapter the spin of the electrons has not been taken into account. Since electrons are fermions, the atomic wave function as well as the wave function of the system (atom plus incident electron), including spins, must be antisymmetric with respect to the exchange of any two electrons. Hence, we should have

$$v(r_1, s_1; r_2, s_2; \ldots; r_i, s_i; \ldots; r_z s_z)$$
$$= -v(r_i, s_i; r_2, s_2; \ldots; r_1, s_1; \ldots; r_z s_z) \qquad (7.4.1)$$

Similar equations are satisfied by the wave function of the system having $Z + 1$ electrons. A single determinant wave function satisfying the above property is given by

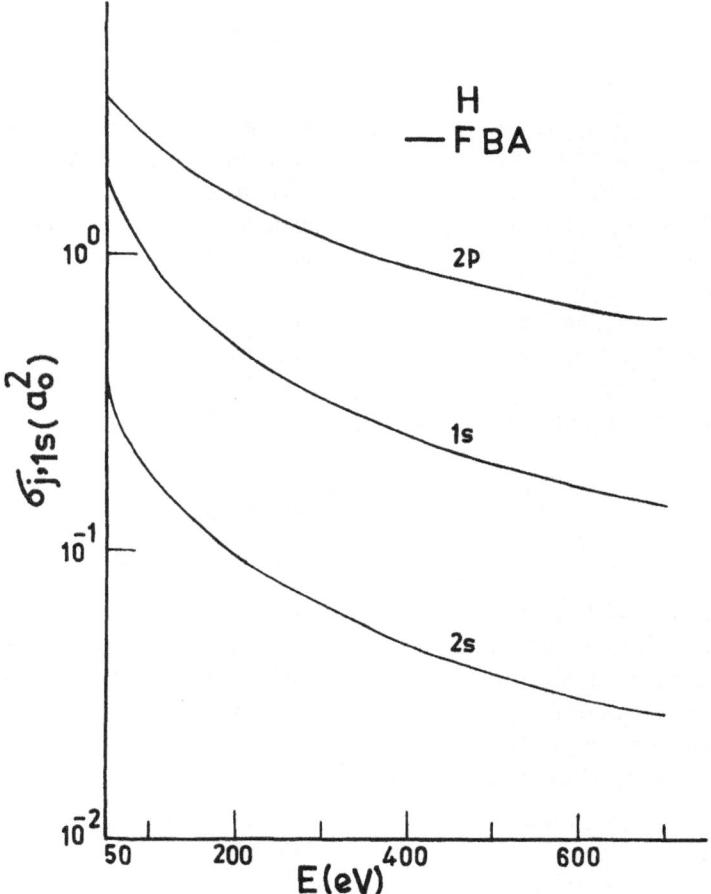

**FIGURE 7.4** Variation of $\sigma_{ji}$ with the energy $E$ of the incident electron: $\sigma_{ji}$ are the total cross sections in the FBA for a transition from $i = 1s$ to $j = 1s$, $2s$, and $2p$ states of the hydrogen atom due to electron impact.

$$\psi = \frac{1}{\sqrt{Z!}} \begin{vmatrix} x_1(r_1, s_1) & x_1(r_2, s_2) & \dots & x_1(r_z, s_z) \\ x_2(r_1, s_1) & x_2(r_2, s_2) & \dots & x_2(r_z, s_z) \\ \dots & \dots & \dots & \dots \\ x_z(r_1, s_1) & x_z(r_2, s_2) & \dots & x_z(r_z, s_z) \end{vmatrix} \qquad (7.4.2)$$

where the $x_i$ are single electron orbitals. The above determinant not only satisfies (7.4.1), but also obeys the Pauli exclusion principle. If any two atomic orbitals are identical, i.e., $x_i = x_j$, then $\psi$ vanishes. Hence, no two electrons can have the same quantum numbers, which is the Pauli exclusion principle.

Let us consider the collision of an electron with a one-electron atom. Then, according to (7.4.2), the initial and final wave functions of the system are given by

$$\psi_{k_i,i} = \frac{1}{\sqrt{2}}[\phi_{k_i}(r_1)\eta(1)v_i(r_2)\chi(2) - \phi_{k_i}(r_2)\eta(2)v_i(r_1)\chi(1)] \qquad (7.4.3a)$$

and

$$\psi_{k_j,j} = \frac{1}{\sqrt{2}}[\phi_{k_j}(r_1)\eta'(1)v_j(r_2)\chi'(2) - \phi_{k_j}(r_2)\eta'(2)v_j(r_1)\chi'(1)] \qquad (7.4.3b)$$

where $\phi_{k_i}$ and $\phi_{k_j}$ are the space orbitals of the projectile before and after the collision. The corresponding spin orbitals are $\eta$ and $\eta'$. Similarly, $v_i$ and $v_j$ are the space orbitals of the target before and after the collision, respectively. The corresponding spin orbitals are $\chi$ and $\chi'$. With the above antisymmetrized wave function, (7.2.10) and (7.2.11) give

$$f_{ji} = -\frac{4\pi^2 m}{\hbar^2}\langle\psi_{k_j,j}|T|\psi_{k_i,i}\rangle \qquad (7.4.4)$$

Use of (7.4.3a) and (7.4.3b) in (7.4.4) gives

$$\begin{aligned}
f_{ji} = -\frac{4\pi^2 m}{\hbar^2}\frac{1}{2}[&\langle\phi_{k_j}(r_1)\eta'(1)v_j(r_2)\chi'(2)|T|\phi_{k_i}(r_1)\eta(1)v_i(r_2)\chi(2)\rangle \\
&+ \langle\phi_{k_j}(r_2)\eta'(2)v_j(r_1)\chi'(1)|T|\phi_{k_i}(r_2)\eta(2)v_i(r_1)\chi(1)\rangle \\
&- \langle\phi_{k_j}(r_1)\eta'(1)v_j(r_2)\chi'(2)|T|\phi_{k_i}(r_2)\eta(2)v_i(r_1)\chi(1)\rangle \\
&- \langle\phi_{k_j}(r_2)\eta'(2)v_j(r_1)\chi'(1)|T|\phi_{k_i}(r_1)\eta(1)v_i(r_2)\chi(2)\rangle]
\end{aligned} \qquad (7.4.5)$$

In the first matrix element of the above equation electron 1 is free and 2 is bound before as well as after the scattering. Similarly, in the second term, electron 2 is always free and 1 is always bound. On the other hand, in the third term, initially electron 1 is free but after the scattering it becomes a bound electron. Similarly, electron 2, which was bound before the scattering, becomes free after the scattering. The same is true for the fourth term, where the electron switches over from the bound (free) to free (bound) orbital. Hence, the first and second terms of (7.4.5) represent direct scattering while the third and fourth terms represent exchange scattering.

Since the $T$ operator does not operate on spin wave functions, we have from (7.4.5)

$$f_{ji} = \tfrac{1}{2} f_d(k_j, k_i)[\langle \eta'(1)\chi'(2)|\eta(1)\chi(2)\rangle + \langle \eta'(2)\chi'(1)|\eta(2)\chi(1)\rangle]$$
$$-\tfrac{1}{2} f_{ex}(k_j, k_i)[\langle \eta'(1)\chi'(2)|\eta(2)\chi(1)\rangle + \langle \eta'(2)\chi'(1)|\eta(1)\chi(2)\rangle] \quad (7.4.6)$$

where the direct scattering amplitude is

$$f_d(k_j, k_i) = -\frac{4\pi^2 m}{\hbar^2}[\langle \phi_{k_j}(r_1)v_j(r_2)|T|\phi_{k_i}(r_1)v_i(r_2)\rangle] \quad (7.4.7a)$$

and the exchange scattering amplitude is given by

$$f_{ex}(k_j, k_i) = -\frac{4\pi^2 m}{\hbar^2}[\langle \phi_{k_j}(r_1)v_j(r_2)|T|\phi_{k_i}(r_2)v_i(r_1)\rangle] \quad (7.4.7b)$$

To evaluate terms like $\langle \eta'|\chi \rangle$, we consider collisions of the spin-up electrons with the unpolarized one-electron atoms $A$. The ensemble of unpolarized atoms is equivalent to a mixture of 50% of $A\uparrow$ and 50% of $A\downarrow$ atoms. Hence, we have the following three types of collisions

$$e\uparrow + A\downarrow \rightarrow e\uparrow + A\downarrow \quad (7.4.8)$$

$$e\uparrow + A\downarrow \rightarrow e\downarrow + A\uparrow \quad (7.4.9)$$

$$e\uparrow + A\uparrow \rightarrow e\uparrow + A\uparrow \quad (7.4.10)$$

The corresponding spin wave functions for the above collisions are given by

| Equation | $\eta$ | $\chi$ | $\eta'$ | $\chi'$ | |
|----------|--------|--------|---------|---------|---|
| (7.4.8) | $\alpha$ | $\beta$ | $\alpha$ | $\beta$ | (7.4.11) |
| (7.4.9) | $\alpha$ | $\beta$ | $\beta$ | $\alpha$ | |
| (7.4.10) | $\alpha$ | $\alpha$ | $\alpha$ | $\alpha$ | |

Putting the above spin functions into (7.4.7), we obtain

$$f_{ji}(k_j, k_i) = f_d(k_j, k_i) \qquad \text{for (7.4.8)}$$
$$= -f_{ex}(k_j, k_i) \qquad \text{for (7.4.9)}$$
$$= (f_d - f_{ex}) \qquad \text{for (7.4.10)}$$

Hence, (7.4.8) and (7.4.9) represent the direct and exchange collisions, respectively. Equation (7.4.10) contains $f_d$ as well as $f_{ex}$; hence, it represents mixed collisions. The differential cross sections for (7.4.8) to (7.4.10) for elastic collisions are $\frac{1}{2}|f_d|^2$, $\frac{1}{2}|f_{ex}|^2$, and $\frac{1}{2}|f_d - f_{ex}|^2$, respectively. Hence, the differential cross section for the sum of the above three types of collisions for elastic scattering is

$$I(\boldsymbol{k}_j, \boldsymbol{k}_i) = \frac{1}{2}\left(|f_d|^2 + |f_{ex}|^2 + |f_d - f_{ex}|^2\right) \tag{7.4.12}$$

The factor $\frac{1}{2}$ on the right-hand side is due to the fact that only half of the atoms participate in each type of collision given by (7.4.8) to (7.4.10). It is easy to see that (7.4.12) is also the differential cross section for collisions of spin-down electrons with unpolarized atoms. Even for collisions of unpolarized electrons with unpolarized atoms the differential cross section is given by (7.4.12) because only 50% of the incident electrons will be involved in the collisions given by (7.4.8) to (7.4.10). Hence, the cross section with $e\uparrow$ electrons will be only one-half that given by (7.4.12). The other half will be contributed by $e\downarrow$ incident electrons. Thus the sum of the two will again be equal to that given by (7.4.12). This equation can also be written as

$$I_{ji}(\boldsymbol{k}_j, \boldsymbol{k}_i) = \frac{3}{4}|f_d - g|^2 + \frac{1}{4}|f_d + g|^2 \tag{7.4.13}$$

where the exchange scattering amplitude is represented by $g(= f_{ex})$.

The above equation can also be obtained from the following simple consideration. Since the system $(e + A)$ has two electrons it will have four types of spin wave functions given by (5.6.1). Of these four wave functions three are symmetric. Hence, the corresponding space wave function has to be antisymmetric, and the scattering amplitude will be $f_d - g$. Similarly for the fourth antisymmetric spin wave function, with a symmetric spatial wave function, the scattering amplitude will be $f_d + g$. Since the weight factors for the symmetric and antisymmetric spin wave functions are $\frac{3}{4}$ and $\frac{1}{4}$, respectively, we get (7.4.13). This equation with $k_j = k_j$ is the same as (5.6.6), obtained for the collision between two unpolarized beams of electrons. However, the expressions for $f_d$ and $g$ in the two cases are different. For electron–atom inelastic collisions (7.4.13) is multiplied by $k_j/k_i$.

In Chapter 4, while discussing the collision of a free electron with a potential, we came across the spin-flip process. But the reason for this process in these two cases is different. In the collisions now being discussed spin-flip takes place because in the system we have two electrons having opposite spins. Loosely, we may say that in the collision the incident electron becomes bound and the atomic electron with opposite spin becomes free, which is detected by the detector. In reaction (7.4.9), the spin–orbit interaction is not considered. $M_S$ and $M_L$ of the system are separately conserved, i.e., $\Delta M_S = \Delta M_L = 0$. On the other hand, the spin-flip process, discussed in Chapter 4, is due to the spin–orbit interaction of the same electron. No second electron is present but the presence of the potential makes the spin–orbit interaction possible. In this case spin-flip changes $\Delta m_s$ by one unit and this is compensated by a corresponding change in $m_l$ in such a manner that $\Delta m_s + \Delta m_l = 0$. Hence, $m_j$ remains unchanged.

The scattering matrix $S(\theta)$ is a $4 \times 4$ matrix. Since without the spin–orbit interaction $M_S(= m_{s1} + m_{s2})$ is equal to $M_S'(= m_{s1}' + m_{s2}')$, $S(\theta)$ for the reactions (7.4.8) to (7.4.10) is a diagonal matrix, whose rows and columns are given by $(M_S', m_{s1}', m_{s2}')$ and $(M_S, m_{s1}, m_{s2})$, respectively. The matrix is

$$
\begin{array}{cccc}
M_s & 1 & 0 & 0 & -1 \\
m_{s1} & \frac{1}{2} & \frac{1}{2} & -\frac{1}{2} & -\frac{1}{2} \\
m_{s2} & \frac{1}{2} & -\frac{1}{2} & \frac{1}{2} & -\frac{1}{2}
\end{array}
$$

$$
\begin{array}{ccc}
M_s' & m_{s1}' & m_{s2}' \\
1 & \frac{1}{2} & \frac{1}{2} \\
0 & \frac{1}{2} & -\frac{1}{2} \\
0 & -\frac{1}{2} & \frac{1}{2} \\
-1 & -\frac{1}{2} & -\frac{1}{2}
\end{array}
\qquad
S(\theta) =
\begin{pmatrix}
f-g & 0 & 0 & 0 \\
0 & f & -g & 0 \\
0 & -g & f & 0 \\
0 & 0 & 0 & f-g
\end{pmatrix}
\qquad (7.4.14)
$$

This $4 \times 4$ matrix breaks up into three submatrices of dimensions $(1 \times 1)$, $(2 \times 2)$, and $(1 \times 1)$. We also have

$$
S(\theta) = f(\theta) - \tfrac{1}{2}(1 + \sigma_1 \otimes \sigma_2) g(\theta) \qquad (7.4.15)
$$

where the outer product of the two Pauli matrices is

$$
\sigma_1 \otimes \sigma_2 = \sigma_{x1} \otimes \sigma_{x2} + \sigma_{y1} \otimes \sigma_{y2} + \sigma_{z1} \otimes \sigma_{z2}
$$

$$
=
\begin{pmatrix}
1 & 0 & 0 & 0 \\
0 & -1 & 2 & 0 \\
0 & 2 & -1 & 0 \\
0 & 0 & 0 & 1
\end{pmatrix}
\qquad (7.4.16)
$$

For the collision of an electron with a helium atom the system has three electrons. Hence, there will be three spatial and three spin orbitals. The antisymmetric wave function of the system will have 9 terms, so the $T$ matrix will have 81 terms. However, the position is considerably simplified if we confine ourselves to elastic scattering of the electrons by the ground state of the helium atom. Since the helium atom will be in the singlet state after the scattering, the incident electron can be exchanged only with the atomic electron that has the identical spin. The spin wave function of these two electrons, considered together, will be symmetric; hence, we are required to have an antisymmetric combination of the scattering amplitude, which will be $f_d - g$.

For a perfect experiment, as discussed in the Sec. 4.5, we have to determine $|f_d|$, $|g|$, and their relative phase $\phi_{rel}$. To obtain these quantities, the unpolarized electrons are scattered by the unpolarized atoms and the differential cross sections $I_{un}$ are measured. This is followed by scattering of the partially polarized spin-up electrons, having $P_e$ as their degree of polarization, by unpolarized atoms. In this experiment $P'_e$ and $P'_A$, the degrees of polarization of the scattered electrons and the recoiled atoms, are measured. Now, (7.4.9) shows that the cross section for the spin-flip process is $\frac{1}{2}|g|^2 P_e$ and the cross section for those collisions, in which spins do not change, from (7.4.8) and (7.4.10), is $\frac{1}{2}(|f_d|^2 + |f_d - g|^2)P_e$. In the incident beam the fraction of unpolarized electrons is $(1 - P_e)$. Half of them behave as $\alpha$ electrons and the other half as $\beta$ electrons. Therefore, the differential cross section for detecting scattered $\alpha$ electrons is

$$I\uparrow = P_e\left[\tfrac{1}{2}|f_d|^2 + \tfrac{1}{2}|f_d - g|^2\right] + \tfrac{1}{2}(1 - P_e)I_{un} \qquad (7.4.17)$$

Similarly, for detecting $\beta$ electrons,

$$I\downarrow = \tfrac{1}{2}P_e|g|^2 + \tfrac{1}{2}(1 - P_e)I_{un} \qquad (7.4.18)$$

As expected

$$I_{un} = I\uparrow + I\downarrow \qquad (7.4.19)$$

By definition, the degree of polarization of the scattered electrons is

$$P'_e = \frac{I\uparrow - I\downarrow}{I_{un}} \qquad (7.4.20)$$

Using (7.4.17) to (7.4.19) in (7.4.20), we get

$$|g|^2 = I_{un}(1 - P'_e/P_e) \qquad (7.4.21)$$

The above equation shows that $P'_e$ is less than $P_e$. Thus the incident beam is partially depolarized by the exchange process but at the same time the recoiled atoms are partially polarized. In reaction (7.4.9), the atom flips its spin. The degree of polarization of the recoiled atoms, obtained from (7.4.8) to (7.4.10), is

$$P'_A = P_e\left(1 - |f_d|^2/I_{un}\right) \qquad (7.4.22)$$

Hence,

$$|f_d|^2 = I_{\text{un}}(1 - P'_A/P_e) \qquad (7.4.23)$$

and

$$|f_d - g|^2 = I_{\text{un}}(P'_e + P'_A)/P_e \qquad (7.4.24)$$

We take $f_d = |f_d|\exp(i\gamma_1)$ and $g = |g|\exp(i\gamma_2)$. Hence, from (7.4.24)

$$\cos(\phi_{\text{rel}}) = \cos(\gamma_1 - \gamma_2) = \frac{1 - (P'_e + P'_A)/P_e}{[(1 - P'_A/P_e)(1 - P'_e/P_e)]^{1/2}} \qquad (7.4.25)$$

We can also obtain $\cos(\phi_{\text{rel}})$ by determining the asymmetry parameter $A(\theta)$, is defined by

$$A(\theta) = \frac{I(\uparrow\downarrow) - I(\uparrow\uparrow)}{I(\uparrow\downarrow) + I(\uparrow\uparrow)} \qquad (7.4.26)$$

where $I(\uparrow\uparrow)$ and $I(\uparrow\downarrow)$ are the cross sections for the parallel and antiparallel orientations of the spins of the electrons and the atoms. Use of (7.4.8) to (7.4.13) gives

$$A(\theta) = \frac{f_d g^* + f_d^* g}{2L_{\text{un}}} = \frac{|f_d||g|\cos(\phi_{\text{rel}})}{I_{\text{un}}} = \frac{|f_d + g|^2 - |f_d - g|^2}{|f_d + g|^2 + 3|f_d - g|^2} \qquad (7.4.27)$$

In general when $P_e$ and $P_A$ are neither parallel nor antiparallel we have (Kesseler, 1985, 1991)

$$I(\theta) = I_{\text{un}}[1 - A(\theta)P_e \cdot P_A] \qquad (7.4.28)$$

A number of measurements of $A(\theta)$ have been made (see Baum et al., 1985, 1988a,b; Fletcher et al., 1985; and Kesseler, 1991).

It is evident that the above measurements yield only $\cos(\phi_{\text{rel}})$ and not the unambiguous $\phi_{\text{rel}}$. To obtain this quantity let us consider $P'_e$ the degree of polarization of the scattered electrons, given by (Kesseler, 1991)

$$P'_e = \frac{\left(1 - |f_d|^2/I_{\text{un}}\right)P_A + \left(1 - |g|^2/I_{\text{un}}\right)P_e + i(f_d^* g - f_d g^*)(P_e \times P_A)/(2L_{\text{un}})}{1 - A(\theta)P_e \cdot P_A} \qquad (7.4.29)$$

From the above equation $(P_e')_n$, the component of $P_e'$ perpendicular both to $P_e$ and $P_A$, is

$$
\begin{aligned}
(P_e')_n &= i \frac{(f_d^* g - f_d g^*) P_e P_A \sin \phi'}{2 I_{un}(1 - A(\theta) P_e P_A \cos \phi')} \\
&= \frac{|f_d| |g| \sin(\phi_{rel}) P_e P_A \sin \phi'}{I_{un}(1 - A(\theta) P_e P_A \cos \phi')}
\end{aligned}
\tag{7.4.30}
$$

where $\phi'$ is the angle between $P_e$ and $P_A$. Thus a measurement of $(P_e')_n$ yields $\sin(\phi_{rel})$ and knowing $\cos(\phi_{rel})$ and $\sin(\phi_{rel})$ an unambiguous value of $\phi_{rel}$ is obtained. Instead of measuring $(P_e')_n$ one may measure $(P_A')_n$. The relationship of $P_A'$ with $|f|^2, |g|^2, P_e,$ and $P_A$ is obtained by interchanging $P_e$ and $P_A$ in (7.4.29). Thus

$$
P_A' = \frac{\left(1 - |f_d|^2/I_{un}\right) P_e + \left(1 - |g|^2/I_{un}\right) P_A - i(f_d^* g - f_d g^*)/2 I_{un}) P_e \times P_A}{1 - A(\theta) P_e \cdot P_A}
\tag{7.4.31}
$$

Such experiments have been performed for elastic scattering, excitation, and ionization collisions (see McClelland et al., 1985, 1987; Baum et al., 1985, 1988a,b; and Kesseler, 1985, 1991).

It may be noted that $|g|$ can also be obtained by the scattering of partially polarized electrons by unpolarized atoms. In this case $P_e'$ from (7.4.29) is

$$
P_e' = \left(1 - |g|^2/I_{un}\right) P_e
\tag{7.4.32}
$$

Hence, a measurement of $P_e'$ enables us to determine $|g|$.

Let us consider (7.4.29) under two extreme cases: (a) $g = 0$ and (b) $f = 0$. For (a) $I_{un} = |f_d|^2$ and $A(\theta) = 0$. Hence, $P_e' = P_e$. This clearly shows that exchange is necessary for polarization transfer. For (b) we have $I_{un} = |g|^2$ and $A(\theta) = 0$. Hence, $P_e' = P_A$. This is due to the fact that in a pure exchange scattering all the scattered electrons come from the target.

$P_A$ and $P_A'$ are measured with a Stern–Gerlach polarimeter. A schematic arrangement of the apparatus, which is kept under ultrahigh vacuum, is shown in Fig. 7.5. The neutral atomic beam passes through a highly inhomogeneous magnetic field $B$ produced by specially designed pole tips. Let the atomic beam be moving along the $x$ direction with a velocity $v$. In Fig. 7.5(a) the plane of the paper is the $x$–$z$ plane. If we look into the atomic beam the cross section of the apparatus is shown in Fig. 7.5(b).

Due to electronic spin the neutral atom behaves like a tiny magnet and has a magnetic moment $\mu$. This $\mu$ interacts with the magnetic field $B$ and a force $F_z$ acts on the atom. The magnitude of $F_z$ is given by

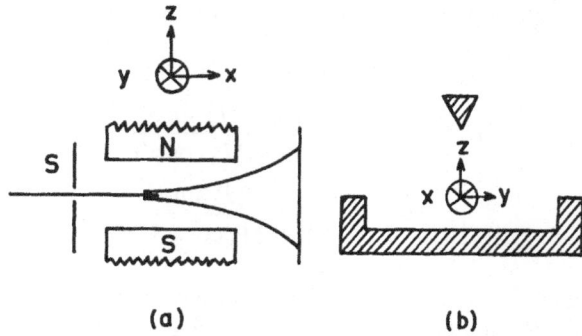

**FIGURE 7.5** The Stern–Gerlach experiment with free atoms having $S = \frac{1}{2}$. The atomic beam travels along the $x$-axis and the inhomogeneous magnetic field is directed along the $z$-axis.

$$F_z = \mu \cos\theta \frac{\partial B_z}{\partial z} \qquad (7.4.33)$$

where $\theta$ is the angle between $\mu$ and $z$. Suppose the atoms are under the influence of the magnetic field for a time $t$. Then the deflection $z$ of the atoms is given by

$$z = \tfrac{1}{2}(F_z/M)t^2 \qquad (7.4.34)$$

where $M$ is the mass of the atom. If the magnetic field acts on the atoms over a length $L$ then $t = L/v$ and

$$z = \frac{\mu \cos\theta}{2M}(L/v)^2 \frac{\partial B_z}{\partial z} \qquad (7.4.35)$$

For a one-electron atom the quantum mechanical values of $\cos\theta$ are only $\pm 1$. These two values correspond to $m_s = \pm\frac{1}{2}$, respectively. Thus the neutral atomic beam splits into two fully polarized atomic beams.

If we perform the Stern–Gerlach experiment with the electrons, then $M$ in (7.4.35) is replaced by $m$, the mass of the electron, and $\mu_z = \mu \cos\theta = \pm\frac{1}{2}\hbar e/mc$. Hence, the angular separation of the two electron beams is

$$\chi = 2z/L = \frac{1}{2c}\frac{\hbar e}{m^2}\frac{1}{L}\frac{\partial B_z}{\partial z}t^2 \qquad (7.4.36)$$

As electrons are charged particles, the Lorentz force also acts on them. The $z$ component of the Lorentz force is $ep_x B_y/mc$, so the angular deflection due to this force is

$$\Phi = \frac{ep_x B_y t^2}{2m^2 cL} \tag{7.4.37}$$

The incident electron beam moving in the $x$ direction will have a certain cross section. Let its width in the $y$ and $z$ directions be $\Delta y$ and $\Delta z$, respectively. Due to $\Delta y$, $\Phi$ will range from $\Phi$ to $\Phi + \Delta \Phi$. The value of $\Delta \Phi$ as obtained from (7.4.37) is

$$\Delta \Phi = \frac{ep_x t^2}{2m^2 cL} \frac{\partial B_y}{\partial y} \Delta y \tag{7.4.38}$$

The magnetic field $B$ satisfies

$$\nabla \cdot B = \partial B_z / \partial z + \partial B_y / \partial y + \partial B_x / \partial x = 0$$

Since $B$ does not vary with $x$ we have

$$\partial B_z / \partial z = -\partial B_y / \partial y \tag{7.4.39}$$

Hence in magnitude

$$\Delta \Phi = \frac{ep_x t^2}{2m^2 cL} \frac{\partial B_z}{\partial z} \Delta y \tag{7.4.40}$$

$$\therefore \quad \chi / \Delta \Phi = \hbar / (p_x \Delta y) \approx \Delta p_y / p_x \tag{7.4.41}$$

where, using the uncertainty principle, we have taken $\Delta y = \hbar / \Delta p_y$. To define a beam we must have $p_x \gg \Delta p_y$. Therefore $\Delta \Phi \gg \chi$. Thus the spread of the electron beam is much greater than the separation produced by the inhomogeneous magnetic field. Therefore we conclude that due to the Lorentz force the Stern–Gerlach apparatus cannot be employed to produce polarized electrons from the unpolarized electron beam. For the same reason we cannot use the Stern–Gerlach experiment to measure $P_e$ and $P_e'$. The polarization of the electron is measured, as discussed in Sec. 4.6, by the Mott detector.

## 7.5   The First-Order Exchange Amplitude

A replacement of the $T$ operator in (7.4.7b) by the interaction energy operator $V$ yields the exchange scattering amplitude $g$ in the *Born–Oppenheimer approximation*. In this approximation $g$ is correct up to first order and for the one-electron atom it is given by

$$g_{ji} = -\frac{m}{2\pi\hbar^2} \int e^{-ik_j \cdot r_2} v_j^*(r_1) V(r_1, r_2) e^{ik_i \cdot r_1} v_i(r_2) dr_1 dr_2 \qquad (7.5.1)$$

However, the Born–Oppenheimer approximation is not found to be a successful approximation, as it suffers from the following defects:

1. If $V$ is replaced by $V + C$, where $C$ is a constant, the value of $g$ changes. This is physically incorrect, because a constant potential produces zero force so the scattering amplitude should not change.

2. The value of $g$ also depends upon the projectile–nucleus interaction $-Z/r_1$. This again seems to be incorrect because the exchange only takes place between identical particles. Hence, the nucleus is not expected to play any role in a first-order approximation.

3. The above approximation also suffers from the post-prior discrepancy. Before the collision, the interaction potential energy is given by

$$V_B = -Ze^2/r_1 + e^2/r_{12}$$

In the collision electrons 1 and 2 are exchanged. Hence, after the collision

$$V_A = -Ze^2/r_2 + e^2/r_{12}$$

It is evident that $V_B \neq V_A$ and that they give different values of $g$.

4. If we break $g$ into partial waves, i.e., $g = \sum_l g_l$, then it is found that $\sigma_l$ due to the $l$th partial wave becomes greater than $4\pi(2l + 1)/k_i^2$. Thus this approximation violates the partial cross section theorem [see (3.9.27)].

The primary reason behind the above defects is the fact that the initial wave function $\exp(ik_i \cdot r_1)v_i(r_2)$ is not orthogonal to the final wave function $\exp(ik_j \cdot r_2)v_i(r_1)$. A number of attempts have been made to remove these defects.

One of the more successful such attempts was made by Ochkur (1964), who expanded $g$ given by (7.5.1) in the power of $k_i^{-1}$ and retained only the leading term, which behaves as $k_i^{-2}$. It is found that the term $-Ze^2/r_2$ gives a contribution that falls faster than $k_i^{-2}$. Thus it is neglected. The electron–electron interaction gives

$$g_{ji} = -\frac{me^2}{2\pi\hbar^2} \int e^{-ik_j \cdot r_2} v_j^*(r_1) \frac{1}{r_{12}} e^{ik_i \cdot r_1} v_i(r_2) dr_1 dr_2 \qquad (7.5.2)$$

The Fourier transform of $1/r_{12}$ is given by

$$\frac{1}{r_{12}} = \frac{1}{2\pi^2} \int \frac{e^{iq\cdot(r_1-r_2)}}{q^2} dq \tag{7.5.3}$$

Putting $q = k_i + p$ into the above equation we get

$$\frac{1}{r_{12}} = \frac{e^{ik_i\cdot r_{12}}}{2\pi^2} \int \frac{e^{iq\cdot r_{12}}}{k_i^2 + p^2 + 2k_i\cdot p} dp \tag{7.5.4}$$

The term $e^{ip\cdot r_{12}}$ oscillates with $p$. Thus the major contribution to the above integral comes from the small values of $p$. Hence, at high energies $p^2 + 2k_i\cdot p$ is neglected in comparison to $k_i^2$, and (7.5.4) reduces to

$$\frac{1}{r_{12}} = \frac{4\pi}{k_i^2} \delta(r_1 - r_2)\exp(ik_i\cdot r_{12}) \tag{7.5.5}$$

Putting (7.5.5) into (7.5.2) and integrating over $r_2$, we get

$$g_{OC} = -\frac{2}{k_i^2 a_0} \int e^{iK\cdot r_1} v_j^*(r_1)v_i(r_1)dr_1 \tag{7.5.6}$$

The above equation gives the exchange scattering amplitude in the Ochkur approximation. It is correct up to $k_i^{-2}$ and is free from the post-prior discrepancy. Further, a constant added to $V$ yields an additional scattering amplitude that falls faster than $k_i^{-2}$ and so it is also neglected. In a good number of cases the values of $g_{OC}$ are found to be reasonable. Hence, it has been employed at all energies. It is easy to see that for inelastic electron–hydrogen atom collisions

$$g_H^{OC} = \frac{K^2}{k_i^2} f_H^{B1} \tag{7.5.7}$$

Use of the above equation in (7.4.13) shows that up to first order in the interaction, the differential cross section for the inelastic collisions of electrons with the hydrogen atom correct up to $k_i^{-2}$ is given by

$$I_{ji}(k_i, k_i) = \frac{k_j}{k_i} \left| f_{ji}^{B1}(k_j, k_i) \right|^2 (1 - K^2/k_i^2 + K^4/k_i^4) \tag{7.5.8}$$

Hence, the integrated cross section in the Born–Ochkur approximation is

$$\sigma_{ji} = \frac{2\pi}{k_i^2} \int_{K_{min}}^{K_{max}} \left| f_{ji}^{B1} \right|^2 F_{ex}(K, k_i) K\, dK \tag{7.5.9}$$

where the exchange factor is

$$F_{ex} = 1 - K^2/k_i^2 + K^4/k_i^4 \qquad (7.5.10)$$

Similarly, for the singlet–singlet excitation of the helium atom we have

$$g_{He}^{OC} = \frac{K^2}{2k_i^2} f_{He}^{B1} \qquad (7.5.11)$$

A factor of two appears in the denominator because out of the two atomic electrons only one has the same spin as the incident electron.

## 7.6 Ionization of Atoms in the First Born Approximation

A single ionizing collision of an electron with an atom $A$ can be represented by

$$e + A \rightarrow e + A^+ + e \qquad (7.6.1)$$

Thus in the final channel we have two free electrons. Further, for a given $E$, the ionization cross section $\sigma_i$ of atoms is fivefold differential: twofold with respect to the scattering angles $(\theta, \phi)$, another twofold with respect to the direction $(\theta_e, \phi_e)$ of the ejected electron, and onefold with respect to the energy $\varepsilon_e$ of the ejected electron. The energy $\varepsilon_e$ varies continuously from zero to $E - I$, $I$ being the ionization potential of the atom. Owing to the above reasons, a quantum mechanical evaluation of $\sigma_i$ is quite involved. Most of the calculations are limited to the FBA and semiempirical methods based on classical binary encounter theory (Grizinski, 1965a,b,c), the FBA, and the Ochkur approximations (Younger and Mark, 1985).

For an ionizing collision, the generalized oscillator strength for a bound–bound transition is to be modified for a bound–free transition. This can be achieved by replacing $\varepsilon_{ex}$ in (7.3.16) by $W$, the energy lost by the incident electron in the ionizing collision, and the matrix element $|\langle j| e^{iK \cdot r} |i\rangle|^2$ by $|\langle k| e^{iK \cdot r} |i\rangle|^2 dk$, where $|k\rangle$ represents the ejected electron and $\hbar k$ is its momentum. Thus (7.3.16) changes to

$$df = \frac{W}{R} \frac{1}{K^2 a_0^2} |\langle k| e^{iK \cdot r} |i\rangle|^2 dk \qquad (7.6.2)$$

To simplify the notation the superscript $G$ of the continuum generalized oscillator strength $df$ (CGOS) is dropped. We note that for a given $K$, $df$ is a dimensionless quantity but it is threefold differential. Now,

$$W = \varepsilon_e + I = Ra_0^2 k^2 + I$$

Hence,

$$dW = 2Ra_0^2 k \, dk$$

Using the above equation in (7.6.2) and integrating over the direction of the ejected electron, we get

$$\frac{df(W, K^2)}{dW} = \frac{kW}{2R^2 a_0^2} \frac{1}{K^2 a_0^2} \int |\langle k|e^{iK \cdot r}|i \rangle|^2 \, d\hat{k} \tag{7.6.3}$$

The above equation gives the CGOS per unit energy loss. Using the ground state hydrogenic wave function, given by (6.11.1), for the initial state and the Coulomb wave for the continuum state, we evaluate the transition matrix element and after somewhat lengthy algebra obtain (Mott and Massey, 1965; Khare et al., 1993; Saksena, 1994)

$$I_s \frac{df(W, Q)}{dW} = \frac{df(\alpha^2, \beta^2)}{d\alpha^2}$$

$$= \frac{2^7(1+\alpha^2)[(1+\alpha^2)/3+\beta^2]}{\left[1+2(\alpha^2+\beta^2)+(\alpha^2-\beta^2)^2\right]^3} F(\beta^2, \alpha^2) \tag{7.6.4}$$

where $I_S = Z_s^2 R$, $\alpha = ka_0/Z_s$, $\beta = Ka_0/Z_s$, $W = I_s(1 + \alpha^2)$, and

$$F(\beta^2, \alpha^2) = \exp\left\{-\frac{2}{\alpha}\arctan[2\alpha/(1+\beta^2-\alpha^2)]\right\}$$

$$\left\{1-\exp\left(-\frac{2\pi}{\alpha}\right)\right\}^{-1} \quad \text{for } a^2 \geq 0 \tag{7.6.5}$$

$$= \exp\left\{-\frac{1}{(-\alpha^2)^{1/2}}\ln\left[\frac{\beta^2+\left[1+(-\alpha^2)^{1/2}\right]^2}{\beta^2+\left[1-(-\alpha^2)^{1/2}\right]^2}\right]\right\} \quad \text{for } \alpha^2 < 0 \tag{7.6.6}$$

We also note that the recoil energy $Q = \beta^2 Z_s^2 R$.

For a one-electron atom, $Z_S = Z$ and $\alpha^2$ is always positive. Figure 7.6 shows the variation of $df(\alpha^2, \beta^2)/d\alpha^2$ for the hydrogen atom as a function of $\ln \beta^2$ for $\alpha^2 = 0.1025$ ($W = 15 \, \text{eV}$) and 2.675 ($W = 50 \, \text{eV}$), respectively. The ionizing collisions corresponding to small values of $\alpha^2$ or $W$ (curve A) are known as soft

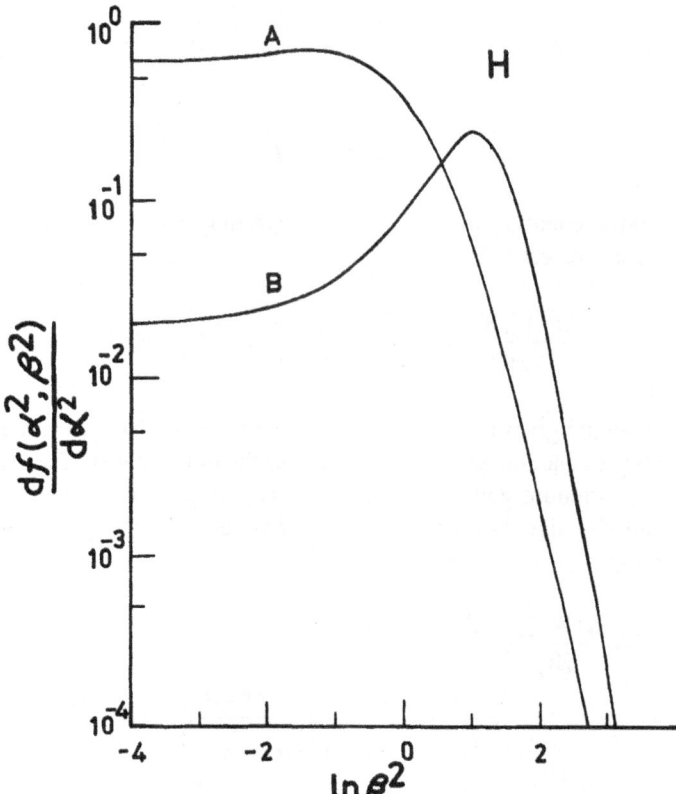

**FIGURE 7.6** Variation of the scaled continuum generalized oscillator strength $df(\alpha^2/\beta^2)/d\alpha^2$ with $\ln \beta^2$ for a hydrogen atom due to electron impact. Curves A and B correspond to the energy loss of 15 and 50 eV, respectively, by the incident electrons.

collisions, whereas those involving large values of $\alpha^2$ are said to be hard collisions. We note that curves $A$ and $B$ of the figure are similar to the curves for $2p$ and $2s$, respectively, of Fig. 7.3.

The above similarity indicates that soft collisions are due to dipole interaction and that hard collisions involve forbidden transitions. The peak in curve $B$ is known as the Bethe peak, and the locus of the Bethe peaks corresponding to different values of $W$ gives rise to the Bethe ridge. As an approximation, the ionization cross section due to hard collisions is evaluated by considering collisions between two free electrons (Sec. 5.6). As discussed in Sec. 6.11, Eq. (7.6.4) can also be utilized to calculate $K$-shell electron impact ionization of multielectron atoms. In this case $Z_S = Z - 0.3$, $W = Ra_0^2\delta^2 + I_S$, $\alpha = \delta a_0/Z_S$, and $I_K < I_S$.

Hence, $\alpha^2$ will be negative for $I_K < W < I_S$ and (7.6.6) is to be used for $F(\beta^2, \alpha^2)$. Further, as there are two $K$-shell electrons, (7.6.4) is to be multiplied by 2. It is evident from the above equation that the scaled continuum generalized oscillator strength $df(\alpha^2, \beta^2)/d\alpha^2$ is the same function of $\alpha^2$ and $\beta^2$ for all the atoms. Hence, $df(\alpha^2, \beta^2)/d\alpha^2$ generates a universal Bethe surface when plotted as a function of $\alpha^2$ and $\beta^2$ (Khare et al., 1993).

Equation (7.6.4) can also be derived by extending (7.3.41) to the continuum states. For the $K$-shell excitation of a multielectron atom, (7.3.41) gives

$$f_{n,1s}(Q) = \frac{2^9 \varepsilon_{ex}}{RZ_s^2} n^{-3} \left[ \frac{1}{3}(1 - 1/n^2) + \beta^2 \right] \frac{\left[ (1 - 1/n)^2 + \beta^2 \right]^{n-3}}{\left[ (1 + 1/n)^2 + \beta^2 \right]^{n+3}} \tag{7.6.7}$$

To make a transition from the $n$th bound state to the continuum state we replace $\varepsilon_{ex}$ by $W$. This shows that $n$ in the above equation is to be replaced by $i/\alpha$. With the above changes $f_{n,1s}$ becomes a differential quantity and for $K$-shell ionization it is given by

$$RZ_s^2 \frac{df(W,Q)}{dW} = 2^8(1+\alpha^2)[\tfrac{1}{3}(1+\alpha^2) + \beta^2] \frac{\left[ (1+i\alpha)^2 + \beta^2 \right]^{-3}}{\left[ (1-i\alpha)^2 + \beta^2 \right]^3}$$
$$\times \left[ \frac{(1+i\alpha)^2 + \beta^2}{(1-i\alpha)^2 + \beta^2} \right]^{i/\alpha} \tag{7.6.8}$$

because with $dn = 1$

$$dW = RZ_s^2 d(\alpha^2) \rightarrow -RZ_s^2 d\left( \frac{1}{n^2} \right) = \frac{2RZ_s^2}{n^3}$$

Further,

$$\frac{\left[ (1+i\alpha)^2 + \beta^2 \right]^{-3}}{\left[ (1-i\alpha)^2 + \beta^2 \right]^3} = \left[ 1 + 2(\beta^2 + \alpha^2) + (\beta^2 - \alpha^2)^2 \right]^{-3}$$

and

$$\left[ \frac{(1+i\alpha)^2 + \beta^2}{(1-i\alpha)^2 + \beta^2} \right]^{i/a} = \exp\left[ \frac{i}{\alpha} \ln\left( \frac{(1+i\alpha)^2 + \beta^2}{(1-i\alpha)^2 + \beta^2} \right) \right]$$

Hence, (7.6.8) reduces to

$$I_s \frac{df(W,Q)}{dW} = df(\alpha^2, \beta^2)/d\alpha^2$$

$$= \frac{2^8(1+\alpha^2)[\frac{1}{3}(1+\alpha^2)+\beta^2]}{\left[1+2(\beta^2+\alpha^2)+(\beta^2-\alpha^2)^2\right]^3} F(\alpha^2,\beta^2) \qquad (7.6.9)$$

where

$$F(\alpha^2,\beta^2) = \exp\left[\frac{i}{\alpha}\ln\frac{(1+i\alpha)^2+\beta^2}{(1-i\alpha)^2+\beta^2}\right] \quad \text{for} \quad \alpha^2 \geq 0 \qquad (7.6.10)$$

For negative values of $\alpha^2$, the above equation changes to

$$F(\alpha^2,\beta^2) = \exp\left\{\frac{-1}{(-\alpha^2)^{1/2}}\ln\left[\frac{\left[1+(-\alpha^2)^{1/2}\right]^2+\beta^2}{\left[1-(-\alpha^2)^{1/2}\right]^2+\beta^2}\right]\right\} \qquad (7.6.11)$$

Let us consider positive values of $\alpha^2$. For this case

$$\ln\left[\frac{(1+i\alpha)^2+\beta^2}{(1-i\alpha)^2+\beta^2}\right] = \ln\left[\frac{1+ix}{1-ix}\right] = 2i\tan^{-1}x \qquad (7.6.12)$$

where

$$x = \frac{2\alpha}{1-\alpha^2+\beta^2} \qquad (7.6.13)$$

Putting (7.6.12) and (7.6.13) into (7.6.10), we obtain

$$F(\alpha^2,\beta^2) = \exp\left[-\frac{2}{\alpha}\arctan\left(\frac{2\alpha}{1-\alpha^2+\beta^2}\right)\right] \qquad (7.6.14)$$

We note that (7.6.9) is twice (7.6.4). The factor of 2 is again due to the fact that there are two $K$-shell electrons in multielectron atoms. However, for $\alpha^2 > 0$ (7.6.14) differs from (7.6.5) by the normalization factor $1 - \exp(-2\pi/\alpha)$ of the Coulomb wave because the bound state wave functions are normalized to unity.

### 7.6.1   The Total Ionization Cross Section

Let us now discuss the evaluation of the total ionization cross section in the plane wave Born approximation (PWBA). We start from (7.3.10), which is for the excitation from the $|i\rangle$ to the $|j\rangle$ state. In this equation $\varepsilon_{ex}$ is replaced by $W$ and $f_{ji}(Q)$ by $df(W, Q)/dW$ and is integrated over $W$ from $I$ to $W_{max}$. Thus, with the help of (7.3.16), the total ionization cross section is

$$\sigma_i = \frac{4\pi R^2 a_0^2}{E} \int_I^{W_{max}} \frac{1}{W} \int_{\ln(Q_-)}^{\ln(Q_+)} \frac{df(W, Q)}{dW} d(\ln Q) dW \qquad (7.6.15)$$

where the maximum and minimum values of the recoil energy are given by

$$Q_{\pm} = \left( \sqrt{E} \pm \sqrt{E - W} \right)^2 \qquad (7.6.16)$$

The maximum value of the energy loss $W_{max}$ in the PWBA is equal to the incident energy $E$. In the plane wave Born–Ochkur approximation (7.6.15) modifies to

$$\sigma_i = \frac{4\pi R^2 a_0^2}{E} \int_I^{W_{max}} \int_{\ln Q_-}^{\ln Q_+} \frac{1}{W} \frac{df(W, Q)}{dW} F_{ex}(E, Q) d(\ln Q) dW \qquad (7.6.17)$$

where the exchange factor as obtained from (7.5.10) is

$$F_{ex}(E, Q) = 1 - Q/E + Q^2/E^2 \qquad (7.6.18)$$

In exchange scattering the bound electron is exchanged with the incident electron. The two electrons (ejected and scattered) are indistinguishable. Hence, out of the two free electrons the faster one is taken to be the scattered electron. Therefore the maximum energy of the ejected electron is taken to be $(E - I)/2$. Thus with exchange the value of $W_{max}$ is $(E - I)/2 + I = (E + I)/2$.

### 7.6.2   The Coulomb Correction

In the PWBA the $\sigma_i$ as given by (7.6.15) are identical for electron and positron impacts. However, there is no exchange scattering in the positron impact. Hence, for positrons $F_{ex} = 1$ and $W_{max} = E$. Thus the $\sigma_i$ as obtained from (7.6.17) are different for the two particles. However, in general, the theoretical ionization cross sections obtained from (7.6.17) do not agree with the experimental data.

An important effect, which should be included in the theory for the calculation of the collision cross section, is the distortion of the plane waves by the atomic field. Thus the higher Born terms, discussed in Chapter 3, have to be evaluated. A simple way to include such an effect within the PWBA and PWBA–Exchange has been proposed by Hippler (1990). The atomic field accelerates the incident electrons but decelerates the positrons. Hence, at the instant of the ionizing collision, the effective kinetic energy of the projectile is different from the incident energy. On the assumption that the acceleration or deceleration takes place only via the Coulomb field of the bare nucleus, the effective distance $r_{\text{eff}}$ at which the ionization takes place is given by

$$r_{\text{eff}} = \frac{\displaystyle\int_{r_{+,-}}^{\infty} R_{nl}(r)rR_{nl}(r)r^2dr}{\displaystyle\int_{r_{+,-}}^{\infty} R_{nl}(r)R_{nl}(r)r^2dr} \tag{7.6.19}$$

where $R_{nl}(r)$ is the radial function of the initial atomic state and $r_-$ is the smallest distance at which the incident positron, after being decelerated by the nucleus, has sufficient energy to knock out an atomic electron; $r_+$ is the similarly defined quantity for the electron impact. However, since electrons are accelerated, $r_+ = 0$, whereas $r_-$ satisfies the following relation:

$$E - \frac{Z'e^2}{r_-} = I_{nl} \tag{7.6.20}$$

Hippler (1990) fixed the value of $Z'$ by taking

$$I_{nl} = \frac{Z'^2 R}{n^2} \tag{7.6.21}$$

For a hydrogen-like atom, (7.6.19) gives

$$r_{\text{eff}} = \frac{a_0}{2Z'}\frac{3n^2 - l(l+1)}{1 + F_{nl}(x)} \tag{7.6.22}$$

where $x = 2Z'r_-/a_0$ For the electron impact $F_{nl}(x)$ is zero but for positron impact it is given by (Khare and Wadehra, 1996)

$$F_{1s}(x) = \frac{x^3}{3}\frac{1}{2 + 2x + x^2}$$

$$F_{2s}(x) = \frac{x^3}{6}\frac{(8 - 5x + x^2)}{8 + 8x + 4x^2 + x^4}$$

and

$$F_{2p}(x) = \frac{x^5}{5} \frac{1}{(24 + 24x + 12x^2 + 4x^3 + x^4)} \tag{7.6.23}$$

At $r_{eff}$ the instant kinetic energy of the projectile is

$$E' = E \pm \frac{Z'e^2}{r_{eff}} = E \pm \frac{hI_{nl}}{1 + F_{nl}(x)} \tag{7.6.24}$$

where + and − correspond to the electron and positron impacts, respectively. It can be easily shown that the value of $h$ for the $K(1^2S_{1/2})$ shell and $L1(2^2S_{1/2})$ sub-shell is $\frac{4}{3}$ but for the $L2(2^2P_{1/2})$ and $L3(2^2P_{3/2})$ subshells its value is $\frac{8}{5}$.

To include the Coulomb correction, $E$ in (7.6.15) and (7.6.17) is replaced by $E'$. For the positron impact $W_{max}$ in (7.6.15) is replaced by $E'$ but for the electron impact $W_{max}$ remains $(E + I_{nl})/2$ in (7.6.17) to avoid nonphysical ionization.

Khare et al. (1993) employed the Coulomb and exchange corrected PWBA to calculate $\sigma_i$ for the $K$-shell ionization of a number of atoms. At low $E$ their theoretical cross sections are in satisfactory agreement with the experimental data but at high $E$ the theory underestimates the cross sections. One of the reasons for this deficiency is the nonrelativistic nature of the theory. For the inner shells the ionization potentials are quite high, so relativistic effects become important. However, it is not essential to solve the Dirac equation. Khare et al. (1994a,b) have shown that a suitable modification of (7.6.17) gives good values of $\sigma_i$ over a wide energy range. According to the well-known relativistic equation, the total energy $E_T$, including the rest mass energy of the projectile, is given by (6.2.1). Before the collision $p_i = \hbar k_i$ and $E_T = E' + mc^2$. During the collision the projectile loses an energy $W$. Hence, $p_j = \hbar k_j$ and $E_T = E' - W + mc^2$. With the help of the above equations we get

$$k_i = \frac{1}{\hbar c}\left[(E' + mc^2)^2 - m^2c^4\right]^{1/2} \tag{7.6.25}$$

and

$$k_j = \frac{1}{\hbar c}\left[(E' - W + mc^2)^2 - m^2c^4\right]^{1/2} \tag{7.6.26}$$

Since $Q_{\pm} = Ra_0^2(k_i \pm k_j)^2$ we get

$$Q_\pm = \frac{1}{2mc^2}\left[\sqrt{E'(E'+2mc^2)} \pm \sqrt{(E'-W)(E'-W+2mc^2)}\right]^2 \quad (7.6.27)$$

Further, for the incident energy $E$

$$E + mc^2 = \frac{m}{\sqrt{1-v^2/c^2}}c^2 \qquad (7.6.28)$$

Thus with the Coulomb correction,

$$v^2 = c^2\left[1 - \frac{1}{\left(1+E'/mc^2\right)^2}\right] \qquad (7.6.29)$$

We also define $E_r = mv^2/2$ and replace $E$ by $E_r$ in (7.6.15).

   To obtain $\sigma_i$ corrected for exchange, Coulomb, and relativistic effects in the PWBA, Khare and Wadehra (1995, 1996) employed (7.6.15) for positron impacts and (7.6.17) and (7.6.18) for electron impacts. In these equations they took $E'$ and $Q_\pm$ given by (7.6.24) and (7.6.27), respectively, and $E$ was replaced by $E_r$. Their calculated inner-shell ionization cross sections are in good agreement with a number of experimental data (see Fig. 7.7) and also with the theoretical cross sections of Scofield (1978), who solved the Dirac equation, up to about 0.2 MeV. For still higher $E$ the above method underestimates the cross sections. To understand the reason for this failure let us consider the electromagnetic interaction between the projectile and the atomic electrons. This interaction can be subdivided into two terms (Fano, 1963). In the Coulomb gauge representation, one term is the unretarded static Coulomb interaction given by $e^2/r_{12}$. The Fourier transform of this interaction is given by (7.5.3), and each Fourier component having wave vector $K$ transfers a momentum $\hbar K$ from the incident electron to the bound electron in the direction of $K$. Hence, it is known as the longitudinal interaction, and is of importance at all velocities. Only this part is included in (7.6.15) and (7.6.17). The other part is the interaction through emission and absorption of virtual photons. The atomic electron absorbs a virtual photon of momentum $\hbar K$ emitted by the projectile. Thus through this interaction a momentum of $\hbar K$ is also transferred from the projectile to the atomic electron. Since the photon field is perpendicular to $K$, this is known as the transverse interaction, and its ionization cross section is given by (Fano, 1963).

$$\sigma_t = -\frac{8\pi a_0^2 R}{mv^2}M^2\left[\ln(1-v^2/c^2)+v^2/c^2\right] \qquad (7.6.30)$$

where $M^2$ is equal to the total dipole matrix squared, measured in the units of $a_0^2$. It is given by

$$M^2 = \int_{I_{nl}}^{W_{max}} \frac{R}{W} \frac{df(W,0)}{dW} dW \tag{7.6.31}$$

The cross section $\sigma_t$ is of importance only at ultrahigh velocities, where $W_{max} \rightarrow \infty$. Further, the longitudinal and transverse interactions are of different parities (Fano, 1963). Hence, the total ionization cross section in the modified PWBA is

$$\sigma_i = \sigma_l + \sigma_t \tag{7.6.32}$$

where $\sigma_l$ is given by (7.6.17) with the Coulomb and the relativistic corrections as discussed above.

Khare and Wadehra (1995, 1996) have calculated $\sigma_i$ for the $K$-shell and three $L$-subshells of a number of atoms due to electron as well as positron impacts over an energy range varying from $I_{nl}$ to $1\,GeV$. For the $K$-shell the expression for $df(W, Q)/dW$ as given by (7.6.4) is utilized. Holt (1969) has derived expressions for the continuum matrix elements for the three $L$-subshells of the hydrogen atom. The same are converted for hydrogen-like atoms by replacing $K$ and $k$ by $K/Z_S$ and $k/Z_S$, respectively. The scaled generalized oscillator strengths for the three $L$-subshells are given by (Khare et al., 1995)

$$\left[\frac{df(\alpha^2, \beta^2)}{d(\alpha^2)}\right]_{L3} = \frac{2^{12}(1+\alpha^2)}{9\left[1+2(\beta^2+\alpha^2)+(\beta^2-\alpha^2)^2\right]^5} F(\alpha^2, \beta^2)$$
$$\times [27\beta^8 - 36\beta^6(1+\alpha^2) + 6\beta^4(19-6\alpha^2-\alpha^4)+$$
$$(4/5)\beta^2(107+98\alpha^2+15\alpha^4)(1+\alpha^2)+(11+3\alpha^2)(1+\alpha^2)^3],$$
$$\tag{7.6.33}$$

$$\left[\frac{df(\alpha^2, \beta^2)}{d\alpha^2}\right]_{L2} = \frac{1}{2}\left[\frac{df(\alpha^2, \beta^2)}{d\alpha^2}\right]_{L3} \tag{7.6.34}$$

and

$$\left[\frac{df(\alpha^2, \beta^2)}{d\alpha^2}\right]_{L1} = \frac{2^{11}(1+\alpha^2)}{3\left[1+2(\alpha^2+\beta^2)+(\beta^2-\alpha^2)^2\right]^5}$$
$$\times [3\beta^{10} - \beta^8(32+11\alpha^2) + 2\beta^6(41+36\alpha^2+7\alpha^4)$$
$$+ 2\beta^4(10-31\alpha^2-20\alpha^4-3\alpha^6)$$
$$+ \frac{1}{5}\beta^2(47-47\alpha^2-35\alpha^4-5\alpha^6)(1+\alpha^2)$$
$$+ (4+\alpha^2)(1+\alpha^2)^4]F(\alpha^2, \beta^2) \tag{7.6.35}$$

where

$$F(\alpha^2, \beta^2) = \exp\left\{-\frac{4}{\alpha}\arctan[(2\alpha)/(\beta^2 + 1 - \alpha^2)]\right\}$$
$$[1 - \exp(-4\pi/\alpha)]^{-1} \quad \text{for} \quad \alpha^2 > 0 \qquad (7.6.36)$$

and

$$F(\alpha^2, \beta^2) = \exp\left[-\frac{2}{\sqrt{-\alpha^2}}\ln\left(\frac{\beta^2 + 1 - \alpha^2 + 2\sqrt{-\alpha^2}}{\beta^2 + 1 - \alpha^2 - 2\sqrt{-\alpha^2}}\right)\right]$$
$$\text{for} \quad \alpha^2 < 0 \qquad (7.6.37)$$

where for L-subshells $\alpha = 2\delta a_0/Z_S$ and $\beta = 2Ka_0/Z_S$. The relationship (7.6.34) is due to the fact that $2^2P_{3/2}$ has four electrons, whereas $2^2P_{1/2}$ has only two. It is also evident from (7.6.33) to (7.6.37) that the scaled CGOS are independent of $Z_S$. Hence, they are the same functions of $\alpha^2$ and $\beta^2$ for all atoms. This provides a scaling relation for the CGOS and we obtain a universal Bethe surface for L-shell ionization similar to that for K-shell ionization given by (7.6.9). According to Slater (1930), the value of the screening parameter $s$ occurring in $Z_S = Z - s$ for the L-subshell is 4.15.

The CGOS given by the above equations reduces to the optical oscillator strength in the limit $\beta^2 \to 0$. Hence, for the three L-subshells

$$\left[\frac{df(\alpha^2, 0)}{d\alpha^2}\right]_{L3} = \frac{2^{12}(11 + 3\alpha^2)}{9(1 + \alpha^2)^6}F(\alpha^2, 0) \qquad (7.6.38)$$

$$\left[\frac{df(\alpha^2, 0)}{d\alpha^2}\right]_{L2} = \frac{1}{2}\left[\frac{df(\alpha^2, 0)}{d\alpha^2}\right]_{L3} \qquad (7.6.39)$$

and

$$\left[\frac{df(\alpha^2, 0)}{d\alpha^2}\right]_{L1} = \frac{2^{11}(4 + \alpha^2)}{3(1 + \alpha^2)^5}F(\alpha^2, 0) \qquad (7.6.40)$$

where

$$F(\alpha^2, 0) = \exp[-(8/\alpha)\arctan\alpha][1 - \exp(-4\pi/\alpha)]^{-1}$$
$$\text{for} \quad \alpha^2 > 0 \qquad (7.6.41)$$

and

$$F(\alpha^2, 0) = \exp\left[-\frac{4}{\sqrt{-\alpha^2}}\ln\left(\frac{1+\sqrt{-\alpha^2}}{1-\sqrt{-\alpha^2}}\right)\right] \quad \text{for} \quad \alpha^2 \leq 0 \qquad (7.6.42)$$

### 7.6.3   The Born–Bethe Ionization Cross Section

Using (7.6.15), we obtain the total ionization cross section $\sigma_l$ due to longitudinal interaction:

$$\sigma_l = \frac{4\pi a_0^2 R}{E_r} \int_I^{W_{\max}} \frac{R}{W} \frac{df(W,0)}{dW}\ln\left(\frac{\overline{Q}}{Q_-}\right) dW \qquad (7.6.43)$$

where $df(W, 0)/dW$ is the continuum optical oscillator strength per unit energy range for the energy loss $W$ and

$$\overline{Q} = Q_- \exp\left[\frac{1}{(df(W,0)/dW)} \int_{\ln Q_-}^{\ln Q_+} \frac{df(W,Q)}{dW} d(\ln Q)\right] \qquad (7.6.44)$$

Equation (7.6.43) is also expressed as

$$\sigma_l^{BB} = \frac{4\pi a_0^2 R}{E_r} M^2 \ln(c_i E_r) \qquad (7.6.45)$$

where $M^2$ is given by (7.6.31) and

$$M^2 \ln(c_i E_r) = \int_I^{W_{\max}} \frac{R}{W} \frac{df(W,0)}{dW}\ln(\overline{Q}/Q_-) dW \qquad (7.6.46)$$

It is rather difficult to determine the value of $\overline{Q}$. Hence $c_i$, known as the Bethe collision parameter, is usually determined with the help of the experimental $\sigma_i$ available at high $E$. The term $\sigma_l^{BB}$ is the Bethe–Born cross section and it includes only longitudinal interaction. Its value is controlled by the dipole interaction. Another Bethe collision parameter $b_{nl}$ for the ionization of the $(nl)$-subshell is defined by

$$b_{nl} = \frac{I_{nl}}{Z_{nl}} \int_0^{W_{\max}} \frac{1}{W} \frac{df(W,0)}{dW} dW \qquad (7.6.47a)$$

where $Z_{nl}$ is the number of electrons in the $(nl)$ subshell of the atom. A comparison of (7.6.46) with (7.6.31) gives

$$b_{nl} = \frac{I_{nl}}{Z_{nl}R} M^2 \tag{7.6.47b}$$

and

$$\sigma_{nl}^{BB} = \frac{4\pi R^2 a_0^2}{E_r} \frac{Z_{nl}}{I_{nl}} b_{nl} \ln(c_{nl}E_r) \tag{7.6.48}$$

for the $(nl)$-subshell. A plot of $y = [\sigma_{nl}I_{nl}E_r/(4\pi R^2 a_0^2 Z_{nl})]$ vs. $\ln E_r$ gives a straight line. This plot is known as the Fano plot. The slope of the line yields $b_{nl}$ and the intersection of the line with the $y$-axis yields $b_{nl}\ln c_{nl}$. Thus using the experimental data for $\sigma_{nl}$, available at large $E$, the Bethe parameters can be determined.

Using (6.10.19) and (7.6.31) in (7.6.45), we obtain the Born–Bethe ionization cross section $\sigma_{nl}^{BB}$ in terms of the photoionization cross section $\sigma_{ph}(W)$:

$$\sigma_{nl}^{BB} = \int_{I_{nl}}^{W_{max}} \frac{dn(W)}{dW} \sigma_{ph}(W)dW \tag{7.6.49}$$

where

$$\frac{dn(W)}{dW} = \frac{1}{2\pi} \frac{c\hbar}{a_0} \frac{1}{E_r} \frac{1}{W} \ln(c_{nl}E_r) \tag{7.6.50}$$

is the number of photons of energy $W$ per unit energy. Thus the interaction of an incident electron of energy $E$ with the atomic electron through the $1/r_{12}$ term is equivalent to the production of photons whose energy $W$ varies continuously, having $E$ as its maximum value. All the photons having $W > I_{nl}$ ionize the atom. These virtual photons have their polarization vectors parallel to $K$. Hence, (7.6.49) gives the ionization cross section due to longitudinal interaction.

Khare and his associates (1995, 1996) have calculated total ionization cross sections due to electron and positron impacts for the $K$-shell and three $L$-subshells ($L1$, $L2$, and $L3$) for a number of atoms. They have employed the plane wave Born approximation with corrections for exchange and Coulomb and relativistic effects. Along with the longitudinal interaction, the contribution of the transverse interaction to the ionization cross section given by (7.6.30) is also included. The energy of the projectile has been varied from $I_{nl}$ to 1 GeV. Their results for the $K$-shell of silver are shown in Fig. 7.7. It is evident from the figure that at low impact energies $\sigma(= \sigma_l + \sigma_t)$ is practically equal to $\sigma_l$. However, at higher values of $E$, $\sigma$ increases with $E$, whereas $\sigma_l$ tends to be constant. A significant difference between $\sigma$ and $\sigma_l$ demonstrates the importance of the transverse interaction at ultrahigh energies. We also noted very good agreement

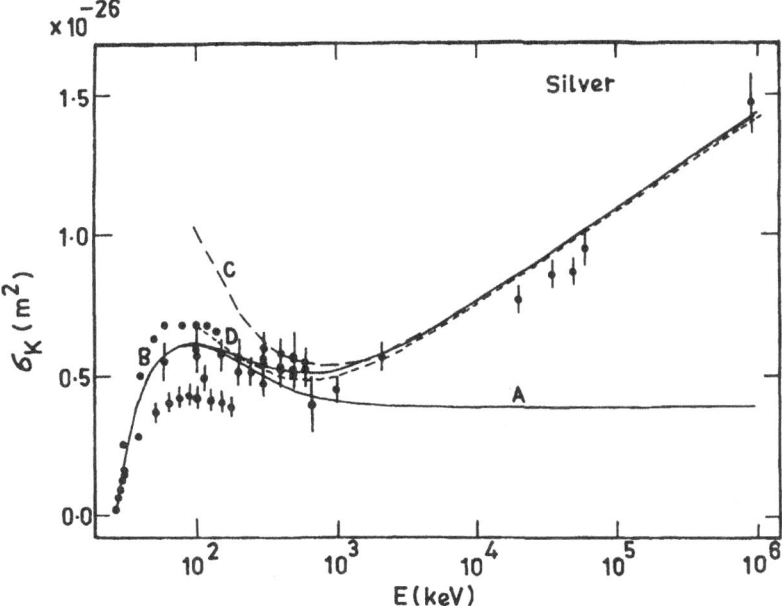

**FIGURE 7.7** Variation of the $K$-shell ionization cross sections of the silver atom with electron impact energy E. Curves A and B represent $\sigma_i$ and $(\sigma_i + \sigma_r)$, respectively, calculated by Khare and Wadehra (1995). Their Born–Bethe cross sections $\sigma^{BB}$ are shown by curve C. Curve D represents the theoretical cross sections of Scofield (1978) and the solid circles with error bars show the experimental data compiled by Long et al. (1990). Reproduced from "$K$-shell ionization of atoms by electron impact," S. P. Khare and J. M. Wadehra, *Phys. Lett. A* 198: 212, 1995, with permission from Elsevier Science.

between $\sigma$ and $\sigma^{BB}$ [obtained from (7.6.53)] for $E > 1$ MeV. The figure also shows a highly satisfactory agreement between the cross sections of Khare and associates and those obtained by Scofield (1978) at high $E$. Finally, a comparison of $\sigma$ with the experimental cross sections (compiled by Long et al., 1990) shows that near the threshold of ionization as well as at ultrarelativistic energies the agreement between the two sets of values is quite good. The theory nicely reproduces the positions of the maximum and minimum in the cross-section curves observed by the experimentalists. For intermediate energies there are considerable differences among the cross sections obtained by different experimental investigators. Khare and Wadehra (1995, 1996) have obtained similar results for a number of other atoms. Figure 7.8 compares the theoretical cross sections for the $L1$-, $L2$-, and $L3$-subshells of gold for electron impact with a number of experimental data.

The agreement between the theoretical cross sections of Khare and Wadehra (1996) and the experimental data is good. Similar results for $K$-shell

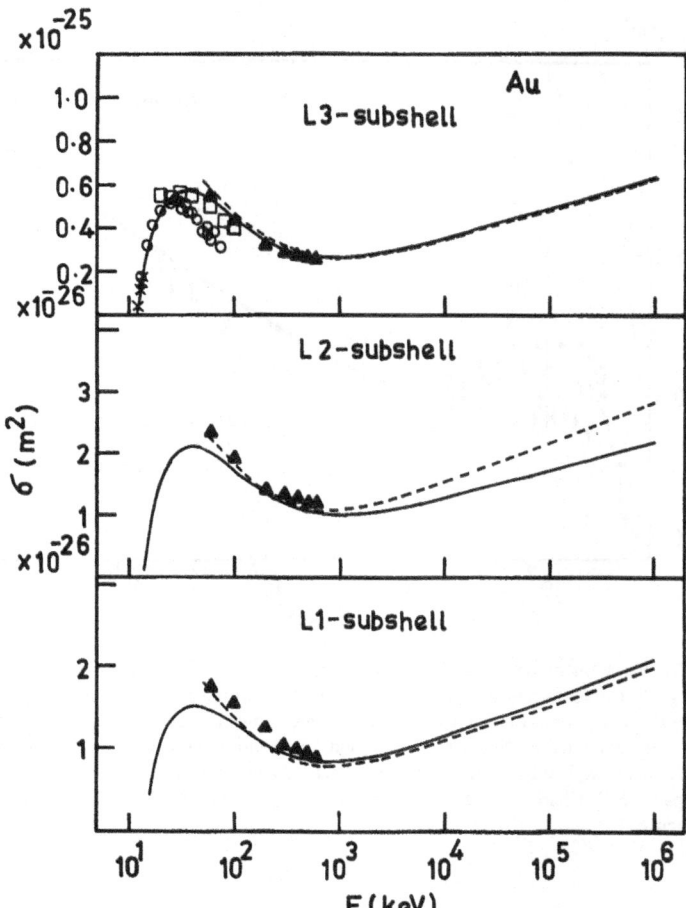

**FIGURE 7.8** Variation of $L1$-, $L2$-, and $L3$- subshell ionization cross sections of the gold atom with energy $E$ of electrons. Theory: —— Khare and Wadehra (1996); ----- Scofield (1978). Experimental: □, Davis (1972); ▲, Palinkas and Schlenk (1980); X, Shima et al. (1981); O Schneider et al. (1993). Reproduced from "$K$-, $L$-, and $M$-shell ionization of atoms by electron and positron impact," S. P. Khare and J. M. Wadehra, *Can. J. Phys.* **74**: 376, 1996, with permission from NRC Research Press, Canada.

and $L$-subshells ionization for a number of other atoms are obtained by Khare and Wadehra. Khare and his associates (1993, 1995) have also determined the values of the collision parameters for the $K$-shell and three $L$-subshells. Their values are shown in Tables 7.1 and 7.2 along with the values of $p_{nl} = (I_{nl}/I_S)$. It is found that whereas $b_{nl}$ decreases with $Z$ $c_{nl}I_{nl}$ increases. Khare and Wadehra (1995, 1996) have shown that $b_{nl}$ and $\ln(c_{nl}I_{nl})$ can be fitted to the following equations:

**Table 7.1** Value of $p_K$ and the Bethe Parameters $b_K$ and $c_K l_K$ for K-Shell Ionization of Atoms in the Plane Wave Born Approximation (Khare et al., 1993)

| Atom | $p_K$ | $b_K$ | $c_K l_K$ |
|------|-------|-------|-----------|
| C    | 0.644 | 0.600 | 6.57      |
| N    | 0.659 | 0.577 | 7.21      |
| O    | 0.660 | 0.576 | 7.30      |
| Ne   | 0.678 | 0.551 | 8.13      |
| Al   | 0.711 | 0.507 | 10.1      |
| Ar   | 0.751 | 0.463 | 13.1      |
| Ni   | 0.798 | 0.418 | 18.1      |
| Ag   | 0.860 | 0.368 | 28.1      |
| Au   | 0.958 | 0.306 | 59.3      |
| H    | 1.00  | 0.283 | 83.0      |

**Table 7.2** Values of $p_{Li}$ and the Bethe Parameters $b_{Li}$ and $c_{Li} l_{Li}$ for the L-Subshells ($i = 1$, 2, 3) of Atoms in the Plane Wave Born Approximation (Khare et al., 1995)

| Atom | $p_{L3}$ | $b_{L3}$ | $c_{L3} l_{L3}$ | $p_{L2}$ | $b_{L2}$ | $c_{L2} l_{L2}$ | $p_{L1}$ | $b_{L1}$ | $c_{L1} l_{L1}$ |
|------|----------|----------|-----------------|----------|----------|-----------------|----------|----------|-----------------|
| Cu   | 0.447    | 0.683    | 7.76            | 0.457    | 0.654    | 8.67            | 0.523    | 0.426    | 16.1            |
| Ag   | 0.537    | 0.480    | 21.4            | 0.565    | 0.433    | 30.5            | 0.606    | 0.365    | 23.3            |
| Sn   | 0.590    | 0.397    | 41.0            | 0.642    | 0.335    | 83.3            | 0.680    | 0.323    | 33.0            |
| Au   | 0.626    | 0.353    | 62.7            | 0.721    | 0.265    | 246             | 0.753    | 0.288    | 46.4            |
| Bi   | 0.635    | 0.343    | 69.9            | 0.743    | 0.249    | 340             | 0.775    | 0.278    | 51.5            |
| H    | 1.0      | 0.133    | 58500           | 1.0      | 0.133    | 58500           | 1.0      | 0.206    | 233             |

$$b_{nl} = a p_{nl}^{-m} \tag{7.6.51}$$

and

$$\ln(c_{nl} I_{nl}) = \alpha_0 + \alpha_1 p_{nl} + \alpha_2 p_{nl}^2 \tag{7.6.52}$$

The values of $\alpha$, $m$, $\alpha_0$, $\alpha_1$, and $\alpha_2$ for different subshells are shown in Table 7.3. With the parameters given by (7.6.51) and (7.6.52), the Born–Bethe cross section, including the transverse component, is

**Table 7.3** Values of $a$, $m$, and $\alpha_i$ for Various Subshells

| Subshell | $a$   | $m$  | $\alpha_1$ | $\alpha_2$ | $\alpha_3$ |
|----------|-------|------|------------|------------|------------|
| $K$      | 0.285 | 1.70 | −9.58      | 26.4       | −13.50     |
| $L1$     | 0.220 | 1.00 | 1.26       | 1.87       | 2.04       |
| $L2$     | 0.158 | 1.80 | −1.24      | 4.17       | 7.19       |
| $L3$     | 0.153 | 1.54 | −1.42      | 5.08       | 6.05       |

$$\sigma_{nl}^{BB} = \frac{4\pi a_0^2 R^2}{E_r I_{nl}} ap^{-m} Z_{nl} [\alpha_1 + \alpha_2 p_{nl} + \alpha_3 p_{nl}^2 + \ln(Er/I_{nl})$$
$$- \ln(1 - v^2/c^2) - v^2/c^2] \tag{7.6.53}$$

As already pointed out, the ionization cross section due to hard collisions is quite often obtained by considering the collision between two free electrons, one moving with energy $E$ and the other at rest. At high $E$ the differential cross section for the transfer of energy $\varepsilon$ from the moving electron to the static electron is given by

$$\frac{d\sigma_i}{d\varepsilon} = \frac{4\pi a_0^2 R^2}{E} \frac{1}{\varepsilon^2} \tag{7.6.54}$$

which is the Rutherford scattering formula.

Considering the two electrons as indistinguishable, the exchange was included by Mott and the above equation changes to

$$\frac{d\sigma_i}{d\varepsilon} = \frac{4\pi a_0^2 R^2}{E} \left[ \frac{1}{\varepsilon^2} - \frac{1}{\varepsilon(E-\varepsilon)} + \frac{1}{(E-\varepsilon)^2} \right] \tag{7.6.55}$$

To apply (7.6.54) and (7.6.55) to the electron impact ionization of atoms one may regard $\varepsilon$ as the energy of the ejected electron. However, the minimum value of $\varepsilon$ is zero. Thus the above equations, which diverge at $\varepsilon = 0$, cannot be utilized to obtain $\sigma_i$. Hence, the Mott formula for an $N$-electron atom is modified to (see Kim and Rudd, 1994)

$$\frac{d\sigma_i}{dW} = \frac{4\pi a_0^2 R^2 N}{E} \left[ \frac{1}{W^2} - \frac{1}{W(E-\varepsilon)} + \frac{1}{(E-\varepsilon)^2} \right] \tag{7.6.56}$$

The integration of Eq. (7.6.56) over $W$ from $I$ to $(E + I)/2$ with $\varepsilon = W - I$ gives

$$\sigma_i^M = \frac{4\pi a_0^2 R^2 N}{E} \left[ \frac{1}{I} - \frac{1}{E} - \frac{\ln(E/I)}{(E+I)} \right] \tag{7.6.57}$$

which is the Mott ionization cross section for hard collisions.

## 7.7   The Second-Order Scattering Term

To obtain the second. Born scattering term $\bar{f}_{ji}^{B2}$, we need the first-order correction $\psi_1$ to the initial wave function of the system. This is given by

$$\psi_1(r, X) = \int G_0^+(r, X; r', X') U(r', X') \varphi_{ki}(r') v_i(X') dr' dX' \qquad (7.7.1)$$

Using (7.2.8) for $G_0^+$, we get

$$\psi_1(r, X) = \lim_{\varepsilon \to 0} S \int \frac{\varphi_{kq}(r) v_p(X) \varphi_{kq}^*(r') v_p^*(X')}{k_p^2 - k_q^2 + i\varepsilon}$$

$$\times U(r', X') \varphi_{ki}(r') v_i(X') dr' dX' dk_q \qquad (7.7.2)$$

It is evident from the above equation that the intermediate atomic states $v_p$ are needed to include the distortion of the initial target wave function $v_i$. Similarly, to represent a distorted plane wave we require the plane waves $\varphi_{kq}(r)$ of the intermediate states. Substitution of (7.2.8) in (7.2.15) with $n = 1$ yields

$$\bar{f}_{ji}^{B2} = -2\pi^2 \lim_{\varepsilon \to 0} S \int \frac{\langle k_j, j|U|k_q, p\rangle\langle k_q, p|U|k_i, i\rangle}{k_p^2 - k_q^2 + i\varepsilon} dk_q \qquad (7.7.3)$$

As shown in the Fig. 7.9, the second Born term is a double scattering term. It should be noted that the intermediate states $|k_q, p\rangle$ do not conserve energy, i.e., $\varepsilon_p + Rk_q^2 a_0^2 \neq \varepsilon_i + Rk_i^2 a_0^2$, although

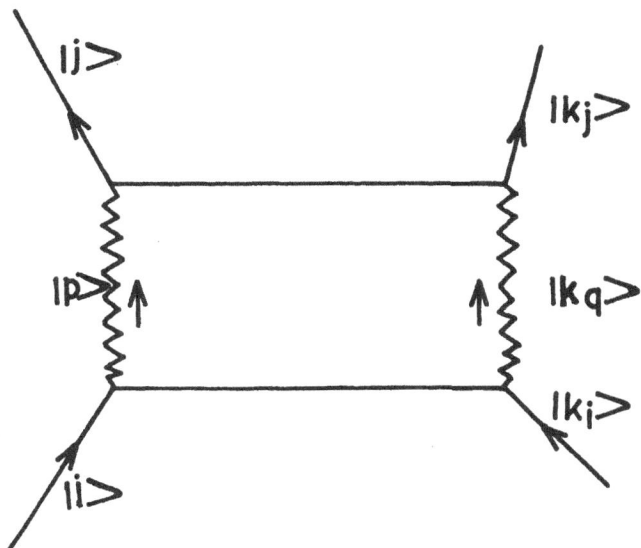

**FIGURE 7.9** Feynman diagram for the second Born amplitude for the scattering of electrons by atoms.

$$\varepsilon_p + Rk_p^2 a_0^2 = \varepsilon_i + Rk_i^2 a_0^2 = \varepsilon_j + Rk_j^2 a_0^2 \qquad (7.7.4)$$

For a given value of $p$ the condition $k_p^2 = k_q^2$ represents a pole in (7.7.3). This is shifted to $k_p^2 + i\varepsilon$ with the help of an infinitesimal quantity $\varepsilon$.

The expression for the second Born term $\bar{f}_{B2}$ (we simplify the notation of $\bar{f}_{ji}^{B2}$ to $\bar{f}_{B2}$) involves a summation over all the bound target states, integration over the continuum target states, and integration over all the projectile states. Hence, an exact evaluation of $\bar{f}_{B2}$, either analytically or numerically, is an involved problem (see Ermolaev and Walters, 1979 and Walters, 1985). Like $\bar{f}_{B2}$ in particle–potential scattering, $\bar{f}_{B2}$ in the electron–atom scattering is also a complex quantity. At high energies and small $K$, Re $\bar{f}_{B2}$ and Im $\bar{f}_{B2}$ vary as $k_i^{-1}$ and $k_i^{-1} \ln k_i$, respectively.

A number of attempts have been made to evaluate $\bar{f}_{B2}$ approximately. Massey and Mohr (1934) were the first to do so. They took $\varepsilon_p = \varepsilon_i$ and thereby $k_p^2 = k_i^2$. Thus the denominator of (7.7.3) becomes independent of $p$. Using the closure relation

$$\underset{p}{S}\langle X|p\rangle\langle p|X'\rangle = \delta(X - X') \qquad (7.7.5)$$

and integrating over $X'$ with the help of the delta function we obtain from (7.7.3) in the *Massey–Mohr approximation*:

$$\bar{f}_{B2M} = \langle j|\bar{f}_{B2P}(k_j, k_i, X)|i\rangle \qquad (7.7.6)$$

where $\bar{f}_{B2P}$ denotes the second-order scattering term for the scattering of the projectile by an atom whose electrons are frozen at $X$. Afterward $\bar{f}_{B2P}$ is unfolded between the atomic states $|i\rangle$ and $|j\rangle$ to obtain $\bar{f}_{B2M}$. It is clear from (7.7.6) that the intermediate atomic states do not appear in the Massey–Mohr approximation. Hence, this approximation completely neglects the effects due to distortion of the target wave function. Since it assumes excitation energies to be zero ($\varepsilon_p = \varepsilon_i$) the imaginary part of $\bar{f}_{B2M}$ diverges in the forward direction for $S$ to $S$ ($l_i = 0$ to $l_j = 0$) excitations. This divergence is due to the absence of any $P$ ($l = 1$) state in the evaluation of the scattering term. Furthermore, Re $\bar{f}_{B2M}(0)$ at large $E$ goes as $k_i^{-2}$ instead of $k_i^{-1}$.

To a great extent the above discrepancies in the Massey–Mohr approximation were removed by Holt and Moiseiwitsch (1968), who proposed a *simplified second Born approximation* (SSBA), in which the first few terms of (7.7.3) are evaluated exactly and for the rest $k_p^2$ is replaced by $k_i^2 - \Delta/(Ra_0^2)$, where $\Delta$ is taken to be the mean excitation energy. The closure is now applied to such terms. Thus in the SSBA

$$\bar{f}_{B2S} = -2\pi^2 \sum_{p=1}^{l} \int \frac{\langle k_j, j|U|k_q, p\rangle\langle p, k_q, |U|k_i, i\rangle}{k_p^2 - k_q^2 + i\varepsilon} dk_q$$

$$-2\pi^2 \left\langle j \left| \int \frac{\langle k_j|U|k_q\rangle\langle k_q|U|k_i\rangle}{k_i^2 - k_q^2 - \Delta/(Ra_0^2) + i\varepsilon} dk_q \right| i \right\rangle$$

$$+2\pi^2 \sum_{p=1}^{l} \int \frac{\langle k_j, j|U|k_q, p\rangle\langle p, k_q|U|k_i, i\rangle}{k_i^2 - k_q^2 - \Delta/(Ra_0^2) + i\varepsilon} dk_q \qquad (7.7.7)$$

The SSBA has been one of the most popular methods of evaluating the second Born term. The choice of $\Delta$ is not unique. Ermolaev and Walters (1979) have discussed the various options.

In many investigations $\Delta$ is so chosen that it reproduces the exact value of the dipole polarizability $\alpha_d$ of the target in the closure approximation. We know that

$$\alpha_d = 2 \underset{p}{S}' \frac{|\langle p|eZ|i\rangle|^2}{\varepsilon_p - \varepsilon_i} \qquad (7.7.8)$$

where $Z$ is the projection of $X$ on the z-axis. If $\varepsilon_p - \varepsilon_i$ is replaced by $\Delta$ then $\bar{\alpha}$, the dipole polarizability in the closure approximation, is given by

$$\bar{\alpha} = \frac{2\langle i|e^2 Z^2|i\rangle^2}{\Delta} \qquad (7.7.9)$$

The integral in the numerator can be easily evaluated. Hence, one takes

$$\Delta = \frac{2\langle i|e^2 Z^2|i\rangle}{\alpha_d} \qquad (7.7.10)$$

where $\alpha_d$ is the experimental value of the dipole polarizability. Jhanwar and Khare (1975) derived the value of $\Delta$ by comparing the total inelastic collision cross section $\sigma_{in}(E)$ obtained in the first Born approximation with that given by the sum rule of Inokuti et al. (1967). Such a procedure gives

$$\Delta = R \exp[L(-1)/S(-1)] \qquad (7.7.11)$$

where

$$S(-1) = \underset{p}{S} \frac{f_{qi}^o}{\varepsilon_{qi}} \qquad (7.7.12)$$

and

$$L(-1) = S_q \frac{f_{qi}^o}{\varepsilon_{qi}} \ln\left(\frac{\varepsilon_{qi}}{R}\right)$$                    (7.7.13)

Using (6.9.1) for $f_{qi}^o$ in (7.7.12), we get

$$S(-1) = \frac{1}{Ra_0^2}\langle i|x^2|i\rangle$$                    (7.7.14)

Thus $S(-1)$ is the ground state property of the target and $L(-1)$ may be computed directly from the optical oscillator strength distribution. Hence $\Delta$ as obtained from (7.7.11) is not an adjustable parameter.

The DCS, which includes $f_{B1}$ and $\bar{f}_{B2}$, is given by

$$I(\theta, \varphi) = |f_{B1}|^2 + 2f_{B1}\,\mathrm{Re}\,\bar{f}_{B2} + (\mathrm{Re}\,\bar{f}_{B2})^2 + (\mathrm{Im}\,\bar{f}_{B2})^2$$                    (7.7.15)

Khare and Shobha (1970, 1971) have suggested a *plane wave approximation* to evaluate $\bar{f}_{B2}$. In this approximation $k_q^2$ in (7.7.3) is replaced by $k_i^2$, and the denominator of (7.7.3) thus becomes independent of $q$. Using the relation

$$\int |k_q\rangle\langle k_q|dk_q = 1$$                    (7.7.16)

we get from (7.7.3)

$$\bar{f}_{PW} = -2\pi^2\langle k_j|U_{ad}|k_i\rangle$$                    (7.7.17)

where the second-order reduced interaction potential energy is

$$U_{ad} = S'_p \frac{\langle j|U|p\rangle\langle p|U|i\rangle}{k_p^2 - k_i^2}$$                    (7.7.18)

The dash over the summation indicates that $p \neq i$ and $j$. It is evident from the above equation that $\bar{f}_{PW}$ is the first Born scattering amplitude due to the second-order polarization potential:

$$V_{ad} = \frac{\hbar^2}{2m}U_{ad}$$

This approximation completely neglects the effects due to the distortion of the wave function of the projectile, but does include the distortion of the target wave function up to first order. The incident electron at $r$ produces an electric field and induces electric multipoles in the atom; i.e., the atom is polarized. This polarized atom produces a polarization potential at $r$. Thus $V_{ad}(r)$ is the second-order polarization potential $V_{dp}(r)$. Since in (7.7.17) the projectile is represented by a plane wave, this approximation is referred to as the plane wave approximation. In the derivation of $U_{ad}(r)$ it is assumed that the projectile is stationary at $r$. Hence, $V_{dp}(r)$ is the adiabatic polarization potential.

To include the nonadiabatic effects we approximate $k_q^2$ by $-\nabla_r^2$ (Khare and Wadehra, 1989) in the denominator of (7.7.3) and take

$$\frac{1}{k_p^2 + \nabla_r^2} = \frac{1}{k_p^2 - k_i^2 + \nabla_r^2 + k_i^2}$$

$$\approx \frac{1}{k_p^2 - k_i^2} - \frac{\nabla_r^2 + k_i^2}{\left(k_p^2 - k_i^2\right)^2} \tag{7.7.19}$$

Using the above equation in (7.7.3), we get

$$\bar{f}_{PW} = -2\pi^2 \langle k_j | U_{dp} + U_{nap} | k_i \rangle \tag{7.7.20}$$

where the nonadiabatic term $U_{nap}$, in its Hermitian form, is given by (Jhanwar et al., 1975)

$$U_{nap} = S' \frac{\nabla_r \langle j|U|p \rangle \cdot \nabla_r \langle p|U|i \rangle}{\left(k_p^2 - k_i^2\right)^2} \tag{7.7.21}$$

Let us evaluate the asymptotic form of $V_{dp}(r)$ for elastic scattering. For the hydrogen atom the interaction energy is

$$V(r) = -\frac{e^2}{r} + \frac{e^2}{|r - X|} \tag{7.7.22}$$

Hence, for large values of $r$ the dipole part of $V(r)$ is given by

$$V(r) \underset{r \to \infty}{\sim} \frac{e^2}{r^2} Z \tag{7.7.23}$$

Putting (7.7.23) into (7.7.18), we obtain

$$V_{ap}(r) \underset{r \to \infty}{\sim} -S' \sum_j \frac{|\langle i|eZ|j \rangle|^2}{\varepsilon_j - \varepsilon_i} \frac{e^2}{r^4} \tag{7.7.24}$$

Now using (7.7.8) yields

$$V_{ad}(r) \underset{r \to \infty}{\sim} -\frac{e^2 \alpha_d}{2r^4} = -\frac{R\alpha_d a_0}{r^4} \qquad (7.7.25)$$

For the interaction potential $V_{dp}$ given by the above equation, the zeroth-order phase shift at small values of $k_i$ is given by (O'Malley et al., 1961; Martyneko et al., 1963)

$$\tan \eta_0 = -a_s k_i - \frac{\pi \alpha_d}{3a_0} k_i^2 - \frac{4\alpha_d a_s}{3a_0} k_i^3 \ln(k_i a_0) + O(k^3) \qquad (7.7.26)$$

where $a_s$ is the scattering length. This equation shows that at low incident energies $\tan \eta_0$ goes to zero at

$$k_i = -\frac{3a_s a_0}{\pi \alpha_d} \qquad (7.7.27)$$

Hence, at the above $k_i$ the cross section becomes a minimum for negative $a_s$, provided that the contribution of the higher partial waves is small. For the above potential we also have

$$\tan \eta_1 = \frac{\pi \alpha_d k_i^2}{15a_0} + A_1 k_i^3 + O(k_i^4) \qquad (7.7.28)$$

Hence, $\tan \eta_1$ vanishes at

$$k_i = -\pi a_d / 15 a_0 A_1 \qquad (7.7.29)$$

Thus the experimental value of $k_i$ at the next minimum, corresponding to $l = 1$, allows an evaluation of $A_1$.

In the FBA

$$\tan \eta_l^{B1} = -\frac{k_i}{Ra_o^2} \int_0^\infty r^2 V_{dp}(r)[j_l(k_i r)]^2 dr \qquad (7.7.30)$$

As the evaluation of $\eta_l^{B1}$ for large values of $l$ by (7.7.30) becomes quite time-consuming, it may be replaced by the semiclassical phase shifts given by (LaBhan and Callaway, 1969)

$$\eta_l^s = -\frac{1}{4k_i a_o^2 R} \int_{r_0}^\infty \frac{r V_{dp}(r)}{r^2 - r_0^2} dr \qquad (7.7.31)$$

In the above equation $r_0$ is equal to $(l + 0.5)/k_i$. For small values of $k_i$ and $V_{dp} = -\alpha_d e^2/2r^4$, the phase shift in the FBA is given by

$$\tan \eta_l^{B1} = \frac{\pi \alpha_d k_i^2}{(2l+3)(2l+1)(2l-1)a_0} \quad \text{for} \quad l > 0 \qquad (7.7.32)$$

The inclusion of the quadrupole term in the expansion of $V(r)$ in the inverse power of $r$ gives

$$V_{dp}(r) \underset{r \to \infty}{\sim} - Ra_0 \left( \frac{\alpha_d}{r^4} + \frac{\alpha_q}{r^6} \right) \qquad (7.7.33)$$

where $\alpha_q$ is the quadrupole polarizability of the atom. The above equations also hold true for multielectron atoms. A similar treatment for the nonadiabatic polarization potential gives

$$V_{nap}(r) \underset{r \to \infty}{\sim} - \frac{6\beta_1}{r^6} Ra_0 \qquad (7.7.34)$$

where $\beta_1$, is the dipole nonadiabatic coefficient of the target (Klienmann et al., 1968), is given by

$$\beta_1 = e^4 a_0 S' \sum_j \frac{|\langle v_i(X)|Z|v_j(X)\rangle|^2}{(\varepsilon_i - \varepsilon_j)^2} \qquad (7.7.35)$$

This coefficient is a measure of the inability of the electric dipole induced in the atom to follow the motion of the incident electron. Hence, up to $r^{-6}$ the reduced interaction polarization energy $U_{pol}$, which includes both adiabatic and nonadiabatic components, is given by

$$U_{pol}(r) \underset{r \to \infty}{\sim} - \frac{1}{a_0} \left( \frac{\alpha_d}{r^4} + \frac{\alpha_q - 6\beta_1}{r^6} \right) \qquad (7.7.36)$$

The above potential does not depend explicitly on the energy of the incident electron and diverges at the origin. Hence, Jhanwar and Khare (1976) proposed a spherically symmetric and energy-dependent Buckingham-type dynamic polarization potential. Their potential is given by

$$V_{dp}(r) = -Ra_0 \left[ \frac{\alpha_d r^2}{(r^2 + d^2)^3} + \frac{\alpha_q r^4}{(r^2 + d^2)^5} \right] \qquad (7.7.37)$$

where $d$ is energy-dependent cut-off parameter, which increases with the incident energy. Hence, for large energies $U_{dp}(r)$ becomes negligible. This is correct physically because a high-velocity incident electron does not have sufficient time to polarize the atom. Further, for small values of $r$, the dipole and the quadrupole part of (7.7.37) vary as $r^2$ and $r^4$, respectively, and tend to zero at the origin. To obtain $d$ we follow Jhanwar and Khare (1975). Using (7.7.37) in (7.7.17) and evaluating the integral, we obtain the second-order scattering term in the plane wave approximation:

$$\bar{f}_{PW}(K) = \frac{\pi}{4da_0} \left\{ \frac{\alpha_d(3 - Kd)}{4} + \frac{5\alpha_q}{64d^2}\left[1 + Kd - \tfrac{2}{3}(Kd)^2 + \tfrac{1}{15}(Kd)^3\right] \right\} \exp(-Kd)$$

(7.7.38)

At large $E$ the cut-off parameter $d$ is large. Hence, in the forward direction

$$\bar{f}_{PW} \approx \frac{3\pi\alpha_d}{16da_0}$$

The above value is equated to that obtained by Byron and Joachain (1974a,b) for the polarized part of the optical Born scattering amplitude in the eikonal approximation. Assuming $\bar{\alpha}$ to be equal to $\alpha_d$, the above comparison gives

$$d = 0.75\frac{k_i a_o^2 R}{\Delta}$$

(7.7.39)

It should be noted that the term $|\operatorname{Im} f_{B2}|^2$ in (7.7.15) is of fourth order in the interaction potential. The term $f_{B1} \operatorname{Re} \bar{f}_{B3}$ is of the same order but is not included in (7.7.15). Hence, to be consistent, the term $|\operatorname{Im} f_{B2}|^2$ is neglected in the plane wave approximation. Thus the differential cross section, including exchange, for the elastic scattering of an electron by a hydrogen atom is given by

$$I_{PW}(\theta, \varphi) = \tfrac{3}{4}(f_{B1} - g)(f_{B1} - g + 2\bar{f}_{PW})$$
$$+ \tfrac{1}{4}(f_{B1} + g)(f_{B1} + g + 2\bar{f}_{PW})$$

(7.7.40)

For a helium atom we have

$$I_{PW}(\theta, \varphi) = (f_{B1} - g)(f_{B1} - g + 2\bar{f}_{PW})$$

(7.7.41)

Jhanwar and Khare (1976) employed the above equation to calculate the DCS for e–He elastic scattering for $E$ varying from 100 to 1000 eV. In Fig. 7.10 the DCS at 200 eV obtained by them are shown along with the experimental data

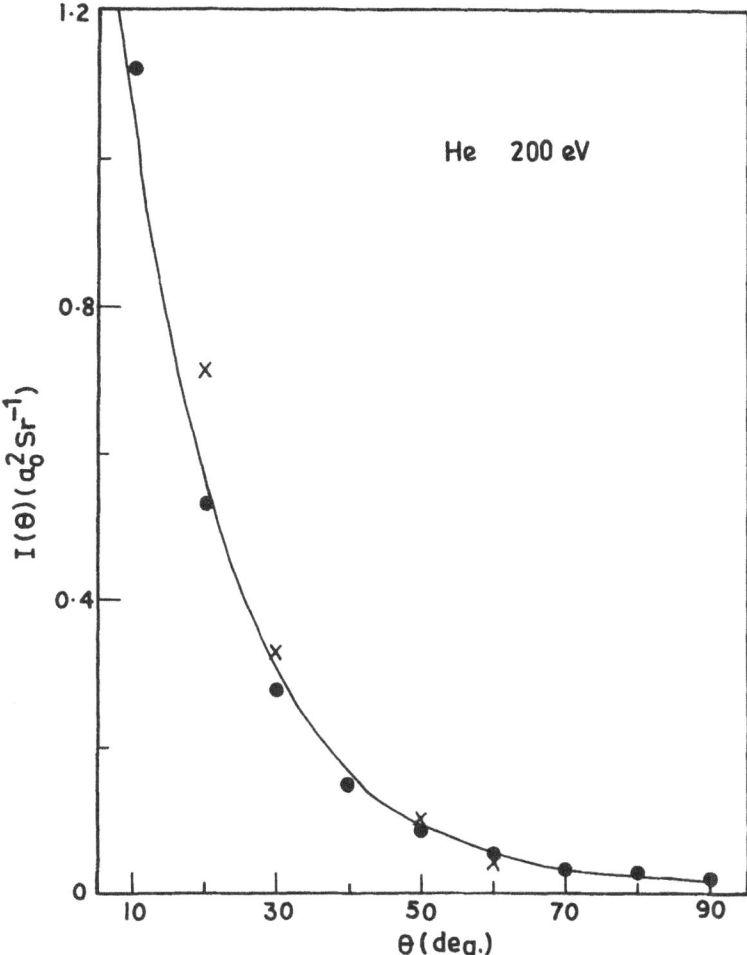

**FIGURE 7.10** Variation of the elastic differential cross sections of the helium atom due to 200-eV electrons in the plane wave approximation (Jhanwar and Khare, 1976). Experimental data: ●, Bromberg (1969, 1974) and X, Crooks and Rudd (1972). The experimental cross sections of Sethuraman et al. (1974) and Jansen et al. (1976) (not shown) are quite close to those of Bromberg et al.

of Jansen et al. (1976), Bromberg (1969, 1974), and Sethuraman et al. (1974). It is found that the FBA–Ochkur approximation underestimates the cross sections at small scattering angles. The inclusion of the polarization effects by $\bar{f}_{PW}$ greatly improves the agreement between theory and experiment. Jhanwar and Khare have noted that for $E \geq 200\,\text{eV}$ their cross sections are within 10% of the experimental values. This indicates the suitability of the dynamic polarization potential given by (7.7.37). This is a long-range potential and is quite important for small-angle scattering. However, $\bar{f}_{PW}$ is purely real, so it cannot be employed to obtain the total collision cross sections through the optical theorem.

The plane wave approximation at intermediate and high $E$ has been successful for the hydrogen atom (Jhanwar et al., 1975) as well as the hydrogen molecule (Gupta and Khare, 1978). However, it overestimates the cross sections for heavier atoms such as neon and argon (Jhanwar et al., 1978; Khare and Kumar, 1978). Hence, it may be concluded that at intermediate $E$ the distortion of the incident plane wave by the atomic field of the light atoms can be ignored but that it becomes quite important for the heavier atoms.

## 7.8 Higher-Order Scattering Terms

A consideration of higher-order Born terms shows that asymptotically they fall faster than $f_{B1}$ and $\bar{f}_{B2}$. For example, for small values of $K$, $\mathrm{Re}\,\bar{f}_{B3}$ falls as $k_i^{-2}$ and $\mathrm{Im}\,\bar{f}_{B3}$ as $k_i^{-3}$ for elastic scattering. Hence, for intermediate and high impact energies ($E \geq 50\,\mathrm{eV}$), the first two terms of the Born series are expected to be sufficient to yield good collision cross sections. However, quite often even for $E \geq 50\,\mathrm{eV}$ it is noted that the cross sections obtained with $f_{B1}$, $\bar{f}_{B2}$, and the first-order exchange amplitude $g$ do not agree with experimental data. Furthermore, sometimes the contribution of the second term to the cross section is quite significant. Both of these observations indicate the need for including the higher terms of the Born series. The evaluation of higher Born terms is extremely difficult. Hence, a number of attempts have been made to evaluate them in an approximate manner and obtain scattering amplitudes correct for all orders of interaction. One such approximation is the Glauber approximation, which we shall discuss now.

### 7.8.1 The Glauber Approximation

The *Glauber approximation* (Glauber, 1959) is an extension of the eikonal approximation of the potential scattering to many-body scattering. In many ways it is similar to the Massey–Mohr approximation. Both these approximations take $k_p^2 = k_i^2$ in the expression of Green's function and use closure, followed by the integration over $X'$. Thus, like the Massey–Mohr approximation, the Glauber approximation (GA) also completely neglects the distortion of the target wave function. In this approximation the scattering amplitude is

$$f_{GA} = -2\pi^2 \langle \varphi_{k_j}(r) v_j(X) | U(r, X) | \psi_p(r, X) v_i(r) \rangle$$

where $\psi_p(r, X)$ is the wave function of the projectile in the eikonal approximation due to a target frozen at $X$. Thus in the Glauber approximation

$$f_{GA} = \langle v_j(X)|f_E(X)|v_i(X)\rangle \qquad (7.8.1)$$

where $f_E(X)$ is the scattering amplitude for the projectile in the eikonal approximation due to a frozen target. It is given by (3.8.16) but now the eikonal phase also depends upon $X$. The similarity between (7.7.6) and (7.8.1) is quite evident. However, whereas $\bar{f}_{PW}$ and $\bar{f}_{B2M}$ are of second order, the eikonal scattering amplitude $f_E(X)$ and $f_G$ include interaction of all orders. The Glauber approximation suffers from the same discrepancies noted earlier for the Massey–Mohr approximation. It also completely neglects the distortion of the target wave function and, hence, $f_G$ for the elastic scattering also diverges in the forward direction.

Like the Born and the eikonal series we also have the Glauber series. The $n^{\text{th}}$ Glauber term is given by

$$\bar{f}_{Gn} = \langle v_j(X)|\bar{f}_{En}(X)|v_i(X)\rangle \qquad (7.8.2)$$

The $n^{\text{th}}$ eikonal scattering term $\bar{f}_{En}(X)$ for a fixed value of $X$ is given by (3.8.24), but now $U$ and the eikonal phase $\zeta$, given by (3.8.15), also depend upon $X$. We have already seen that $\bar{f}_{En}$ is alternatively purely real and imaginary; hence, the $\bar{f}_{Gn}$ follow the same trend. A detailed comparison between $\bar{f}_{En}$ and $\bar{f}_{Gn}$ was made by Byron and Joachain (1973a, b, 1977a)

### 7.8.2 The Eikonal Born Series

Since the elastic scattering amplitude $f_G$ diverges in the forward direction, the Glauber approximation cannot be employed to obtain the integrated elastic cross section $\sigma_{el}$ and the total collision cross section $\sigma_T$. However, even at intermediate energies, scattering terms higher than second order are required to explain the experimental data. At the same time evaluation of $\bar{f}_{B3}$, $\bar{f}_{B4}$, etc., is very difficult. It is relatively easier to evaluate $\bar{f}_{G3}$, $\bar{f}_{G4}$, etc. Hence, quite often the higher-order terms are included through $\bar{f}_{Gn}$.

A careful analysis of $\bar{f}_{G2}$ ($\theta = 0$) [see (7.8.15)] shows that the divergence of $f_{Gn}$ is due to its second term $\bar{f}_{G2}$, which is purely imaginary but agrees very nicely with Im $\bar{f}_{B2}$ for intermediate and large values of $K$ even for low values of $E$. A similar agreement is expected between $\bar{f}_{G3}$ and Re $\bar{f}_{B3}$. Both of them are finite at all $K$ and fall as $k_i^{-2}$ for large $E$. Hence, Byron and Joachain (1973a,b) proposed the *eikonal Born series* (EBS) method, in which the direct scattering amplitude is given by

$$f_{EBS} = f_{B1} + \bar{f}_{B2} + \bar{f}_{G3} \qquad (7.8.3)$$

Not only is $f_{EBS}$ free from the divergence at $K = 0$ but it also includes distortion of the target wave function up to first order through $\bar{f}_{B2}$. Asymptotically, the above scattering amplitude is correct up to $k_i^{-2}$. The exchange scattering amplitude $g$ in the Ochkur approximation also falls as $k_i^{-2}$. Hence, the DCS in the EBS method for electron–hydrogen-atom scattering is given by

$$I_{\text{EBS}}(\theta) = \tfrac{3}{4}|f_{\text{EBS}} - g|^2 + \tfrac{1}{4}|f_{\text{EBS}} + g|^2 \qquad (7.8.4)$$

### 7.8.3    The Modified Glauber Approximation

In the EBS method the scattering amplitude is truncated at the order of $k_i^{-2}$, which seems quite arbitrary. It is desirable to include higher-order terms. Such a scattering amplitude is given by the *modified Glauber approximation* (*MGA*), proposed by Byron and Joachain (1975) (see also Gien, 1976). $f_G - \bar{f}_{G2}$ is free from the divergence and includes terms of all orders except second order. To include the term of this order as well $\bar{f}_{B2}$ is added to $f_G - \bar{f}_{G2}$. Thus in the MGA the direct scattering amplitude is given by

$$f_{\text{MG}} = f_G - \bar{f}_{G2} + \bar{f}_{B2} \qquad (7.8.5)$$

Like $f_{\text{EBS}}$ the amplitude $f_{\text{MG}}$ is also free of divergence and includes the effects due to polarization of the target through the real part of $\bar{f}_{B2}$. A comparison of (7.8.3) and (7.8.5) shows that

$$f_{\text{MG}} = f_{\text{EBS}} + \sum_{n=4}^{\infty} \bar{f}_{Gn} \qquad (7.8.6)$$

Hence, the MGA should be regarded as a better approximation in comparison with the EBS method.

Since $\bar{f}_{Gn}$ is real for odd values of $n$ and imaginary for even values of $n$, we may write

$$
\begin{aligned}
f_{\text{MG}} &= \operatorname{Re} f_{\text{MG}} + i \operatorname{Im} \bar{f}_{\text{MG}} \\
&= \operatorname{Re} \bar{f}_{B2} + \sum_{n=1}^{\infty} \bar{f}_{Gn} + i\left( \operatorname{Im} \bar{f}_{B2} + \sum_{n'=4}^{\infty} \bar{f}_{Gn} \right)
\end{aligned}
\qquad (7.8.7)
$$

where the integers $n$ and $n'$ increase by two units, i.e., $n = 1, 3, 5, \ldots$ and $n' = 4, 6, 8, \ldots$, etc. If we consider positron scattering, all the odd terms of the Glauber and Born series change their sign but the even terms do not. Hence,

$$\operatorname{Re} f_{\text{MG}}^+ = -\operatorname{Re} f_{\text{MG}}^- + 2\operatorname{Re} \bar{f}_{B2} \qquad (7.8.8)$$

and

$$\operatorname{Im} f_{MG}^+ = \operatorname{Im} f_{MG}^- \tag{7.8.9}$$

where the $f_{MG}^{+-}$ are the scattering amplitudes for the positron and the electron impacts, respectively. In the absence of exchange, the differential cross sections $I_{MG}(\theta, \varphi)$ for the positron and the electron impact are related by

$$I_{MG}^+ = I_{MG}^- - 4\operatorname{Re} f_{MG}^- \operatorname{Re} f_{B2}^- + 4(\operatorname{Re} f_{B2}^-)^2 \tag{7.8.10}$$

It has been noted (Gien, 1977a,b) that for the elastic scattering of electrons by a hydrogen atom, $\operatorname{Re} f_{MG}$ and $\operatorname{Re} \bar{f}_{B2}$ are rather large in the intermediate energy range. Hence, $I_{MG}^+$ is much smaller than $I_{MG}^-$. The total collision cross section $\sigma_T$ depends only on $\operatorname{Im} \bar{f}_{MG}(K = 0)$, so according to (7.8.9), the two total cross sections $\sigma_T^{+,-}$ are identical.

The inclusion of the exchange scattering amplitude $g$, obtained through the Ochkur approximation in (7.8.7), changes only the real part of the scattering amplitude for the electron impact. Hence, Eqs. (7.8.8) and (7.8.10) are modified but (7.8.9) does not change and we still have $\sigma_T^+ = \sigma_T^-$. Thus as far as the total collision cross sections are concerned the modified Glauber approximation does not differentiate between electron and positron impacts.

Let us now consider elastic scattering of the electrons and positrons in the Glauber approximation by a hydrogenic atom, represented by

$$v_i(r) = \sqrt{\frac{\lambda^3}{8\pi}} \exp(-\lambda r) \tag{7.8.11}$$

where $\lambda = 2Z/a_0$. For small values of the momentum transfer $K$ we have (Jhanwar et al., 1982a)

$$f_G(K) \underset{K \to 0}{\sim} -\frac{4ik_i}{\lambda^2}\left[\eta^2 \ln\left(\frac{K^2}{\lambda^2}\right) + \tfrac{5}{2}\eta^2 + \frac{i\eta}{1 - i\eta} + J(\eta)\right] \tag{7.8.12}$$

where

$$J(\eta) = \sum_{m=1}^{\infty}\left[\frac{(-i\eta)_m}{m!}\right]^2 \frac{1}{1 + m - i\eta} - \sum_{m=2}^{\infty}\left[\frac{(-i\eta)_m}{m!}\right]^2 \frac{1}{1 - m} \tag{7.8.13}$$

and $\eta = -q/k_i a_0$; $q = \pm 1$ correspond to positron and electron scattering, respectively; $(a)_m$ is the Pochhammer symbol and is equal to

$$(a)_m = a(1 + a)(2 + a) \cdots (m - 1 + a) = \Gamma(a + m)/\Gamma(a) \tag{7.8.14}$$

with $(a)_0 = 1$. An expansion of $f_G(K)$ in powers of $\eta$ yields

$$f_G(K) \underset{K \to 0}{\sim} \frac{4k_i\eta}{\lambda^2} - \frac{4ik_i}{\lambda^2}\eta^2\left[\ln\left(\frac{K^2}{\lambda^2}\right) + \frac{1}{2}\right] + 0(\eta^3) \qquad (7.8.15)$$

As expected, the first term of the above equation is purely real and represents the first Born (Glauber) term. The second term, which is purely imaginary, is $\bar{f}_{G2}$. It diverges logarithmically in the forward direction. The rest of the terms containing higher powers of $\eta$ are free from divergence. Hence, the divergence of $f_G(0)$ is due to its second term, so in the forward direction,

$$f_G(0) - \bar{f}_{G2}(0) = -\frac{4ik_i}{\lambda^2}\left[\frac{i\eta}{1-i\eta} + 2\eta^2 + J(\eta)\right] \qquad (7.8.16)$$

A number of studies have been carried out using the modified Glauber approximation to investigate collisions of charged particles with various atoms and molecules (see Khare and Vijaishri, 1988).

### 7.8.4    The Unitarized Eikonal Born Series

To obtain (3.8.6) from (3.8.5) we have taken

$$\frac{1}{p^2 + 2p \cdot k_i - i\varepsilon} = \frac{1}{2p \cdot k_i - i\varepsilon} \qquad (7.8.17)$$

If we include the next term of the expansion we get

$$\frac{1}{p^2 + 2p \cdot k_i - i\varepsilon} = \frac{1}{2p \cdot k_i - i\varepsilon} - \frac{p^2}{(2p \cdot k_i - i\varepsilon)^2} \qquad (7.8.18)$$

The second term of the above equation gives rise to the Wallace phase correction to the eikonal phase $\zeta(b, k_i, X)$ represented by (3.8.15) (Wallace, 1973). This correction changes (7.8.1) to

$$f_W(k_j, k_i) = \langle j|f_{EW}(X)|i\rangle \qquad (7.8.19)$$

where

$$f_{EW}(X) = \frac{k_i}{2\pi i}\int db\,\exp(iK \cdot b)\{\exp[i(k_i a_0)^{-1}\zeta(b, k_i, X)$$
$$+ i(k_i a_0)^{-3}\zeta_W(b, k_i, X)] - 1\} \qquad (7.8.20)$$

Like $f_G$, the Wallace scattering amplitude, $f_W$ can also be expanded in a series. However, difficulties arise in the evaluation of the Wallace terms $\bar{f}_{Wn}$ (with

$n \geq 4$) of the electron–atom scattering amplitude. This led Byron et al. (1982) to define the *unitarized eikonal Born series* in the following manner:

$$f_{\text{UEBS}} = \tilde{f}_W - \tilde{f}_{W2} + \tilde{f}_{B2} \qquad (7.8.21)$$

where $\tilde{f}_W$ is again obtained from (7.8.19), except that in the evaluation of $f_{EW}$ $(X)$ the phase term $\exp[i(k_i a_0)^{-3}\zeta_W(b, k_i, X)]$ in (7.8.20) is replaced by $1 + i(k_i a_0)^{-3} \zeta_W(b, k_i, X)$. Byron et al. (1985) have utilized the UEBS method to obtain the cross sections for the elastic and inelastic scattering of electrons and positrons by atomic hydrogen at intermediate and high energies. Their results for the elastic scattering are shown in Table 7.5.

## 7.8.5   The Schwinger Variational Principle and the Fredholm Integral Equation

Equation (3.7.8) for the scattering amplitude $[f_{pn}]$ is valid for the scattering of charged particles by atoms and molecules provided that we use (7.2.15), instead of (3.4.11) for $\bar{f}_{Bn}$. In this section we give an alternative derivation of (3.7.8). For electron–atom scattering in which the atom makes a transition from $v_i(X)$ to $v_j(X)$, the Fredholm integral equation (3.4.17) becomes

$$f^{ji}(k_j, k_i) = f_B^{ji}(k_j, k_i) + \frac{1}{2\pi^2} S_m \int \frac{f_{B1}^{jm}(k_j, k)f^{mi}(k, k_i)}{k^2 - k_m^2 - i\varepsilon} dk \qquad (7.8.22)$$

where $f^{mi}(k, k_i)$ is the exact scattering amplitude for the transition from the initial state $|i, k_i\rangle$ to the intermediate state $|m, k\rangle$. Iterating the above equation $p$ times we obtain

$$f^{ji}(k_j, k_i) = f_{Bp}^{ji}(k_j, k_i) + \frac{1}{2\pi^2} S_m \int \frac{\bar{f}_{Bp}^{jm}(k_j, k)f^{mi}(k, k_i)}{k^2 - k_m^2 - i\varepsilon} dk \qquad (7.8.23)$$

where $\bar{f}_{Bp}^{jm}$ is the $p^{\text{th}}$ Born term and $f_{Bp}^{ji}$ is the scattering amplitude in the $p^{\text{th}}$ Born approximation. We note that (7.8.23) is exact. To obtain an approximate solution we take

$$f^{mi}(k, k_i) = \lambda_{p,n} f_{Bn}^{mi}(k, k_i) \qquad (7.8.24)$$

for all values of $m$ (including $m = i$ and $j$), where $n$ is an integer and $\lambda_{p,n}$ is a complex multiplying factor. Putting (7.8.24) into (7.8.23), we obtain

$$f^{ji}(k_j, k_i) = f_{Bp}^{ji} + \lambda_{p,n}\left(\bar{f}_{B(p+1)}^{ji} + \bar{f}_{B(p+2)}^{ji} + \cdots \bar{f}_{B(p+n)}^{ji}\right)$$
$$= f_{Bp}^{ji} + \lambda_{p,n}\left(f_{B(p+n)}^{ji} - f_{Bp}^{ji}\right) \qquad (7.8.25)$$

Taking $m = j$ and $k = k_j$ in (7.8.24), we obtain

$$f^{ji}(k_j, k_i) = \lambda_{p,n} f_{Bn}^{ji}(k_j, k_i) \tag{7.8.26}$$

Both the above equations are approximate solutions of the exact equation (7.8.23). Equating them we get

$$\lambda_{p,n} = \frac{f_{Bp}^{ji}}{f_{Bn}^{ji} + f_{Bp}^{ji} - f_{B(p+n)}^{ji}} \tag{7.8.27}$$

Use of (7.8.27) in either (7.8.25) or (7.8.26) yields (3.7.8). As expected, for $p = n = 1$ we get $[f_{11}]$, given by (3.7.9). For $n = 1$ and $p = 2$ we have

$$[f_{12}] = \frac{f_{B1} f_{B2}}{f_{B1} - \bar{f}_{B3}} \tag{7.8.28}$$

Khare and Lata (1984, 1985) replaced Re $\bar{f}_{B3}$ with $\bar{f}_{G3}$ and neglected Im $\bar{f}_{B3}$, which falls faster than $k_i^{-2}$. Thus for the direct scattering amplitude they took

$$f_d = \frac{f_{B1} f_{B2}}{f_{B1} - \bar{f}_{G3}} \tag{7.8.29}$$

To calculate $f_d$ we require $f_{B1}$, $\bar{f}_{B2}$, and $\bar{f}_{G3}$. The expression for $f_{B1}$ is well known, and $\bar{f}_{B2}$ is calculated in the SSBA of Holt and Moiseiwitsch (1968). $\bar{f}_{G2}$ and $\bar{f}_{G3}$ for hydrogenic atoms are as follows (Yates, 1974):

$$\bar{f}_{G2} = \frac{-iq^2}{k_i} \left[ \frac{4(K^2 + 2\lambda^2)}{(K^2 + \lambda^2)^2} \ln\left( \frac{K\lambda}{K^2 + \lambda^2} \right) + \frac{2(\lambda^2 - K^2)}{(\lambda^2 + K^2)^2} \right] \tag{7.8.30}$$

and

$$\bar{f}_{G3} = \frac{-q^3}{2k_i^3 \lambda^2} \frac{1}{x^3} \frac{\partial}{\partial x} \left( \frac{x^4}{1+x^2} \right) \left\{ 4\left[ \ln\left( \frac{1+x^2}{x} \right) \right]^2 + \frac{\pi^2}{3} - 2A(x) \right\} \tag{7.8.31}$$

where

$$A(x) = 2(\ln x)^2 + \frac{\pi^2}{6} + \sum_{n=1}^{\infty} \frac{(-x^2)^n}{n^2} \qquad \text{for } x < 1$$

$$= -\sum_{n=1}^{\infty} \frac{(-1/x^2)^n}{n^2} \qquad \text{for } x > 1 \tag{7.8.32}$$

and $x = K/\lambda$. As expected, (7.8.30) tends to the second term of (7.8.15) for small values of $K$.

Khare and Lata (1984, 1985) employed the Ochkur approximation to obtain $g$. They calculated the real part of the forward elastic scattering amplitudes, total collision cross sections (with the help of the optical theorem), and differential cross sections for the elastic scattering of electrons and positrons by hydrogen and helium atoms and hydrogen molecules. For $e$–$H$ collisions the real and imaginary parts of the forward scattering amplitude are given by (Gerjoy and Krall, 1960)

$$\mathrm{Re}\, f(E, 0) = \mathrm{Re}[f_d(E, 0) - \tfrac{1}{2} g(E, 0)] \qquad (7.8.33)$$

and

$$\mathrm{Im}\, f(E, 0) = \mathrm{Im}[f_d(E, 0) - \tfrac{1}{2} g(E, 0)] \qquad (7.8.34)$$

On the other hand, for helium,

$$\mathrm{Re}\, f(E, 0) = \mathrm{Re}[f_d(E, 0) - g(E, 0)] \qquad (7.8.35)$$

and

$$\mathrm{Im}\, f(E, 0) = \mathrm{Im}[f_d(E, 0) - g(E, 0)] \qquad (7.8.36)$$

Since in the Ochkur approximation the exchange scattering amplitude $g$ is real, we have

$$\mathrm{Im}\, f(E, 0) = \mathrm{Im}\, f_d(E, 0) \qquad (7.8.37)$$

Das and his associates (Das and Biswas, 1980, 1981; Das and Saha, 1981) took $n = p = 1$ in (7.8.27) and thus obtained

$$\lambda_{11} = \frac{f_{B1}}{f_{B1} - \bar{f}_{B2}} \qquad (7.8.38)$$

They then rationalized the above equation and integrated the numerator and denominator over all the scattering angles to obtain

$$\mathrm{Re}\, \lambda_{11} = \frac{\int f_{B1}(f_{B1} - \mathrm{Re}\, \bar{f}_{B2}) d\Omega}{\int \left[ (f_{B1} - \mathrm{Re}\, \bar{f}_{B2})^2 + (\mathrm{Im}\, \bar{f}_{B2})^2 \right] d\Omega} \qquad (7.8.39)$$

**Table 7.4(a)** The Differential Cross Section $I(\theta)$ (in $10^{-21} m^2/Sr$) for the Elastic Collision of 100-eV Electrons with Hydrogen Atom[a]

| $\theta$ (deg) | Theory | | | | Experiment | |
|---|---|---|---|---|---|---|
| | UEBS | SVP | MGA | EBS | V | W |
| 0 | 2.3 + 1[b] | 2.32 + 1 | — | 2.32 + 1 | — | — |
| 10 | 6.7 | 6.38 | 6.38 | 6.74 | — | — |
| 20 | 2.4 | 2.35 | 2.42 | 2.48 | 3.27 | 3.08 |
| 30 | 1.0 | 1.08 | 1.11 | 1.11 | 1.47 | 1.42 |
| 40 | 4.8 − 1 | 5.34 − 1 | 5.46 − 1 | 5.42 − 1 | 7.19 − 1 | 8.06 − 1 |
| 60 | 1.5 − 1 | 1.60 − 1 | 1.63 − 1 | 1.69 − 1 | 2.07 − 1 | 2.02 − 1 |
| 80 | 6.4 − 2 | 6.30 − 2 | 6.44 − 2 | 7.19 − 2 | 8.93 − 1 | 8.26 − 2 |
| 100 | 3.4 − 2 | 3.39 − 2 | 3.22 − 2 | 3.97 − 2 | 4.53 − 2 | 4.34 − 2 |
| 120 | 2.2 − 2 | 1.90 − 2 | 1.99 − 2 | 2.64 − 2 | 2.99 − 2 | 2.57 − 2 |
| 140 | 1.7 − 2 | 1.35 − 2 | 1.40 − 2 | 2.02 − 2 | — | 1.82 − 2 |
| 160 | 1.4 − 2 | 1.11 − 2 | 1.16 − 2 | 1.74 − 2 | — | — |
| 180 | 1.3 − 2 | 1.04 − 2 | 1.09 − 2 | 1.66 − 2 | — | — |

[a]SVP, MGA, and EBS, Lata (1984). In the SVP method $n = 1$ and $p = 2$ were taken; UEBS, Byron et al. (1982, 1985); W, Williams (1975); V, van Wingerden et al. (1977).
[b]$A(\pm B) \equiv A \times 10^{\pm B}$.

and

$$\text{Im}\,\lambda_{11} = \frac{\int f_{B1}\,\text{Im}\,\bar{f}_{B2}\,d\Omega}{\int\left[\left(f_{B1} - \text{Re}\,\bar{f}_{B2}\right)^2 + \left(\text{Im}\,\bar{f}_{B2}\right)^2\right]d\Omega} \qquad (7.8.40)$$

Thus in the Das method, we get from (7.8.25)

$$f_d = f_{B1} + (\text{Re}\,\lambda_{11} + i\,\text{Im}\,\lambda_{11})(\text{Re}\,\bar{f}_{B2} + i\,\text{Im}\,\bar{f}_{B2}) \qquad (7.8.41)$$

where $\text{Re}\lambda_{11}$ and $\text{Im}\lambda_{11}$ are given by (7.8.39) and (7.8.40), respectively. Since $f_{B1}$ for $e^-$ and $e^+$ scattering are of opposite signs, $\text{Re}\lambda_{11}$, $\text{Im}\lambda_{11}$, and thus $f_d$ are different for the electron and positron collisions.

Tables 7.4 to 7.9 show theoretical differential cross sections $I(\theta)$ and total collision cross sections $\sigma_T$ obtained by Khare and associates, Joachain and associates, and Dewangan and Walters in the intermediate energy range for the collision of electrons and positrons with hydrogen and helium atoms. These investigators have employed different theoretical methods. Experimental data are also given for comparison. According to the tables $I(\theta)$ and $\sigma_T$ in the EBS method are higher than those obtained in the MGA, the SVP, and the UEBS methods. This shows that the effect of the higher-order terms ($n > 3$) is to reduce the cross sections. The differences between the cross sections obtained by the latter three

**Table 7.4(b)** Same as Table 7.4(a) but for 200 eV

| $\theta$ (deg) | Theory | | | | Experiment | |
|---|---|---|---|---|---|---|
| | UBES | SVP | MGA | EBS | V | W |
| 0 | 1.6 + 1 | 1.54 + 1 | — | 1.54 + 1 | — | — |
| 10 | 3.1 | 3.36 | 3.02 | 3.05 | — | — |
| 20 | 1.1 | 1.09 | 1.10 | 1.11 | 1.61 | 1.17 |
| 30 | 4.2 − 1 | 4.23 − 1 | 4.25 − 1 | 4.25 − 1 | 5.60 − 1 | 4.81 − 1 |
| 40 | 1.7 − 1 | 1.80 − 1 | 1.81 − 1 | 1.83 − 2 | 2.16 − 1 | 1.98 − 1 |
| 60 | 4.5 − 2 | 4.56 − 2 | 4.65 − 2 | 4.84 − 2 | 6.63 − 2 | 5.23 − 1 |
| 80 | 1.8 − 2 | 1.71 − 2 | 1.75 − 2 | 1.93 − 2 | 2.80 − 2 | 2.40 − 2 |
| 100 | 9.0 − 3 | 8.42 − 2 | 8.73 − 2 | 1.00 − 2 | 1.15 − 2 | 1.15 − 2 |
| 120 | 5.6 − 3 | 5.09 − 3 | 5.32 − 3 | 6.41 − 3 | — | 7.61 − 3 |
| 140 | 3.9 − 3 | 3.64 − 3 | 3.86 − 3 | 4.79 − 3 | — | 4.98 − 3 |
| 160 | 3.4 − 3 | 2.97 − 3 | 3.16 − 3 | 4.00 − 3 | — | — |
| 180 | 3.1 − 3 | 2.80 − 3 | 2.97 − 3 | 3.78 − 3 | — | — |

approximations are small: $\sigma_T(e^-)$ and $\sigma_T(e^+)$ are identical in the EBS and MGA methods, and they are nearly the same in the SVP and the UEBS methods.

We see a qualitative agreement between the theoretical results and the experimental data but, in general, the theoretical methods have a tendency to underestimate the cross sections. According to the experimental data of Kauppila et al. (1981), $\sigma_T(e^-)$ and $\sigma_T(e^+)$ for the helium atom are very nearly the same for $E \geq 200$ eV. However, these two cross sections obtained theoretically by Byron and Joachain (1977b) in the optical model (OM) and Dewangan and

**Table 7.4(c)** Same as Table 7.4(a) but for 400 eV

| $\theta$ (deg) | Theory | | | | Experiment |
|---|---|---|---|---|---|
| | UBES | SVP | MGA | EBS | W |
| 0 | 1.09 + 1 | 1.07 + 1 | — | 1.07 + 1 | — |
| 10 | 1.75 | 1.74 | 1.75 | 1.75 | — |
| 20 | 4.67 − 1 | 4.76 − 1 | 4.64 − 1 | 4.64 − 1 | 5.48 − 1 |
| 30 | 1.41 − 1 | 1.43 − 1 | 1.44 − 1 | 1.44 − 1 | 1.73 − 1 |
| 40 | 5.26 − 2 | 5.29 − 2 | 5.32 − 2 | 5.37 − 2 | 5.76 − 2 |
| 60 | 1.23 − 2 | 1.22 − 2 | 1.24 − 2 | 1.28 − 2 | 1.23 − 2 |
| 80 | 4.53 − 3 | 4.51 − 3 | 4.59 − 3 | 4.84 − 3 | 4.39 − 3 |
| 100 | 2.25 − 3 | 2.22 − 3 | 2.30 − 3 | 2.49 − 3 | 2.56 − 3 |
| 120 | 1.37 − 3 | 1.34 − 3 | 1.40 − 3 | 1.54 − 3 | 1.69 − 3 |
| 140 | 9.85 − 4 | 9.71 − 4 | 1.02 − 3 | 1.14 − 3 | 1.42 − 3 |
| 160 | 8.14 − 4 | 8.00 − 4 | 8.42 − 4 | 9.49 − 4 | — |
| 180 | 7.64 − 4 | 7.53 − 4 | 7.92 − 4 | 8.95 − 4 | — |

**Table 7.5** Same as 7.4(a) but for Positron Collisions at $E = 100$ and $200\,\text{eV}$

| $\theta$ (deg) | UBES | SVP | MGA | EBS |
|---|---|---|---|---|
| $E = 100\,\text{eV}$ | | | | |
| 0 | 6.4 | 7.22 | — | — |
| 10 | 2.7 | 3.02 | 2.85 | 3.22 |
| 20 | 1.2 | 1.22 | 1.17 | 1.26 |
| 30 | 5.6 − 1 | 5.54 − 1 | 5.12 − 1 | 5.34 − 1 |
| 40 | 2.7 − 1 | 2.73 − 1 | 2.43 − 1 | 2.55 − 1 |
| 60 | 7.8 − 2 | 8.70 − 2 | 7.47 − 2 | 8.81 − 2 |
| 80 | 3.0 − 2 | 3.75 − 2 | 3.30 − 2 | 4.51 − 2 |
| 100 | 1.6 − 2 | 2.02 − 2 | 1.84 − 2 | 2.85 − 2 |
| 120 | 1.0 − 2 | 1.29 − 2 | 1.21 − 2 | 2.09 − 2 |
| 140 | 7.3 − 3 | 1.07 − 2 | 9.51 − 3 | 1.70 − 2 |
| 160 | 6.2 − 3 | 7.84 − 3 | 7.75 − 3 | 1.51 − 2 |
| 180 | 5.9 − 3 | 7.44 − 3 | 7.36 − 3 | 1.45 − 2 |
| $\theta$ (deg) | UBES | SVP | MGA | EBS |
| $E = 200\,\text{eV}$ | | | | |
| 0 | 4.5 | 4.90 | — | 4.90 |
| 10 | 2.0 | 2.06 | 2.03 | 2.09 |
| 20 | 7.8 − 1 | 7.67 − 1 | 7.53 − 1 | 7.58 − 1 |
| 30 | 3.1 − 1 | 2.97 − 1 | 2.85 − 1 | 2.88 − 1 |
| 40 | 1.3 − 1 | 1.28 − 1 | 1.23 − 1 | 1.25 − 1 |
| 60 | 3.4 − 2 | 3.44 − 2 | 3.30 − 2 | 3.58 − 2 |
| 80 | 1.3 − 2 | 1.35 − 2 | 1.32 − 2 | 1.51 − 2 |
| 100 | 6.4 − 3 | 6.86 − 3 | 6.80 − 3 | 8.28 − 3 |
| 120 | 3.9 − 3 | 4.23 − 3 | 4.25 − 3 | 5.46 − 3 |
| 140 | 2.8 − 3 | 3.05 − 3 | 3.13 − 3 | 4.17 − 3 |
| 160 | 2.4 − 3 | 2.52 − 3 | 2.57 − 3 | 3.50 − 3 |
| 180 | 2.3 − 3 | 2.37 − 3 | 2.43 − 3 | 3.30 − 3 |

**Table 7.6** Total Collision Cross Section $\sigma_T$ (in $10^{-21}\,\text{m}^2$) for the Collision of Electrons and Positron with Hydrogen Atoms[a]

| $E$ (eV) | UBES | | SVP | | MGA | EBS | H |
|---|---|---|---|---|---|---|---|
| | $e^-$ | $e^+$ | $e^-$ | $e^+$ | $e^{-,+}$ | $e^{-,+}$ | $e^-$ |
| 50 | — | — | 27.4 | 27.9 | 28.3 | 33.2 | 28.8 |
| 100 | 19.7 | 19.1 | 18.9 | 19.1 | 19.2 | 20.6 | 19.2 |
| 200 | 11.8 | 11.7 | 11.7 | 11.7 | 11.7 | 12.1 | 11.7 |
| 300 | 8.68 | 8.59 | 8.20 | 8.20 | 8.56 | 8.73 | 8.56 |
| 400 | 6.86 | 6.83 | 6.80 | 6.80 | 6.80 | 6.91 | 6.80 |
| 500 | — | — | 5.68 | 5.68 | 5.68 | 5.74 | — |

[a] UEBS, Byron et al. (1982, 1985); SVP, Khare and Prakash (1985); MGA and EBS, Jhanwar et al. (1982a); H, de Heer et al. (1977).

**Table 7.7(a)** The Differential Cross Section $I(\theta)$ (in $10^{-21}m^2/$ Sr) for the Elastic Collision of 100-eV Electrons with Helium Atoms[a]

| $\theta$ (deg) | Theory | | Experiment | | |
|---|---|---|---|---|---|
| | SVP | EBS | SR | KV | J |
| 0 | 9.82 | 12.18 | — | — | — |
| 10 | 5.99 | 7.67 | — | 6.88 | 4.76 |
| 20 | 3.22 | 3.95 | — | 3.76 | 2.61 |
| 30 | 1.77 | 2.01 | 1.58 | 2.07 | 1.54 |
| 40 | 1.04 | 1.10 | 9.37 − 1 | 1.28 | 9.23 − 1 |
| 60 | 4.11 − 1 | 4.11 − 1 | 3.81 − 1 | 6.01 − 1 | — |
| 80 | 1.87 − 1 | 2.05 − 1 | 2.00 − 1 | 3.29 − 1 | — |
| 100 | 1.03 − 1 | 1.32 − 1 | 1.42 − 1 | 2.20 − 1 | — |
| 120 | 6.46 − 2 | 1.00 − 1 | 1.17 − 1 | 1.72 − 1 | — |
| 140 | 4.65 − 2 | 8.60 − 2 | 1.07 − 1 | 1.43 − 1 | — |
| 160 | 3.83 − 2 | — | — | — | — |
| 180 | 3.58 − 2 | — | — | — | — |

[a] SVP, Khare and Lata (1985) with $n = 1$ and $p = 2$; J, Jansen et al. (1976); SR, Sethuraman et al. (1974); KV, Kurepa and Vuskovic (1975).

Walters (1977) in the distorted wave second Born approximation (DWSBA) continue to differ even at the higher impact energies.

We conclude this chapter by noting that, in principle, the scattering amplitude correct to any order in the interaction potential can be evaluated. However, the higher Born terms are very difficult to calculate. Hence, as we have seen, practically all the methods employ only the first and second Born terms. The

**Table 7.7(b)** Same as Table 7.7(a) but for 200 eV. B Represents Experimental Data of Bromberg (1974)

| $\theta$ (deg) | Theory | | Experiment | | | |
|---|---|---|---|---|---|---|
| | SVP | EBS | SR | KV | J | B |
| 0 | 7.64 | 8.85 | — | — | — | — |
| 10 | 3.33 | 3.75 | — | 3.45 | 3.02 | 3.13 |
| 20 | 1.53 | 1.64 | — | 1.61 | 1.48 | 1.47 |
| 30 | 7.78 − 1 | 8.06 − 1 | 7.36 − 1 | 8.18 − 1 | 7.86 − 1 | 7.72 − 1 |
| 40 | 4.14 − 1 | 4.31 − 1 | 4.39 − 1 | 4.52 − 1 | 4.23 − 1 | 4.25 − 1 |
| 60 | 1.38 − 1 | 1.52 − 1 | 1.51 − 1 | 1.61 − 1 | — | 1.56 − 1 |
| 80 | 5.71 − 2 | 7.2 − 2 | 6.91 − 2 | 7.6 − 2 | — | 7.36 − 2 |
| 100 | 2.88 − 2 | 4.4 − 2 | 3.90 − 2 | 4.5 − 2 | — | 4.06 − 2 |
| 120 | 1.75 − 2 | 3.2 − 2 | 2.74 − 2 | 3.4 − 2 | — | — |
| 140 | 1.25 − 2 | 2.6 − 2 | 1.90 − 2 | 2.8 − 2 | — | — |
| 160 | 1.02 − 2 | — | — | — | — | — |
| 180 | 9.54 − 3 | — | — | — | — | — |

**Table 7.7(c)**  Same as Table 7.7(a) but for 500 eV. EBS Values are taken from Lata (1984) and O Represents Experimental Data of Oda et al. (1972)

| $\theta$ (deg) | Theory | | Experiment | | | |
|---|---|---|---|---|---|---|
| | SVP | EBS | SR | J | B | O |
| 0 | 5.37 | 5.66 | — | — | — | — |
| 10 | 1.59 | 1.62 | — | 1.54 | 1.61 | 1.64 |
| 20 | 6.27 – 1 | 6.32 – 1 | — | 6.27 – 1 | 6.32 – 1 | 6.38 – 1 |
| 30 | 2.50 – 1 | 2.54 – 1 | 2.63 – 1 | 2.61 – 1 | 2.56 – 1 | 2.58 – 1 |
| 40 | 1.09 – 1 | 1.12 – 1 | 1.16 – 1 | 1.15 – 1 | 1.15 – 1 | 1.20 – 1 |
| 60 | — | 3.19 – 2 | 3.11 – 2 | — | 3.16 – 2 | 3.25 – 2 |
| 80 | 1.07 – 2 | 1.34 – 2 | 1.07 – 2 | — | 1.25 – 2 | 1.33 – 2 |
| 100 | 5.26 – 3 | 7.36 – 3 | 4.98 – 3 | — | 6.44 – 3 | 5.88 – 3 |
| 120 | 3.16 – 3 | 4.87 – 3 | 2.60 – 3 | — | — | 2.55 – 3 |
| 140 | 2.25 – 3 | 3.72 – 3 | 2.04 – 3 | — | — | — |
| 160 | 1.85 – 3 | 3.19 – 3 | — | — | — | — |
| 180 | 1.73 – 3 | 3.02 – 3 | — | — | — | — |

higher terms are evaluated through the Glauber approximation, which completely neglects the effects due to the distortion of the target wave function. Thus the EBS method, the MGA, the UEBS, etc., consider the effect due to the distortion of the target wave function only up to the second order through $\bar{f}_{B2}$. Joachain (1990) has reviewed some of these methods.

At low impact energies the distortion of the target wave function becomes quite significant. Hence, none of the methods discussed in the present chapter are

**Table 7.8**  The Differential Cross Section $I(\theta)$ (in $10^{-21}m^2/Sr$) for the Elastic Collision of Positrons with Helium Atoms[a]

| $\theta$ (deg) | $E = 100$ eV | | $E = 200$ eV | |
|---|---|---|---|---|
| | SVP | EBS | SVP | EBS |
| 0 | 1.75 | 3.47 | 1.67 | 2.19 |
| 10 | 1.21 | 2.20 | 1.26 | 1.45 |
| 20 | 7.44 – 1 | 1.21 | 6.91 – 1 | 7.22 – 1 |
| 30 | 4.42 – 1 | 6.46 – 1 | 3.55 – 1 | 3.50 – 1 |
| 40 | 2.63 – 1 | 3.72 – 1 | 1.88 – 1 | 1.85 – 1 |
| 60 | 1.02 – 1 | 1.97 – 1 | 6.42 – 2 | 7.86 – 2 |
| 80 | 5.01 – 2 | 1.65 – 1 | 2.91 – 2 | 5.01 – 2 |
| 100 | 2.99 – 2 | 1.52 – 1 | 1.60 – 2 | 3.75 – 2 |
| 120 | 2.08 – 2 | 1.42 – 1 | 1.04 – 2 | 3.05 – 2 |
| 140 | 1.62 – 2 | 1.33 – 1 | 7.67 – 3 | 2.66 – 2 |
| 160 | 1.39 – 2 | 1.28 – 1 | 6.41 – 3 | 2.45 – 2 |
| 180 | 1.33 – 2 | 1.26 – 1 | 6.04 – 3 | 8.00 – 3 |

[a] SVP and EBS represent theoretical cross sections of Khare and Lata (1985), obtained in the Schwinger variational principle and the eikonal Born series methods, respectively.

**Table 7.9** (a) The Total Collision Cross Section $\sigma_T$ (in $10^{-21}$ m$^2$) for $e^{-,+}$—He Collisions[a]

(a) Theoretical results

| $E$ (eV) | SVP $e^{-,+}$ | EBS $e^{-,+}$ | MGA $e^{-,+}$ | DWSBA $e^-$ | DWSBA $e^+$ | OM $e^-$ | OM $e^+$ |
|---|---|---|---|---|---|---|---|
| 50 | 10.4 | 16.5 | 7.11 | — | — | — | — |
| 80 | 10.5 | 14.4 | 10.1 | — | — | — | — |
| 100 | 9.88 | 12.8 | 9.88 | 15.3 | 10.6 | 17.2 | 11.1 |
| 200 | 7.05 | 8.09 | 7.25 | 8.95 | 7.42 | 9.43 | 7.50 |
| 300 | 5.46 | 5.99 | 5.57 | 6.49 | 5.71 | 6.66 | 5.74 |
| 400 | 4.48 | 4.79 | 4.56 | 5.12 | 4.67 | 5.21 | 4.67 |
| 500 | 3.81 | 4.03 | 3.86 | 4.25 | 3.95 | 4.31 | 3.97 |
| 700 | 2.94 | 3.08 | 2.97 | 3.22 | 3.05 | 3.25 | — |
| 1000 | 2.21 | 2.29 | 2.24 | 2.37 | 2.29 | — | — |

[a] SVP and EBS, Khare and Lata (1985); MGA, Jhanwar et al. (1982a); DWSBA, Dewangan and Walters (1977); OM, Byron and Joachain (1977b).

(b) Experimental data

| $E$ (eV) | Brenton et al. (1977) $e^+$ | Twomey et al. (1977) $e^+$ | Dalba et al. (1980) $e^-$ | Blaauw et al. (1980) $e^-$ | Kauppila et al. (1981) $e^-$ | Kauppila et al. (1981) $e^+$ |
|---|---|---|---|---|---|---|
| 50 | 11.1 (at 49 eV) | 10.6 | — | 17.3 | 17.3 | 11.2 |
| 80 | — | — | — | 12.5 | — | — |
| 100 | 9.15 | 9.40 | 11.6 | 11.1 | 11.1 | 10.2 |
| 200 | 7.22 | 6.41 | 7.64 | 7.22 | 7.14 | 7.00 |
| 300 | 5.71 | 4.76 | 5.65 | 5.54 | 5.46 | 5.40 |
| 400 | 4.66 | 3.86 | 4.51 | 4.59 | 4.53 | 4.53 |
| 500 | 3.69 | 3.86 | 3.78 | 3.78 | 3.81 | 3.83 |
| 700 | 2.91 | 2.63 | 2.83 | 2.91 | — | — |
| 1000 | 2.10 | 8.68 − 1 | 2.07 | — | — | — |

suitable at low $E$, as they give reasonable results only at intermediate and high energies. A crude lower limit for these methods may be taken as 50 eV. This limit moves to higher values of $E$ as the nuclear charge of the target increases.

## Questions and Problems

7.1 The helium atom in its ground state is represented by

$$\psi(r_2, r_3) = v(r_2)v(r_3)$$

with

**Table 7.10** The Real Part of the Forward Scattering Amplitude for the Elastic Scattering of Electron by Hydrogen Atom in the EBS Method (Lata, 1984)

| $E$ (eV) | $f(\theta = 0)$ (in $a_0$) |
|----------|----------------------------|
| 50       | 3.106                      |
| 100      | 2.394                      |
| 200      | 1.937                      |
| 300      | 1.747                      |
| 400      | 1.638                      |
| 500      | 1.565                      |

$$v(r) = N\left(e^{-Zr/a_0} + ce^{-2Zr/a_0}\right)$$

The variational parameters $Z$ and $C$ are equal to 1.4558 and 0.6, respectively. Show that $N^2$ is equal to $0.7012/a_0^3$.

7.2 Use the above wave function for the helium atom to obtain an expression for the scattering amplitude $f_{B1}(K)$ in the FBA for the elastic scattering of 200-eV electrons by a helium atom. Show that the values of $f_{B1}(K)$ in the forward and backward directions are $0.7879a_0$ and $6.592 \times 10^{-2}a_0$, respectively.

7.3 Use (7.7.10) to calculate $\Delta$ (in Rydberg units) for the hydrogen atom. Take $\alpha_d = 4.5a_0^3$.

7.4 Derive (7.7.38) and use it with (7.7.39) to obtain $\bar{f}_{PW}(K)$ in the forward and backward directions for the scattering of 200-eV electrons by helium atoms. For this atom $\alpha_d = 1.395a_0^3$, $\alpha_q = 2.327a_0^5$, and $\Delta = 35.373$ eV. Also calculate the values of $|1 + f_{dp}(\theta)/f_{B1}(\theta)|^2$ for both directions. Take the values of $f_{B1}(\theta)$ from problem 2. Comment on the importance of the polarization potential in electron–atom collisions.

7.5 Represent the space part of the helium atom by the following wave functions

$$\psi(1^1 S_0) = v_{1s}(r_2)v_{1s}(r_3)$$

and

$$\psi(2^1 S_0) = \frac{1}{\sqrt{2}}[v_{1s}(r_2)v_{2s}(r_3) + v_{1s}(r_3)v_{2s}(r_1)]$$

where $v(r)$ are one-electron hydrogenic orbitals. Use the above wave functions to derive an expression for the differential cross section $I(K) \, dK$ in the FBA for the excitation of the helium atom from $1^1 S_0$ to $2^1 S_0$ due to electron impact.

7.6 Derive an expression for the generalized oscillator strength for the excitation of the helium atom from $1^1S_0$ to $2^1P_1$ ($M_L = 0$) due to electron impact. Represent the helium atom by the same wave function as taken in problem 5 but replace the $\nu_{2s}(r)$ orbital with the $\nu_{2p}$ ($m = 0$) orbital. From the derived GOS obtain the optical oscillator strength and compare it with (6.9.5).

7.7 Derive (7.6.4) and check the values of $df(\alpha^2, \beta^2)/d(\alpha^2)$ given in Fig. 7.6 for $W = 15\,\mathrm{eV}$.

7.8 Helium atoms are excited from $1^1S_0$ to $2^3P$ ($M_S = 1$) by polarized electrons having $m_s = \frac{1}{2}$. Show that this is not possible in the FBA. Obtain an expression for the scattering amplitude for this excitation in the Ochkur approximation. Represent the helium atom, in the initial and final states, by a suitable combination of the hydrogenic orbitals with proper symmetrization.

7.9 Compare and contrast the Born series with the Glauber series. Derive (7.8.12) and (7.8.15). Of the FBA and the EBS method, which one is expected to give better results. Give reasons for your answer.

7.10 Use (7.8.33) and (7.8.34) along with Tables 7.6 and 7.10 to calculate $\bar{f}_{B2}(0)$ for the elastic scattering of electrons by hydrogen atoms in the EBS approximation. With the help of the calculated values of $\bar{f}_{B2}(0)$ and (3.7.8) obtain $[f_{np}]$ and, hence, $\sigma_T$ for $p = n = 1$ in the energy range 50 to 500 eV. Compare your calculated values with those given in Table 7.6 and comment.

# Collision of Electrons with Atoms: The Differential Approach

## 8.1 Introduction

In the last chapter we discussed the integral approach to study electron–atom collisions. These methods are suitable only at intermediate and high energies. At low impact energies the distortion of the wave functions of the target and the projectile due to the interaction potential becomes quite important. To take a proper account of this distortion we have to solve the Schrödinger differential equation. Quite often, even in the intermediate energy range, the differential approach is used to explain the experimental data.

As pointed out earlier, an electron–atom collision is a many-body problem, and an exact solution of the Schrödinger differential equation is not yet possible. In this chapter we shall develop a number of approximate methods to solve this differential equation and obtain the phase shifts. These phase shifts are used to obtain the scattering amplitudes and the differential, integrated, and total collision cross sections.

## 8.2 The Basic Differential Equation

Let us consider the collision of an electron with an atom having $Z$ electrons. As discussed earlier, electrons are fermions so the atomic wave functions as well as the wave function of the system (atom + incident electron) must be antisymmetric with respect to the exchange of any two electrons. Under steady state conditions the system satisfies the following time-independent Schrödinger equation:

$$H \psi(r, s; X, S) = E_T \psi(r, s; X, S) \qquad (8.2.1)$$

where $r$ and $s$ are the space and spin coordinates of the incident electron, respectively. The space and spin coordinates of all the atomic electrons are represented by $X$ and $S$, respectively. The total energy $E_T$ is a conserved quantity. To start with, we ignore the antisymmetrization condition of the wave function (i.e., neglect the exchange) and also the influence of the spin on the collision. Then (8.2.1) reduces to (7.2.1), and (7.2.2) to (7.2.6) are also satisfied.

To solve (7.2.1), we expand $\Psi(r, X)$ in a complete set formed by the eigenfunctions $v_n(X)$ of the atom. Hence,

$$\psi(r, X) = \underset{n}{S} F_n(r) v_n(X) \qquad (8.2.2)$$

where the expansion coefficient $F_n(r)$ is the wave function of the scattered electron when the atom is in its $n^{\text{th}}$ excited state. $F_n(r)$ is different from the plane wave $\phi_{kn}(r)$ due to the presence of the interaction energy $V$ [given by (7.2.6)] in the Hamiltonian of the system. We use (8.2.1) and (7.2.5) in (8.2.2) and obtain

$$\underset{n}{S}(\varepsilon_n + V + H_e) F_n(r) v_n(X) = \underset{n}{S}\left(\varepsilon_n + \frac{\hbar^2 k_n^2}{2m}\right) F_n(r) v_n(X)$$

or

$$\underset{n}{S}(V + H_e) F_n(r) v_n(X) = \underset{n}{S} \frac{\hbar^2 k_n^2}{2m} F_n(r) v_n(X) \qquad (8.2.3)$$

We now multiply (8.2.3.) by $v_p^*(X)$ from the left and integrate over $X$. Using the orthogonal property of $v_n(X)$, we get

$$-\frac{\hbar^2}{2m} \nabla^2 F_p(r) + \underset{n}{S} \int v_p^*(X) V(r, X) v_n(X) dX \, F_n(r) = \frac{\hbar^2}{2m} k_p^2 F_p(r)$$

or

$$(\nabla^2 + k_p^2) F_p(r) = \frac{2m}{\hbar^2} \underset{n}{S} \langle p|V|n\rangle F_n(r) \qquad (8.2.4)$$

It is evident from the above equation that $F_p(r)$ is different from a plane wave due to the presence of the interaction energy $V$. Further to obtain $F_p(r)$ we must know $F_n(r)$ for all possible values of $r$. Since the target has an infinite number of eigenstates,(8.2.4) represents an infinite number of coupled differential equations. Thus finding an exact solution of $F_p(r)$ is an impossible task.

To investigate elastic scattering we reduce the many-body problem to a one-body problem by assuming the existence of an optical potential defined by

$$V_{op}(r)F_i(r) = \sum_n S\langle i|V|n\rangle F_n(r) \qquad (8.2.5)$$

where $v_i(X)$ is the initial atomic state. Putting(8.2.5) into(8.2.4) we get the following one-body differential equation:

$$[\nabla^2 + k_i^2 - U_{op}(r)]F_i(r) = 0 \qquad (8.2.6)$$

where the reduced optical interaction energy $U_{op}(r)$ is equal to $2mV_{op}(r)/\hbar^2$. Equation (8.2.6) describes the scattering of an electron of mass $m$ (much smaller than the mass of target) by the optical potential. In general, in the absence of exchange symmetry, the optical potential is noncentral, energy dependent, and complex. The inclusion of exchange symmetry makes it nonlocal as well. Its construction in an exact form is again an impossible task. Hence, quite often an approximate spherically symmetric form of $U_{op}(r)$ is utilized and the one-body differential equation given by (8.2.6) is solved by following the method discussed in Chapter 3 under the boundary conditions (3.9.4) and (3.9.5) to find the elastic scattering phase shifts $\eta_l$. The scattering amplitudes and the differential and integrated cross sections are obtained from these phase shifts. The total collision cross section $\sigma_T$ is also arrived at by using the optical theorem. We note once again that $\sigma_{el}$ is different from $\sigma_T$ only if $U_{op}(r)$ is complex. In that case the phase shifts are also complex and (3.9.19), (3.9.23), and (3.9.24) are utilized to get $\sigma_{el}$, the absorption cross section $\sigma_{ab}$, and $\sigma_T$, respectively.

Let us now discuss some of the approximate differential approaches to obtain $F_p(r)$.

## 8.3 The Static Field Approximation

In the *static field approximation* (SFA), all the matrix elements $V_{in}$ occurring in (8.2.5) are assumed to be zero except $V_{ii}$. Thus coupling of the initial target state $|i\rangle$ to all other target states, given by $|n\rangle$, is neglected. Hence, (8.2.6) reduces to

$$[\nabla^2 + k_i^2 - U_{SF}(r)]F_i(r) = 0 \qquad (8.3.1)$$

For an atom, having $Z$ electrons the interaction energy $V$ is given by (7.2.6). Hence,

$$U_{SF}(r) = U_{ii}(r) = \frac{2m}{\hbar^2} \left\langle i \left| -\frac{Ze^2}{r} + \sum_{l=1}^{z} \frac{e^2}{|r - r_l|} \right| i \right\rangle \qquad (8.3.2)$$

Thus the reduced interaction energy $U_{SF}$ in the SFA is equal to the average value of the interaction potential energy multiplied by $2m/\hbar^2$. In this case the optical potential is equal to the static potential. To evaluate $U_{SF}(r)$ let us represent the target wave function $v_i(X)$ in the Hartree self-consistent field approximation and take

$$v_i(X) = \pi_{q=1}^{z} \varphi_q(r_q) \qquad (8.3.3)$$

where the one-electron atomic orbitals $\varphi_q(r_q)$ are orthonormal. They are the solutions of the coupled Hartree self-consistent field equations,

$$(H_q - \varepsilon_{iq})\varphi_q(r_q) = 0$$

where

$$H_q = -\frac{\hbar^2}{2m}\nabla_q^2 - \frac{Ze^2}{r_q} + e^2 \sum_{k \neq q} \int \frac{|\varphi_k(r_k)|^2}{|r_k - r_q|} dr_k \qquad (8.3.4)$$

With the help of the above equations we obtain

$$U_{SF}(r) = \frac{2me^2}{\hbar^2}\left[ -Z/r + \sum_{q=1}^{z} \int \frac{|\varphi_q(r_q)|^2}{|r - r_q|} dr_q \right] \qquad (8.3.5)$$

The term $U_{SF}(r)$ is very simple in form. It behaves like $-Z/r$ at short distances and falls off exponentially beyond a distance of the order of the size of the atom. Thus the static potential is a short-range potential. At short distances the static potential is quite strong. Hence, it plays a very important role in large-angle scattering. The SFA includes the effect of the distortion of the projectile's wave function in the evaluation of the collision cross sections to all orders of the interaction potential. However, it completely neglects the effects due to the distortion of the target wave function, the exchange symmetry, and the loss of flux due to simultaneous inelastic scattering (absorption effect). In general, the static potential is real and spherically symmetrical; hence, the method of partial waves discussed in Sec. 3.9 becomes applicable to calculate $\eta_l$, $I(\theta)$, and $\sigma_{el}$.

Now

$$\frac{1}{|r - r_q|} = \frac{1}{r_>} + \sum_{l=1}^{\infty} \sum_{m=-l}^{+l} \frac{4\pi}{2l+1} \frac{r_<^l}{r_>^{l+1}} Y_{lm}^*(\hat{r}) Y_{lm}(\hat{r}_q) \qquad (8.3.6)$$

and

$$\int \varphi_q^*(r_q) \frac{1}{r_>} \varphi_q(r_q) dr_q = \frac{1}{r} + \int_r^{\infty} \int_0^{4\pi} \varphi_q^*(r_q) \left( -\frac{1}{r} + \frac{1}{r_q} \right) \varphi_q(r_q) r_q^2 dr_q d\Omega_q \quad (8.3.7)$$

Using the above two equations in (8.3.5) we get for the electron–atom scattering

$$U_{SF}(r) = \frac{2}{a_0} \sum_{q=1}^{Z} \int_r^{\infty} |\varphi_q(r_q)|^2 \left( -\frac{1}{r} + \frac{1}{r_q} \right) r_q^2 dr_q d\Omega_q$$

$$+ \frac{2}{a_0} \sum_{q=1}^{Z} \sum_{l=1}^{\infty} \sum_{m=-l}^{+l} \frac{4\pi}{2l+1} \int |\varphi_q(r_q)|^2 \frac{r_<^l}{r_>^{l+1}} Y_{lm}^*(\hat{r}) Y_{lm}(\hat{r}_q) dr_q \qquad (8.3.8)$$

where for the electron–atom collision $\hbar^2/me^2$ has been replaced by $a_0$. Let us evaluate $U_{SF}(r)$ for the ground state of hydrogenic atoms. The atom is represented by only one orbital, given by (6.11.1). Since $v_i(X) = \varphi_1(r_1)$ is spherically symmetric, the second term of (8.3.8) is zero and

$$\int_r^{\infty} |\varphi_1(r_1)|^2 \left( -\frac{1}{r} + \frac{1}{r_1} \right) r_1^2 dr_1 d\Omega_1 = \frac{4Z_s^3}{a_0^3} \int_r^{\infty} e^{-2Z_s r_1/a_0} \left( -\frac{1}{r} + \frac{1}{r_1} \right) r_1^2 dr_1$$

We use the standard integral

$$\int_a^b r^n e^{-\lambda r} dr = \frac{n!}{\lambda^{n+1}} \sum_{l=0}^{n} \frac{\lambda^l r^l}{l!} e^{-\lambda r} \Big|_b^a \qquad (8.3.9)$$

in the above equation to evaluate it and put the result in (8.3.8) to get

$$U_{SF}(r) = -\frac{2}{a_0} \left( \frac{1}{r} + \frac{Z_s}{a_0} \right) \exp(-2Z_s r/a_0) \qquad (8.3.10)$$

As expected $U_{SF}$ is real. It falls exponentially at large values of and goes to $-\infty$ at the origin. The negative sign of (8.3.10) shows that the interaction between

**Table 8.1** Values of the Parameters Required in (8.3.12) for a Few Light Atoms (Strand and Bonham, 1964)

| Atom | $a_{\gamma 1}$ | $a_{\gamma 2}$ | $a_{\lambda 1}$ | $a_{\lambda 2}$ | $b_{\gamma 1}$ | $b_{\gamma 2}$ | $b_{\lambda 1}$ | $b_{\lambda 2}$ |
|------|------|------|------|------|------|------|------|------|
| C  | 1.3391 | −0.3391 | 1.7315 | 13.713 | −2.7379 | −2.0444 | 4.718 | 8.333 |
| N  | 1.3521 | −0.3521 | 2.0249 | 15.700 | −3.0744 | −2.3369 | 5.671 | 9.960 |
| O  | 1.2806 | −0.2806 | 2.2376 | 18.263 | −3.0715 | −1.9710 | 6.803 | 11.548 |
| Fe | 1.2538 | −0.2538 | 2.4796 | 20.644 | −3.2697 | −1.8073 | 7.971 | 13.392 |
| N  | 1.2464 | −0.2464 | 2.7385 | 22.850 | −3.5467 | −1.7746 | 9.129 | 15.381 |

the projectile and the target in the SFA is attractive. Hence, the phase shifts are positive and real. They are obtained by solving (3.9.17) under the boundary conditions (3.9.18) with $U(r) = U_{SF}$.

The helium atom in its ground state can be represented by

$$\nu(r_1, r_2) = \varphi_1(r_1)\varphi_2(r_2) \tag{8.3.11}$$

where $\varphi_i(r)$ is again given by (6.11.1). But the value of $Z_s$, as obtained from the variational principle, is 1.27. Hence, with (8.3.11), the static potential for the ground state of the helium atom is twice that of (8.3.10) but with $Z_s = 1.27$. The static field of any multielectron atom can be obtained without much difficulty. Using Hartree–Fock target wave functions (see Weissbluth, 1978), Strand and Bonham (1964) have shown that $U_{ii}(r)$ for the ground state of an atom with $2 \leq Z \leq 18$ is given by

$$U_{SF}(r) = -\frac{4Z}{a_0 r}\left[\sum_{i=1}^{2} a_{\gamma i}\exp(-a_{\lambda i}r/a_0) + \frac{r}{a_0}\sum_{j=1}^{2} b_{\gamma j}\exp(-b_{\lambda j}r/a_0)\right] \tag{8.3.12}$$

The values of the parameters $a_{\gamma i}$, $a_{\lambda i}$, $b_{\gamma j}$, and $b_{\gamma j}$ for a few atoms, as given by Strand and Bonham (1964), are shown in Table 8.1.

## 8.4  The Two-State Approximation

Let us consider collision of electrons with a hypothetical atom that has only two eigenstates $\nu_i(X)$ and $\nu_j(X)$. Under these circumstances, (8.2.4) gives rise to two coupled differential equations:

$$(\nabla^2 + k_i^2)F_i(r) = \frac{2m}{\hbar^2}[V_{ii}(r)F_i(r) + V_{ij}(r)F_j(r)] \tag{8.4.1}$$

and

$$(\nabla^2 + k_j^2)F_j(r) = \frac{2m}{\hbar^2}[V_{ji}(r)F_i(r) + V_{jj}(r)F_j(r)] \qquad (8.4.2)$$

If initially the atom was in the $i^{th}$ state, then $F_i(r)$ is the wave function of the elastically scattered electron and $F_j(r)$ is the wave function of that scattered electron which during collision has excited the atom from the $i^{th}$ to the $j^{th}$ state. As before, asymptotically, $F_i(r)$ is a linear combination of a plane wave and an outgoing spherical wave. However, $F_j(r)$, which is created by the collision, is represented asymptotically only by an outgoing spherical wave. Hence,

$$F_i(r) \underset{r \to \infty}{\sim} A\left[e^{ik_i \cdot r} + \frac{e^{ik_i r}}{r} f_{ii}(\theta, \varphi)\right] \qquad (8.4.3)$$

and

$$F_i(r) \underset{r \to \infty}{\sim} A\frac{e^{ik_j r}}{r} f_{ji}(\theta, \varphi) \qquad (8.4.4)$$

The DCS for the elastic and the inelastic collisions are given by $|f_{ii}|^2$ and $k_j/k_i$ $|f_{ji}|^2$, respectively. For the spherically symmetric $V_{ii}$, $V_{ij}$, and $V_{ji}$, the method of the partial waves is employed and two coupled equations for the initial and final states are solved simultaneously to obtain the required cross sections.

Let us consider collisions where the incident energy $E$ is less than the excitation energy $\varepsilon_{ex} = \varepsilon_j - \varepsilon_i$. In this situation real excitation is not possible and $F_j(r)$ given by (8.4.4) must go to zero as $r \to \infty$. This can be achieved by replacing $k_j$ by $i\mu_j$, where $i = \sqrt{-1}$, $\mu_j$ is a real positive quantity and

$$\frac{2m}{\hbar^2}\varepsilon_{ex} - k_i^2 = \mu_j^2 = -k_j^2 \qquad (8.4.5)$$

Here the $i^{th}$ channel is open and $j^{th}$ channel is closed. To examine the effect of the $j^{th}$ channel on the elastic scattering we approximate $\nabla^2$ by $-k_i^2$ and neglect the term $V_{jj}F_j$ in (8.4.2). Thus, approximately,

$$F_j(r) = -\frac{2m}{\hbar^2}\frac{V_{ji}(r)}{k_i^2 + \mu_j^2} F_i(r) \qquad (8.4.6)$$

Putting (8.4.6) into (8.4.1), we obtain

$$[\nabla^2 + k_i^2 - U_{SF}(r) - U_{pol}(r)]F_i(r) = 0 \qquad (8.4.7)$$

$U_{\text{pol}}(r)$ is the second-order reduced polarization potential. It is given by

$$U_{\text{pol}}(r) = -\left(\frac{2m}{\hbar^2}\right)^2 \frac{|V_{ji}(r)|^2}{k_i^2 + \mu_j^2} = \frac{2m}{\hbar^2} \frac{|V_{ji}(r)|^2}{\varepsilon_i - \varepsilon_j} \qquad (8.4.8)$$

and is always negative. Since $U_{\text{pol}}(r)$ is independent of $E$, $(\hbar^2/2m)U_{\text{pol}}(r)$ is known as the reduced adiabatic polarization potential energy. In the dipole approximation it is equal to the $j^{\text{th}}$ term of (7.7.24). Due to the presence of the closed $j^{\text{th}}$ channel the electric field of the incident electron induces multipoles in the target atom, i.e., the atom is polarized. Hence, the real polarization potential is due to the virtual excitations of the atom.

For electron–atom collisions both $U_{\text{SF}}$ and $U_{\text{pol}}$ are attractive but for positron scattering $U_{\text{SF}}$ is repulsive while $U_{\text{pol}}$ is attractive. Hence, they oppose each other.

For $E \rangle \varepsilon_{\text{ex}}$ the elastic and the inelastic scattering take place simultaneously. As before, the electric field of the incident electron will polarize the atom. Furthermore, due to real excitations, some of the electrons will leave the $i^{\text{th}}$ channel and appear in the $j^{\text{th}}$ channel with energy $E - \varepsilon_{\text{ex}}$ and momentum $\hbar k_j$. Hence, the second-order potential will now be complex. The real part $V_{\text{pol}}$, which is due to virtual excitations, is the polarization potential. The imaginary part $V_{\text{ab}}$ is due to real excitations and is known as the absorption potential. Hence, now

$$V_{\text{op}}(r) = V_{\text{SF}} + V_{\text{pol}} + iV_{\text{ab}} \qquad (8.4.9)$$

where $V_{\text{ab}}$ is real. To investigate the elastic scattering one is now required to solve (8.2.6) with $U_{\text{op}}(r) = (2m/\hbar^2) V_{\text{op}}(r)$. This approximation for the elastic scattering is known as the *static field-polarization-absorption approximation* (*SFPAA*). Both $V_{\text{pol}}$ and $V_{\text{ab}}$ are of second order. However, the former is of long range whereas the latter is of short range. Due to their presence the DCS are changed by a substantial amount.

Since in our present model there is only one inelastic channel, $\sigma_{\text{ab}}$ is equal to $\sigma_{\text{inel}}$. We note that $V_{\text{ab}}$, which arises due to inelastic scattering, affects elastic scattering by its presence in $V_{\text{op}}(r)$. In other words, the effect of the inelastic scattering on the elastic scattering is governed by $V_{\text{ab}}$.

## 8.4.1 Resonance in Elastic Scattering

Let us again consider the case when $E$ is less than $\varepsilon_{\text{ex}}$. Equation (8.4.2) then changes to

$$[\nabla^2 - \mu_j^2 - U_{jj}]F_j(r) = U_{ji}F_i(r) \qquad (8.4.10)$$

If $U_{jj}$ is strong and attractive then for certain values of $\lambda^2$ we have

$$H\chi = \lambda^2 \chi \qquad (8.4.11)$$

with

$$H = -\hbar^2/2m \, \nabla^2 + V_{jj} \qquad (8.4.12)$$

For negative values of $\lambda^2$ the eigenfunctions $\chi$ represent bound states. Let such values of $\lambda^2$ be $E_1, E_2, \ldots, E_n \ldots$. At these incident energies the incident electron gets attached to the target and forms a complex. This complex exists for a brief period and then decays again into the initial state of the target and a free-electron of energy $E$. Near the energy $E_1, E_2, \ldots, E_n \ldots$ due to the formation of bound states the elastic cross sections become quite large and give rise to resonances. These are known as the Feshbach resonances and have been experimentally observed in electron–atom collisions. For the resonances to occur the incident energy $E$ should be less than but close to $\varepsilon_{\mathrm{ex}}$.

Let us consider collisions between electrons and the ground state $(1s)$ hydrogen atom. The excitation of the atom is possible only if $E > 10.204\,\mathrm{eV}$. Below this energy is the region of pure elastic collisions. As discussed above, for $E$ slightly less than $10.204\,\mathrm{eV}$, doubly excited state $H^{-**}$ $(2s, nl)$ may be formed at some specified energies. The eigenenergies of these bound states are embedded into the continuum states of $H^-(1s, E)$, consisting of $H(1s)$ and a free electron of energy $E$. The doubly excited states of $H^-$ are unstable and decay into $H(1s)$ and $e(E)$. This decay represents a transition from a bound to a continuum state. However, such transitions are radiationless, occur at specified energies, and produce Feshbach resonances. The doubly excited states of $H^-$ are known as autoionizing states. O'Malley and Geltman (1965) have calculated eight autoionizing states of $H^{-**}$ lying between 9.559 and 10.203 eV. Three such states are shown in Fig. 8.1.

The elastic scattering of electrons by a helium ion also produces resonances. The reaction is

$$e + \mathrm{He}^+(1s) \to \mathrm{He}^{**}(2s, 2p) \Big\langle \begin{array}{l} \mathrm{He}^+(1s) + e \\ \mathrm{He}^*(1s, 2p) + h\nu \end{array} \qquad (8.4.13)$$

In this case we have an additional decay channel corresponding to the production of a neutral singly excited helium atom and a photon. However, the probability of autoionization is much higher than that of radiative decay. This is

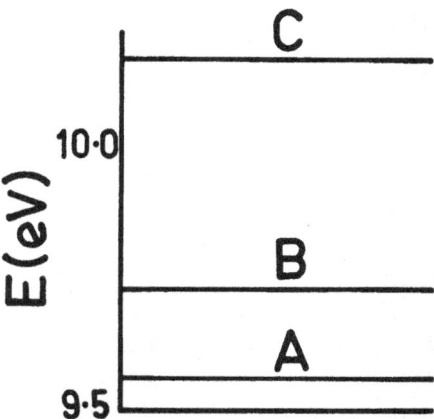

**FIGURE 8.1** Three autoionizing states A, B, and C of H lying between 9.559 and 10.204 eV. The energies of A, B, and C, as calculated by O'Malley and Geltman (1965), are 9.559, 9.127, and 10.149 eV, respectively.

confirmed by the observation that the spectral line has a width much greater than expected for a radiative emission line.

The electron–helium collision process can also form doubly excited states of helium:

$$e + \mathrm{He} \rightarrow e + \mathrm{He}^{**} \qquad (8.4.14)$$

The formation of these doubly excited states produces resonances in the energy loss spectrum of the incident electron. Silvermann and Lassetre (1964) and Simpson (1964) detected two peaks in the energy loss spectrum of 500-eV electrons scattered by helium at 60.0 and 63.5 eV above the ground state of helium. These experiments confirmed the existence of doubly excited helium atoms that decay by autoionization. Burke (1968) discussed the theory of resonances, and experimental details are provided by Golden (1978).

We have discussed resonances between the first two states of atoms. Such resonances have also been noticed below the excitation threshold of $n = 3, 4,$ . . . excited states.

### 8.4.2   Weak Coupling

Let us consider the situation when $U_{ii}$ and $U_{jj}$ are large but the coupling terms $U_{ij}$ and $U_{ji}$ are small. Then, approximately, (8.4.1) and (8.4.2) reduce to

$$[\nabla^2 + k_1^2 - U_{ii}]F_i(\mathbf{r}) = 0 \qquad (8.4.15)$$

and

$$[\nabla^2 + k_j^2 - U_{jj}]F_j(r) = U_{ji}(r)F_i(r) \tag{8.4.16}$$

respectively, the excitation of the atom taking place due to the coupling term $U_{ji}(r)F_i(r)$. The asymptotic solution of (8.4.16) is given by

$$F_j(r) \underset{r \to \infty}{\sim} -2\pi^2 \frac{e^{ik_j r}}{r} \int U_{ji}(r')F_i(r')\chi_j(r', \pi - \Theta)dr' \tag{8.4.17}$$

where $\chi_j$ is the solution of the following homogeneous equation:

$$(\nabla^2 + k_j^2 - U_{jj})\chi_j = 0 \tag{8.4.18}$$

For the scattering angles $(\theta, \varphi)$ the angle $\Theta$ is given by

$$\cos\Theta = \cos\theta \cos\theta' + \sin\theta \sin\theta' \cos(\phi - \phi') \tag{8.4.19}$$

In this approximation the differential cross section is (Mott and Massey, 1965)

$$I_{ji}(\theta, \varphi)d\Omega = \frac{k_j}{k_i} 16\pi^4 \left(\frac{m}{\hbar^2}\right)^2$$

$$\times \left| \int V(r', X)v_i(X)v_j(X)F_i(r', \theta')\chi_j(r', \pi - \Theta)dr'dX \right|^2 \tag{8.4.20}$$

If we represent $F_i$ and $\chi_j$ by plane waves, the above equation gives the DCS in the first Born approximation. Equation (8.4.15) shows that $F_i$ represents the motion in the mean field of the initial state but that $\chi_j$ is due to the mean field of the excited state. This method is known as the distorted wave method.

## 8.5   The Static Field and Polarization Approximation

For atoms having an infinite number of eigenstates, (8.4.8) is modified to

$$U_{\text{pol}} = \frac{2m}{\hbar^2} S' \frac{|V_{ji}(r)|^2}{\varepsilon_i - \varepsilon_j} \tag{8.5.1}$$

where $j \neq i$. This equation can also be obtained using first-order time-independent perturbation theory. According to this theory, for an atom $A$ perturbed by a first-order perturbing potential $V$, we have (Schiff, 1968)

$$(H_A - \varepsilon_i)v_i = 0 \tag{8.5.2}$$

$$(H_A - \varepsilon_i)v_i^1 = (\varepsilon_i^1 - V)v_i \tag{8.5.3}$$

$$(H_A - \varepsilon_i)v_i^2 = (\varepsilon_i^2 - V)v_i^1 + \varepsilon_i^2 v_i \tag{8.5.4}$$

where $H_A$ is the unperturbed Hamiltonian of the atom with eigenenergy $\varepsilon_i$ and eigenfunction $v_i$; $v_i^{1,2}$ are first- and second-order corrections to the eigenfunction; and similarly $\varepsilon_i^{1,2}$ correspond to first- and second-order corrections to the eigen-energy $\varepsilon_i$.

It is easy to verify that if we replace $v_i^{1,2}$ by $v_i^{1,2} + c_i^{1,2}v_i$, in the left-hand sides of (8.5.3) and (8.5.4), where $c_i^{1,2}$ are arbitrary multiplying factors, these equations do not change. Hence, this process does not affect the evaluation of the $v_i^{1,2}$ in terms of their lower-order wave functions. We choose $c_i^{1,2}$ in such a way that the $v_i^{1,2}$ are orthogonal to $v_i$. Now from (8.5.3)

$$\langle i|H_A - \varepsilon_i|v_i^1\rangle = \langle i|\varepsilon_i^1 - V|i\rangle$$

Hence,

$$\varepsilon_i^1 = \langle i|V|i\rangle \tag{8.5.5}$$

Similarly from (8.5.4)

$$\varepsilon_i^2 = \langle i|V|v_i^1\rangle \tag{8.5.6}$$

where we have used $\langle i|v_i^1\rangle = 0$. To obtain an expression for $v_i^1$ we expand it in terms of a complete set formed by $v_j$ and take

$$v_i^1 = \underset{j\neq i}{S}\, a_j v_j \tag{8.5.7}$$

The term $j = i$ is excluded because $v_i^1$ is orthogonal to $v_i$. Using the above equation in (8.5.3) and the orthogonality relation, we get

$$a_j = -\frac{\langle j|V|i\rangle}{\varepsilon_j - \varepsilon_i} \tag{8.5.8}$$

Hence,

$$v_i^1 = \underset{j\neq i}{S}\, \frac{\langle j|V|i\rangle}{\varepsilon_i - \varepsilon_j} v_j \tag{8.5.9}$$

and

$$\varepsilon_i^2 = S_{j \neq i} \frac{|V_{ji}|^2}{\varepsilon_i - \varepsilon_j} \tag{8.5.10}$$

A comparison of the above equation with (8.5.1) shows that $\varepsilon_i^2$ is the second-order polarization energy.

Let us now consider the polarization of a ground state hydrogen atom by a slowly moving incident electron. We assume that the electron is moving so slowly that as far as the polarization is concerned it can be regarded as stationary at $r$ (adiabatic approximation). Hence, the perturbing potential is due only to the potential energy of the incident electron and is given by

$$V(r, X) = -\frac{e^2}{r} + \frac{e^2}{|r - X|} = e^2 \sum_{l=1}^{\infty} \frac{r_<^l}{r_>^{l+1}} P_l(\cos\theta) \tag{8.5.11}$$

where $\theta$ is the angle between $r$ and $X$, the position vector of the atomic electron. For a meaningful concept of the polarization of the atom by the incident electron we should have $r > X$. Hence,

$$V(r, X) = e^2 \sum_{l=1}^{\infty} \frac{X^l}{r^{l+1}} P_l(\cos\theta) \tag{8.5.12}$$

With the above form of $V$ and the spherically symmetrical ground state wave function $v_i$, it is easy to see that the matrix element $V_{ji}$ will be zero if $v_j$ is also spherically symmetric. Hence, in an expansion of $v_i^1$ [given by (8.5.9)] in the complete set of $P_l(\cos\theta)$ we exclude the $l = 0$ term and take

$$v_i^1 = \sum_{l=1}^{\infty} f_l(r, X) P_l(\cos\theta) \tag{8.5.13}$$

Putting the above equation into (8.5.3), we obtain

$$\sum_{l=1}^{\infty} \left( -\frac{\hbar^2}{2m} \nabla^2 - \frac{e^2}{X} - \varepsilon_i \right) f_l(r, X) P_l(\cos\theta) = (\varepsilon_i^1 - V) v_i(X) \tag{8.5.14}$$

In the above equation we use (2.6.6), (2.6.8), and (2.6.2) to get

$$\left( \frac{d^2}{dX^2} + \frac{2}{X} \frac{d}{dx} - \frac{l(l+1)}{X^2} + \frac{2}{a_0 X} - \frac{1}{a_0^2} \right) f_l(r, X) = \frac{2X^l}{r^{l+1} a_0} v_i(X) \tag{8.5.15}$$

The solution of the above differential equation is (Sternheimer, 1954; Dalgarno and Lewis, 1955)

$$f_l(r, X) = -\frac{1}{r^{l+1}}\left(\frac{a_0 X^l}{l} + \frac{X^{l+1}}{l+1}\right)v_i(X)$$

Hence,

$$v_i^1(r, X) = -\sum_{l=1}^{\infty}\frac{1}{r^{l+1}}\left(\frac{a_0 X^l}{l} + \frac{X^{l+1}}{l+1}\right)P_l(\cos\theta)v_i(X) \qquad (8.5.16)$$

Thus from (8.5.6)

$$\varepsilon_i^2 = -e^2\sum_{l=1}^{\infty}\frac{(l+2)(2l+1)!}{l2^{l+1}}\frac{a_0^{2l+1}}{r^{2l+2}} \qquad (8.5.17)$$

It should be noted that in the evaluation of $\langle i|V|v_i^1\rangle$ we have taken the range of $X$ from 0 to $\infty$. But at the same time $r$ is taken to be greater than $X$. Hence, $\varepsilon_i^2$, given by (8.5.17), is its asymptotic value. The dipole ($l = 1$) component of $\varepsilon_i^2$ is

$$(\varepsilon_i^2)_{\mathrm{dp}} \underset{r\to\infty}{\to} -\frac{\alpha_d}{r^4}a_0 R \qquad (8.5.18)$$

because the dipole polarizability $\alpha_d$ of the hydrogen atom is $4.5a_0^3$ and the Rydberg energy $R$ is $e^2/2a_0$. As expected, the above equation agrees with the first term of (7.7.33). Since (8.5.18) diverges at $r = 0$, various empirical forms of $V_{\mathrm{dp}}(r)$, which go to (8.5.18) asymptotically and reduce to zero at the origin, are utilized in the calculations. One such form is given by (7.7.37).

In the *static field-polarization approximation (SFPA)*, the optical potential is the sum of the spherically symmetric static field and the polarization potential. The radial equation in this approximation is

$$\left(\frac{d^2}{dr^2} + k^2 - U_{\mathrm{SF}}(r) - U_{\mathrm{pol}}(r) - \frac{l(l+1)}{r^2}\right)f_l(r) = 0 \qquad (8.5.19)$$

To obtain phase shifts the above equation is solved numerically under the boundary conditions given by (3.9.18). In many calculations $U_{\mathrm{pol}}(r)$, given by (7.7.37), has been utilized. The calculated phase shifts are employed to obtain the differential cross sections and the integrated elastic cross sections $\sigma_{\mathrm{el}}$ in the SFPA. For real $U_{SF}$ and $U_{pol}$ the phase shifts are also real. Hence, the total collision cross section $\sigma_T$ obtained by the application of the optical theorem is equal to $\sigma_{el}$.

So far we have considered only the unsymmetrized wave functions. Thus the exchange effect is neglected. This effect will be considered in the next section.

## 8.6   The Static Field and Exchange Approximation

We take $\Psi$ of (8.2.1) in a completely antisymmetric form as demanded by Fermi–Dirac statistics. For simplicity we consider scattering of the electrons by the hydrogen atom and take the space part of the wave function of the system $e + H$ as

$$\psi^\pm(r_1, r_2) = S_n[F_n^\pm(r_1)\nu_n(r_2) \pm F_n^\pm(r_2)\nu_n(r_1)] \qquad (8.6.1)$$

It is evident that $\Psi^+$ is symmetric with respect to the exchange of the incident electron and the atomic electron. Hence, the spin wave function associated with $\Psi^+$ must be antisymmetric. Similarl $\Psi^-$ is antisymmetric; hence, the associated spin wave function will be symmetric. Since for a two-electron system there are three symmetric spin wave functions and only one antisymmetric spin wave function [see (5.6.1)], $\Psi^\pm$ describes singlet and triplet scattering, respectively. Now with the help of (8.2.1) and, (8.2.2), from (8.6.1) we obtain

$$S_n\left(-\frac{\hbar^2}{2m}\nabla_1^2 - \frac{e^2}{r_1} + \frac{e^2}{r_{12}} - E_T + \varepsilon_n\right)F_n^\pm(r_1)\nu_n(r_2)$$
$$= \mp S_n\left(-\frac{\hbar^2}{2m}\nabla_2^2 - \frac{e^2}{r_2} + \frac{e^2}{r_{12}} - E_T + \varepsilon_n\right)F_n^\pm(r_2)\nu_n(r_1) \qquad (8.6.2)$$

Multiplying both the sides by $\nu_p^*(r_2)$ and integrating over $r_2$ yields

$$(\nabla_1^2 + k_p^2)F_p^\pm(r_1) = S_n\left(U_{pn}(r_1)F_n^\pm(r_1) \mp \frac{2m}{\hbar^2}\int \kappa_{pn}(r_1, r_2)F_n^\pm(r_2)dr_2\right) \qquad (8.6.3)$$

where the kernel $\kappa_{pn}$ is equal to

$$\kappa_{pn}(r_1, r_2) = \nu_p^*(r_2)\left(E_T - \varepsilon_p - \varepsilon_n - \frac{e^2}{r_{12}}\right)\nu_n(r_1) \qquad (8.6.4)$$

Equation (8.6.3) represents an infinite number of coupled integro-differential equations and obtaining their exact solutions is again impossible. A comparison of this equation with (8.2.4) shows that the symmetry consideration has added one more term to the differential equation in which the coordinates of the two electrons are exchanged: the exchange term. To evaluate $F_p$ at $r_1$ we require $F_n$ at all possible value of $r_2$. Thus it is nonlocal in nature.

Let us go back to the one-state approximation. We replace $p$ by the initial channel $i$ and neglect all the terms in (8.6.3) for which $n \neq i$. The resultant equation

$$(\nabla_1^2 + k_i^2)F_i^{\pm}(r_1) = U_{ii}(r_1)F_i^{\pm}(r_1) \mp \frac{2m}{\hbar^2}\int \kappa_{ii}(r_1, r_2)F_i^{\pm}(r_2)dr_2 \qquad (8.6.5)$$

with

$$\kappa_{ii}(r_1, r_2) = v_i^*(r_2)\left(E_T - 2\varepsilon_i - \frac{e^2}{r_{12}}\right)v_i(r_1) \qquad (8.6.6)$$

is the scattering equations in the *static field exchange approximation (SFEA)*. A partial wave expansion of $F_i^{\pm}(r_1)$ given by

$$F_i^{\pm}(r_1) = \sum_l i^l(2l+1)e^{i\eta_{ii}^{\pm}} \sin \eta_{ii}^{\pm} \frac{\varphi_{ii}^{\pm}(r_1)}{r_1} P_l(\cos\theta) \qquad (8.6.7)$$

yields the following one-dimensional differential equation for the $l^{th}$ partial wave

$$\left[\frac{d^2}{dr^2} + k_i^2 - U_{ii}(r_1) - \frac{l(l+1)}{r_1^2}\right]\varphi_{ii}^{\pm} = \mp \frac{2m}{\hbar^2}\int \kappa_{ii}^l(r_1, r_2)\varphi_{ii}^{\pm}(r_2)dr_2 \qquad (8.6.8)$$

with

$$\kappa_{ii}^l(r_1, r_2) = \frac{4\pi}{2l+1} r_1 r_2 v_i(r_1)v_i(r_2)(E_T - 2\varepsilon_i - e^2 r_<^l/r_>^{l+1}) \qquad (8.6.9)$$

In the above derivation it is assumed that the atom is in the $S(L = 0)$ state, which is certainly true for the ground state of the hydrogen atom. Equation (8.6.8) is solved numerically for $\varphi_{ii}^{\pm}$ under the boundary conditions

$$\varphi_{ii}^{\pm}(r_1 = 0) = 0$$

and

$$\varphi_{il}^{\pm}(r_1) \underset{r_1 \to \infty}{\sim} \frac{1}{k_i} \sin(k_i r_1 - l\pi/2 + \eta_{il}^{\pm}) \tag{8.6.10}$$

The phase shifts $\eta_{il}^{\pm}$ are utilized in (3.9.13) to obtain the scattering amplitudes $f_{el}^{\pm}(\theta)$. Considering the relative weight of the singlet and triplet spin states, the differential cross section for elastic scattering of electrons by unpolarized ground state hydrogen atoms is given by

$$I_{el}(\theta) = \tfrac{3}{4}|f^-(\theta)|^2 + \tfrac{1}{4}|f^+(\theta)|^2 \tag{8.6.11}$$

In Fig. 8.2 the phase shifts $\eta_0^{\pm}$ and $\eta_0^{SF}$ are shown as functions of the incident energy $E$. $\eta_0^{SF}$ is found to be zero at $E = 0$. On the other hand, at this energy both $\eta_0^{\pm}$ are equal to $\pi$. This value is in accordance with the Lavinson theorem, given by (3.9.31), because $H^-$ has only one bound state. Thus as expected, the SFEA is superior to the *SFA*. With the increase in $E$ the phase shift $\eta_0^{SF}$ increases, attains a maximum, and then falls with further increase in $E$. In contrast, both $\eta_0^{\pm}$ continuously decrease with an increase in $E$. $\eta_0^-$ is greater than $\eta_0^+$ and $\eta_0^{SF}$ lies between the two for $E$ greater than about 40 eV. At large $E$ the three curves merge into one another. This shows that the exchange effect is of importance at low $E$. Physically, the probability of exchange is also expected to be appreciable for those velocities of the incident electron that are of the same order as the velocity of the atomic electron.

### 8.6.1  Local Exchange Potential

To obtain a solution of an integro-differential equation is much more difficult than to solve a differential equation, so there have been a number of attempts to convert (8.6.5) into

$$[\nabla_1^2 + k_i^2 - U_{SF}(r_1) - U_{ex}^{\pm}(r_1)]F_i^{\pm}(r_1) = 0 \tag{8.6.12}$$

with $U_{ex}^{\pm}(r_1)$ as a local exchange term. For a spherically symmetrical potential, the above differential equation is reduced to a one-dimensional equation [similar to (8.6.8)] but with the additional local term $U_{ex}^{\pm}(r_1)$. These equations are solved numerically and the $\eta_i^{\pm}$ are obtained.

We follow the treatment of Vanderpoorten (1975) to obtain local $U_{ex}^{\pm}$. Let us define

$$J(r_1) = \int v_i^*(r_2)\left(E_T - 2\varepsilon_i - \frac{e^2}{r_{12}}\right)F_i^{\pm}(r_2)v_i(r_1)dr_2$$

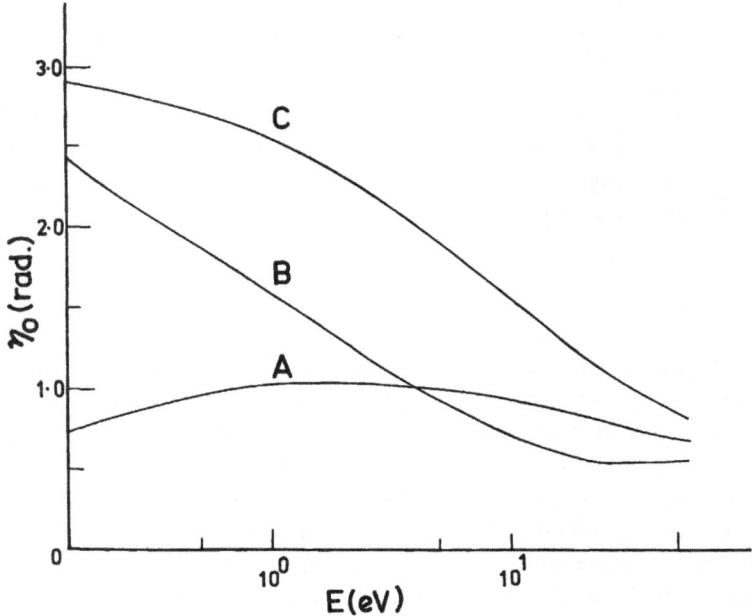

**FIGURE 8.2** Variation of the $l = 0$ phase shift $\eta_0$ for elastic scattering of a electrons of a hydrogen atom with the impact energy E. Curve A represents phase shifts in the static field approximation. Singlet and triplet phase shifts in the static field-exchange approximation are represented by the curves B and C, respectively.

Since the functions $v_i(r_2)$ and $F_i(r_2)$ represent, respectively, bound and continuum states, of the same Hamiltonian, they should be orthogonal to each other. Hence,

$$j(r_1) = -e^2 v_i(r_1) \int v_i^*(r_2) \frac{1}{r_{12}} F_i(r_2) dr_2 \qquad (8.6.13)$$

Now taking $F_i(r_2) = F_i(r_1 - r_{12})$ and using the Taylor's expansion, we get

$$F_i(r_2) = F_i(r_1) + (-r_{12} \cdot \nabla_F) F_i(r_i) + \frac{(-r_{12} \cdot \nabla_F)^2}{2!} F_i(r_i) + \cdots$$

$$= \exp(-r_{12} \cdot \nabla_F) F_i(r_i)$$

Similarly,

$$v_i^*(r_2) = \exp(-r_{12} \cdot \nabla_v) v_i^*(r_i)$$

Hence,

$$J(r_1) = e^2 \nu_i(r_1) \int dr_{12} \frac{1}{r_{12}} \exp[-r_{12} \cdot (\nabla_F + \nabla_\nu)] \nu_i^*(r_1) F_i(r_1)$$

$$= e^2 \nu_i(r_1) \frac{4\pi}{|\nabla_F + \nabla_\nu|^2} \nu_i^*(r_1) F_i(r_1) \qquad (8.6.14)$$

On the assumption that $\nu_i$ is a much smoother function of $r_1$ than $F_i$, we neglect the gradient $\nabla_\nu$ in (8.6.14). Thus,

$$U_{ex}^\pm(r_1) F_i(r_1) = \mp \frac{2m}{\hbar^2} \int \kappa_{ii}(r_1, r_2) F_i^\pm(r_2) dr_2$$

$$= \mp \frac{8\pi e^2 m}{\hbar^2} |\nu_i(r_1)|^2 \frac{1}{|\nabla_1|^2} F_i^\pm(r_1) \qquad (8.6.15)$$

Using (8.6.12) we approximate $\nabla_1^2$ by $U_{ii}(r_1) + U_{ex}^\pm(r_1) - k_i^2$ in the above equation and obtain the following quadratic equation for $U_{ex}^\pm$:

$$U_{ex}^\pm(r_1) = \mp \frac{8\pi e^2 m}{\hbar^2} |\nu_i(r_1)|^2 \frac{1}{U_{ii}(r_1) + U_{ex}^\pm(r_1) - k_i^2} \qquad (8.6.16)$$

Solving the above equation under the physical condition $U_{ex}^\pm(r_1) \to 0$ as $k_i \to \infty$, we get

$$2U_{ex}^\pm(r_1) = k_i^2 - U_{ii}(r_1) - \left\{ [k_i^2 - U_{ii}(r_1)]^2 \mp \frac{32\pi e^2 m}{\hbar^2} |\nu_i(r_1)|^2 \right\}^{1/2} \qquad (8.6.17)$$

At large $k_i$ the reduced local exchange potential energy tends to

$$U_{ex}^\pm(r_1) = \pm \frac{8\pi e^2 m}{\hbar^2 k_i^2} |\nu_i(r_1)|^2 \qquad (8.6.18)$$

In the first Born approximation (8.6.18) gives the Ochkur exchange amplitude. Hence, $(\hbar^2/2m) U_{ex}^\pm(r_1)$ is known as the Ochkur exchange potential (energy).

For multielectron atoms the (8.6.17) modifies to (Riley and Truhlar, 1975)

$$V_{ex}(r_1) = \frac{1}{2}(E - V_{st}) - \frac{1}{2}\left\{ (E - V_{SF})^2 + \alpha^2 \right\}^{1/2} \qquad (8.6.19)$$

where

$$\alpha^2 = \frac{8\pi e^2 \hbar^2}{m} \sum_{i=1}^{n_0} N_i |v_i|^2 \qquad (8.6.20)$$

In the above equation $n_0$ is the number of different single-particle spatial states $v_i$ occupied in the target. $N_i$ is a positive or negative constant depending on the state of the target. For example, for targets with doubly occupied spatial orbitals, $N_i$ is $+\frac{1}{2}$.

### 8.6.2   The Polarized Orbital Method

Temkin and Lemkin (1961) considered polarization of the atom along with the exchange and the static field interactions in the investigation of elastic scattering of electrons by hydrogen atoms. The incident energy $E$ was taken to be less than $\varepsilon_{ex}$ so that no real excitation of the atom is possible but the atom is polarized due to virtual excitations. They employed the adiabatic one-state approximation and considered the distortion of the target wave function by the incident electron only up to first order. Thus, in their method, which is known as the *polarized orbital method* (POM), the wave function of the system (H + e) for singlet and triplet scattering is given by

$$\Psi^{\pm}(r, X) = F^{\pm}(r)[v_i(X) + v_{pol}(r, X)] \pm F^{\pm}(X)[v_i(r) + v_{pol}(X, r)] \qquad (8.6.21)$$

where $v_{pol}$ is the dipole ($l = 1$) term of $v_i'$, given by (8.5.16), and $F$ is the wave function of the scattered electron. Due to the presence of $v_{pol}$ in (8.6.21), the scattering equation (8.6.5) modifies to

$$(\nabla^2 + k^2)F^{\pm}(r) = [U_{SF}(r) + U_{pol}(r, X)]F^{\pm}(r)$$

$$\mp \frac{2m}{\hbar^2} \int [\kappa(r, X) + \kappa_{ep}(r, X)]F^{\pm}(X)dX \qquad (8.6.22)$$

where $U_{pol}$ is the reduced second-order dipole polarization interaction energy and the kernel $\kappa_{ep}$ arises due to exchange and polarization. $U_{pol}(r)$ is evaluated from the dipole term of (8.5.6) but the upper limit of $X$ is taken to be $r$. Hence,

$$U_{pol} = -\frac{8}{3(ra_0)^4} \int_0^r (a_0 X^4 + \tfrac{1}{2}X^5)\exp(-2X/a_0)dX$$

Using (8.3.9), we finally obtain

$$U_{pol}(r) = -\frac{\alpha_d}{a_0 r^4}[1 - p(r)\exp(-2r/a_0)] \qquad (8.6.23)$$

where the dipole polarizability is equal to $4.5a_0^3$ and

$$p(r) = \frac{4}{27}\left(y^5 + \frac{9}{2}y^4 + 9y^3 + \frac{27}{2}y^2 + \frac{27}{2}y + \frac{27}{4}\right) \qquad (8.6.24)$$

In the above equation $y$ is equal to $r/a_0$. Since $U_{SF}$ and $U_{pol}$ are spherically symmetric, a partial wave expansion is carried out and the phase shifts $\eta_l^{\pm}$ for each partial wave are obtained by numerical integration of the resulting integro-differential equation. Temkin and Lemkin (1961) considered $l = 0$, 1, and 2 and varied $ka_0$ from 0 to 0.8. A comparison of the scattering lengths $a_S$ and the phase shifts obtained in the POM and SFE approximations shows appreciable differences. For example the values of $a_S^{\pm}$ in the SFE approximation are $8.10a_0$ and $2.36a_0$, respectively. They change to $5.7a_0$ and $1.7a_0$, respectively, when calculated in the POM. The latter values are quite close to the variational values of Schwartz (1961), which are regarded as almost exact. Similarly at $ka_0 = 0.75$, the values of $\eta_2^{\pm}$ (in radians), in the SFEA, are $-0.0176$ and $0.0555$, respectively. The corresponding values given by the POM are $0.0627$ and $0.112$. As expected, the relative change in $\eta_l$ increases with $l$. Since the higher values of $l$ contribute significantly to the DCS in the forward direction, the polarization potential play an important role in the evaluation of the DCS at small scattering angles. It is noted that the effect of $\kappa_{ep}$ on the phase shifts is quite small. In the *exchange-adiabatic approximation* $\kappa_{ep}$ is neglected.

One of the early calculations to investigate the effect of the polarization potential on differential cross sections for elastic scattering of electrons by helium atoms in the intermediate energy range was carried out by Khare and Moise-witsch (1965). They solved the radial equation (8.5.19) under the proper boundary conditions numerically and thus obtained $\eta_{SFP}$, the phase shifts in the static field and polarization approximation. To obtain $U_{SF}$ they represented the helium atom by the following wave function of Green et al. (1954):

$$v(r_1, r_2) = \phi(r_1)\phi(r_2) \qquad (8.6.25)$$

with

$$\phi(r) = N[\exp(-zr) + c\exp(-2zr)] \qquad (8.6.26)$$

where $N$ is the normalization constant, and the variational parameters $Z$ and $c$ are equal to 1.4558 and 0.6, respectively. The polarization potential $U_{pol}(r)$ is given by (8.6.23) but with

$$p(r) = 1 + 2(Zr/a_0) + 2(Zr/a_0)^2 + \tfrac{4}{3}(Zr/a_0)^3 + \tfrac{2}{3}(Zr/a_0)^4 \qquad (8.6.27)$$

and $\exp(-2r/a_0)$ of (8.6.23) being replaced by $\exp(-2Zr/a_0)$. The cut-off parameter $Z$ is taken to be same as that given by the variational method in (8.6.26), i.e., 1.4558. The dipole polarizability of the helium atom is $1.39a_0^3$ (Sternheimer, 1957). Khare and Moiseiwitsch (1965) included the effect of exchange in the following approximate manner. For $l = 0$ and 1 they took $\eta_{SFE}$ evaluated by Morse and Allis (1933) from the exact numerical solution of the appropriate static field-exchange integro-differential equation. To this phase shift ($\eta_{SFP} - \eta_{SF}$) was added to obtain the approximate value of $\eta_{SFPE}$, the phase shift in the static field-polarization-exchange approximation. The phase shifts $\eta_{SF}$ in the static field alone was again obtained from (8.5.19) with $U_{pol} = 0$. For the higher partial waves ($l \geq 1$) the effect of exchange was included through the first-order exchange approximation of Bell and Moiseiwitch (1963). The calculation of Khare and Moisewitsch clearly demonstrated that, as expected, the inclusion of the polarization potential sharply increases the DCS at small scattering angles and gives much better agreement between theory and experiment.

In a number of investigations carried out in the intermediate energy range following one-channel differential equation:

$$\left[\frac{d^2}{dr^2} + k^2 - U_{SF}(r) - U_{pol}(r) - U_{ex}^l(r) - \frac{l(l+1)}{r^2}\right] f_l(r) = 0 \qquad (8.6.28)$$

with the local exchange potential $U_{ex}^l(r)$ being solved to obtain the DCS and $\sigma_{el}$ for multielectron atoms.

At low impact energies only the $l = 0$ partial wave is of importance. If $V_{op}(r)$ is so strong that at $E = E_0$, $\eta_0$ is equal to $\pi$, then $\sigma_0$ from (3.9.26) is zero. Since $\sigma_l ( l > 0 )$ is negligible, $\sigma_{el}$ at $E_0$ is quite small. Hence, a curve of $\sigma_{el}$ vs. $E$ shows a dip at $E_0$. This dip in the cross section at small $E$ is known as the Ramsauer–Townsend effect and has been observed for a number of targets. For Ar, Kr, and Xe the dips are observed at $E = 0.37\,\text{eV}$, $0.60\,\text{eV}$, and $0.65\,\text{eV}$, respectively.

Temkin (1962) developed a nonadiabatic model for the $S$ partial wave (the orbital angular momentum $L$ of the electron–hydrogen atom system being zero). The initial investigations employing the above model were carried out in the elastic, inelastic, and ionization energy domains of the hydrogen atom (Temkin, 1962; Kyle and Temkin, 1964). However, these investigations faced severe rounding off problem. Poet (1978) overcame this difficulty by using a very fine energy grid and obtained a resonant structure for $ka_0 < 1$. Since then this model is known as the *Temkin-Poet model*. Bhatia et al. (1993) developed an *ab initio* method to calculate $\sigma_{el}$ and $\sigma_T$ for $S$ wave $e$–H scattering. They considered $ka_0$

up to 2 and thus also included the ionization domain. Their cross sections are in excellent agreement with the more accurate cross sections of Callaway and Oza (1984). Recently Temkin et al. (1998a,b), extended the Temkin–Poet model for $L > 0$ partial waves. The approximation developed by these authors is known as the *generalized exchange approximation (GEA)*. This approximation gives resonances, differentiates between singlet and triplet scattering for all $L$, and contains inelastic and ionization channels.

## 8.7   The Static Field, Exchange, Polarization, and Absorption Approximation

In the *state field, exchange, polarization, and absorption approximation (SFPEAA)* a complex optical potential given by

$$V_{op}(r) = V_{SF}(r) + V_{pol}(r) + V_{ex}(r) + iV_{ab}(r) \qquad (8.7.1)$$

is used. These four potentials approximately account for the dynamics of the collision process. All the components of $V_{op}(r)$ are taken in the spherically symmetric form and the method of partial waves is utilized to obtain the phase shifts. Due to complex $V_{op}(r)$, the phase shifts are also complex and the real and imaginary parts of the scattering matrix $S_l$ are given by

$$\mathrm{Re}S_l = \exp(-2\beta_l)\cos(2\alpha_l) \qquad (8.7.2)$$

$$\mathrm{Im}S_l = \exp(-2\beta_l)\sin(2\alpha_l) \qquad (8.7.3)$$

where the complex phase shift $\eta_l = \alpha_l + i\beta_l$.

The construction of the absorption potential is not straightforward. One of the guiding factors is that $V_{op}$ should yield $\sigma_T$ in agreement with the experimental data. A number of $V_{ab}$ have been proposed. According to McCarthy et al. (1977)

$$V_{ab}(r) = \frac{W \times 15.015 \times 10^4}{(E - V_{SF} - V_{ex})^2} \sum_l (2l+1)r^2 v_1^2(r) \qquad (8.7.4)$$

where $v_l(r)$ is the single-orbital radial wave function. Summation over $l$ includes only those single-particle states that contribute most to the ionization cross section of the atom. For example, for the argon atom McCarthy et al. (1977) considered only the $3p$ orbitals. According to Furness and McCarthy (1973), the inclusion of the $3s$ orbital produces only slight modification in the shape of the absorption potential curve, and the cross sections are not very sensitive to the shape.

Another absorption potential has been proposed by Reitan (1981). To construct this potential, the target wave functions are given in terms of the exponential density functions fitted to the statistical Thomas–Fermi distribution. With such density functions the absorption potential is given by

$$V_{ab} = -\frac{2Z}{rk_i}\phi_{ab}(t)R \tag{8.7.5}$$

where

$$\phi_{ab}(t) = A_1\alpha_1 \exp(-2.9\alpha_1 t) + \left(\frac{\bar{\delta}}{\alpha_1}\right)^2 \left[K_0(\bar{\delta}t) + \frac{K_1(\bar{\delta}t)}{\bar{\delta}t}\right]\exp[-2.3/(\alpha_1 t)] \tag{8.7.6}$$

$$A_1 = 2(1-\ln 2) - \frac{1}{4}\left(\frac{\bar{\delta}}{a_1}\right)^2\left[1 - 2\ln\left(\frac{\bar{\delta}}{2\alpha_1}\right)\right] \tag{8.7.7}$$

$$t = \frac{r\,Z^{1/3}}{0.8853\alpha_0}, \qquad \bar{\delta} = \frac{243\Delta}{137k_i Z^{1/3}e^2}$$

and $\alpha_1$ is the one-term fit to the Thomas–Fermi distribution. $K_0$ and $K_1$ are modified Bessel functions.

A third absorption potential is the nonempirical potential derived from a quasi-free scattering model by Staszewska et al. (1983); it is given by

$$V_{ab}(r) = -\frac{\rho(r)}{a_0^2}(T_{loc}R)^{1/2}\frac{8\pi}{5k_i^2 k_F^3}H(x)(A_1 + A_2 + A_3) \tag{8.7.8}$$

with

$$k_F = \left[3\pi^2\rho(r)\right]^{1/3}$$
$$A_1 = 5k_F^3 a_0^3 R/\Delta, \qquad A_2 = -k_F^3(5k_i^2 - 3k_F^2)a_0\big/(k_i^2 - k_F^2)^2$$
$$A_3 = \frac{2H(y)y^{5/2}}{(k_i^2 - k_F^2)^2 a_0^4}, \qquad y = (1/R)[(2k_F^2 - k_i^2)a_0^2 R + \Delta]$$
$$x = [(k_i^2 - k_F^2)a_0^2 R + \Delta]/R \tag{8.7.9}$$

where $H(x)$ and $H(y)$ are Heavyside unit step functions and $T_{loc}$ is the total kinetic energy.

## Comparison of Theoretical and Experimental Crosssection

Let us consider the results obtained by the application of the above approximations to investigate the elastic scattering of electrons and positrons by the inert gases neon and argon. Jhanwar et al. (1978) used the SF, SFP, and SFPE approximations to investigate electron–neon elastic collisions. Khare and Kumar (1978) employed the SF and SFPE approximations for electron–argon collisions. Lata (1984) added the absorption potential of Reitan [given by (8.7.5)] to the real potential of Jhanwar et al. to investigate the electron–neon collision. The same absorption potential was added by Khare et al. (1986) to the real potential of Khare and Kumar (1978) for electron–argon collisions. In all the above investigations the energy of the projectile was varied from 100 to 1000 eV and the targets were represented by the Hartree wave functions given by Sheorey (1969). Jhanwar et al. and Khare and Kumar employed the local exchange potential given by (8.6.19) with $V_{SF}$ being replaced by $V_{SF} + V_{dp}$. To evaluate $V_{dp}$, they used (7.7.37) with d given by (7.7.39) and solved (8.2.6) with $V_{op} = V_{st} + V_{ex} + V_{dp}$ for the first $M$ partial waves for a reasonably high value of $M$. The higher partial waves $(l > M)$ are effectively due to the long-range potential and are small. Hence, their contribution was included through the FBA.

Thus in the investigations of Jhanwar et al. (1978),

$$f(\theta) = \frac{1}{k_i} \sum_{l=0}^{M} (2l+1) e^{i\eta_l} \sin \eta_l P_l(\cos\theta) + f_{dp}^{B1}(\theta)$$

$$- \frac{1}{k_i} \sum_{l=0}^{M} (2l+1) \eta_l^{B1} P_l(\cos\theta) \qquad (8.7.10)$$

where $f_{dp}^{B1}$ and $\eta_l^{B1}$ are due to a long-range polarization potential in the FBA.

The theoretical differential cross sections for the elastic scattering of 100-eV electrons by neon and argon atoms are shown in the Figs. 8.3 and 8.4. The theoretical results of Byron and Joachain (1977b) and Joachain et al. (1977), who employed an optical model (OM) with complex potential, are also included. The experimental data are shown for comparison. The failure of the plane wave approximation (which is quite evident from the figures) shows the importance of the distortion of the wave function of the projectile. This distortion, to all orders, is included in the SFA but it completely neglects the distortion of the target (polarization effect). Due to this shortcoming, the SFA underestimates $I(\theta)$ at low $\theta$. In the backward direction (large $\theta$) agreement between the SFA cross sections and the experimental data is satisfactory. For electron–neon scattering $I(\theta)$ in the SFA falls to a deep minimum, which is partially supported by the experimental data.

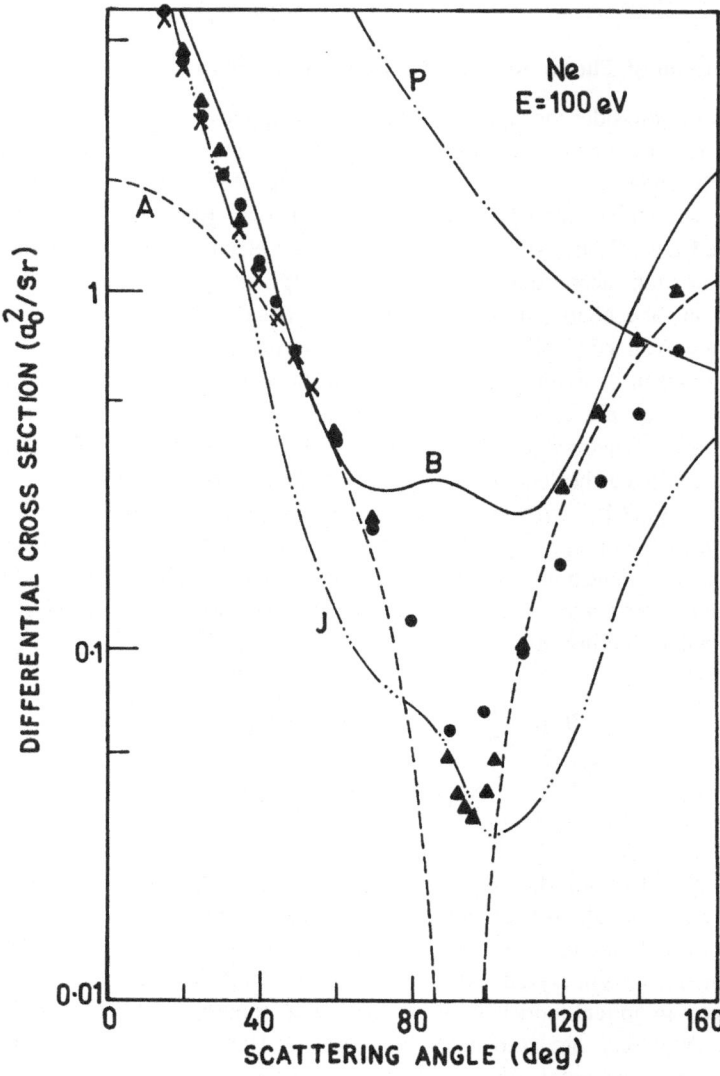

**FIGURE 8.3** Differential cross sections for 100-eV electrons elastically scattered by neon atoms. Curves A, B, and P are obtained in SF, SFPE, and PW (Jhanwar et al., 1978) approximations, respectively. Curve J represent cross sections obtained by Byron and Joachain (1977b) with an OM complex potential. Experimental data, ●, Gupta and Rees (1975a); ▲, Williams and Crowe (1975); X, Jansen et al. (1976). Reproduced from "Elastic scattering of electrons on Ne atom at intermediate energies," Jhanwar, B. L., Khare S. P., and Kumar Jr. A, *J. Physics B* **11**, 887, 1978, with permission from the Institute of Physics Publishing Ltd. (U.K.). The curve J was added.

**Table 8.2** Total Elastic Cross Sections $\sigma_{el}$ (in $10^{-21}\,m^2$) for Electron (Positron)-Neon Scattering. The Numbers within the Brackets are for Positron Impact, Calculated by Lata (1984)

| E (eV) | Theory | | | | | Experiment | |
|---|---|---|---|---|---|---|---|
| | SF | SFPE | SFPEA | OM | DWSBA | de Heer | Jansen |
| | Jhanwar et al. | | Lata | Byron and | Dewangan | et al. | et al. |
| | (1978) | | (1984) | Joachian | and Walters | (1979) | (1976) |
| | | | | (1977b) | (1977) | | |
| 100 | 16.7 | 36.3 | 16.7 | 18.0 | — | 24.4 | 23.8 |
| | | | (4.22) | | | | |
| 150 | 12.5 | 24.4 | 14.3 | — | — | 17.3 | 16.2 |
| | | | (4.13) | | | | |
| 200 | 10.6 | 17.5 | 12.0 | 12.2 | 15.4 | 14.0 | 14.0 |
| | | | (3.60) | | | | |
| 300 | 8.54 | 11.5 | 8.53 | 9.67 | 11.5 | 11.2 | 10.7 |
| | | | (3.87) | | | | |
| 400 | 7.33 | 8.95 | 6.95 | 8.18 | 9.41 | 8.97 | 8.98 |
| | | | (3.78) | | | | |
| 500 | 6.49 | 7.50 | 6.07 | 7.21 | 8.09 | 7.61 | 7.70 |
| | | | (3.69) | | | | |
| 750 | 5.09 | 5.57 | 4.75 | — | — | — | 6.24 |
| | | | (3.25) | | | | |
| 1000 | 4.23 | 4.50 | 3.96 | — | 5.01 | 4.48 | 4.98 |
| | | | (2.90) | | | | |

On the other hand, for the electron–argon system the theoretical minimum at $\theta \approx 125°$ is in accordance with the experimental data. The figures show that for neon the inclusion of $V_{ex}$ and $V_{dp}$ increases the values of $I(\theta)$ at all angles. For the argon atom $I(\theta)$ also increases with the inclusion of this potential, except near the minimum, where a decrease in $I(\theta)$ is noted. The agreement with the SFPEA cross section is good at low $\theta$ but at the higher values of $\theta$, the theory overestimates the cross sections. The OM results of Joachain et al. are also in good accord with the experimental data at low $\theta$ but underestimation is noted at higher values of $\theta$. According to Table 8.2 inclusion of $V_{ex}$ and $V_{dp}$ potentials with $V_{SF}$ increases the value of $\sigma_{el}$. With the increase of E the importance of $V_{ex}$ and $V_{dp}$ decreases. The inclusion of $V_{ab}$ decreases the cross sections for both the atoms.

According to Table 8.2 inclusion of $V_{ex}$ and $V_{dp}$ potentials with $V_{SF}$ increases the value of $\sigma_{el}$. With an increase in $E$ the importance of $V_{ex}$ and $V_{dp}$ decreases. Tables 8.2 and 8.3 show that the inclusion of $V_{ab}$ decreases cross sections for both atoms.

The decrease in $\sigma_{el}$, due to the inclusion of $V_{ab}$, is found to be valid even at energies just above the threshold of the formation of positronium (Brown and Humperston, 1985). The OM cross sections are slightly higher than those of Lata (1984) and Khare et al. (1986). The cross sections in the distorted wave second Born approximation (DWSBA) of Dewangan and Walters (1977) are still higher.

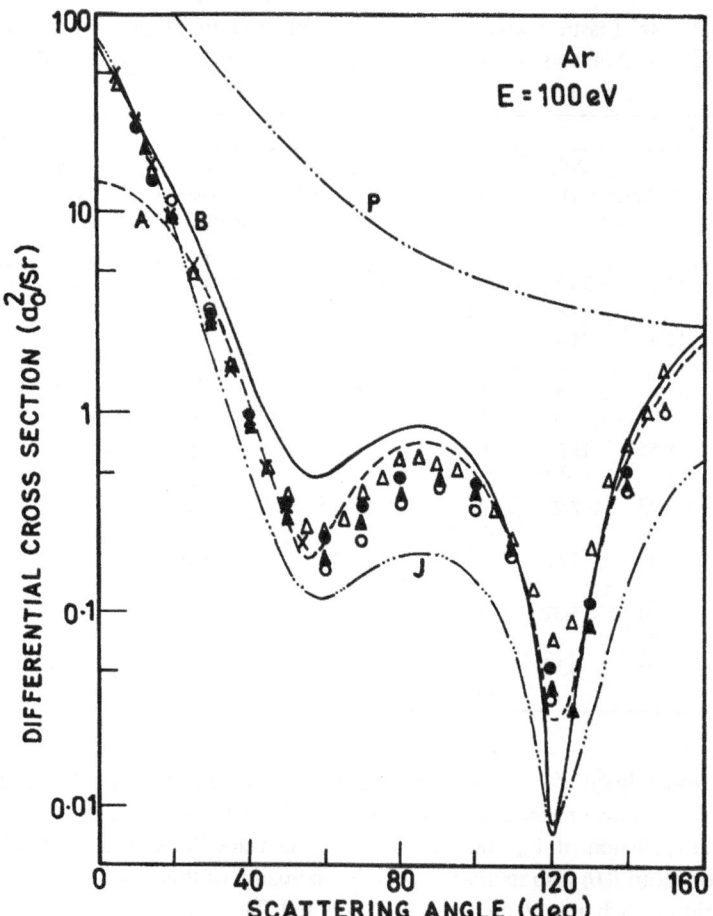

**FIGURE 8.4** Same as Fig. 8.3 but for argon atoms. The theoretical curves A, B, and P are in the SF, SFPE, and PW approximations (Khare and Kumar, 1978), respectively. Curve J represents cross section of Joachain et al. (1977) obtained with an OM complex potential. Experimental data: ●, Gupta and Rees (1975b); ▲, Dubois and Rudd (1975); O, Williams and Willis (1975); Δ, Vuskovic and Kurepa (1976); X, Jansen et al. (1976). Reproduced from "Elastic scattering of electrons by argon atoms," Khare and Kumar *Pramana* **10**, 63, 1978, with permission from the Indian Academy of Sciences.

In general the $\sigma_{el}$ in SFPEA and those given by the OM potential are lower than the experimental data. The underestimation by the theories decreases with an increase in $E$.

Tables 8.4 to 8.7 compare theoretical $\sigma_T$ with the experimental data. For the neon atom $\sigma_T$ in the SFPEA are lower than those given by the OM potential

**Table 8.3** Total Elastic Cross Sections $\sigma_{el}$ (in $10^{-21}\,\text{m}^2$) for Electron (Positron)-Argon Scattering. The Numbers within the Brackets Are for Positron Impact, Calculated by Lata (1984)

| $E$ (eV) | Theory | | | Experiment | | |
|---|---|---|---|---|---|---|
| | SFPA Khare-Kumar (1978) | SFPEA Khare et al. (1986) | OM Joachain et al. (1977) | Duboi and Rudd (1975) | Jansen et al. (1976) | de Heer et al. (1979) |
| 100 | 64.3 (16.6) | 34.3 (15.7) | 42.6 | 42.6 | 38.1 | 48.5 |
| 150 | 43.8 (18.4) | 25.8 (14.8) | — | — | 33.6 | 37.9 |
| 200 | 34.5 (18.5) | 21.6 (14.3) | 30.2 | 28.8 | 30.2 | 32.0 |
| 300 | 26.2 (17.5) | 17.8 (13.5) | 24.6 | — | 25.6 | 24.6 |
| 400 | 22.0 (16.2) | 15.7 (12.7) | 21.4 | — | 22.4 | 21.1 |
| 500 | 19.3 (14.9) | 14.4 (11.9) | 19.2 | — | 20.0 | 18.8 |
| 750 | — | 12.2 (10.5) | — | — | — | 14.9 |
| 1000 | 12.9 (10.9) | 10.7 (9.23) | 13.2 | — | — | 12.7 |

and the DWSBA. The agreement between the theoretical and the experimental cross sections is fairly good for $E$ greater than about 300 eV (see Table 8.4). For the argon atom the SFPEA cross sections are greater than the OM cross sections at low $E$ but the reverse is the case at the higher $E$ (see Table 8.5). The theoretical cross sections are again in fair accord with the experimental data for

**Table 8.4** Total Collision Cross Sections $\sigma_T$ (in $10^{-21}\,\text{m}^2$) for Electron-Neon Scattering

| $E$ (eV) | Theory | | | Experiment | | |
|---|---|---|---|---|---|---|
| | OM Byron and Joachain (1977b) | DWSBA Dewangan-Walters (1977) | SFPEA Lata (1984) | de Heer et al. (1979) | Wagenaar and de Heer (1980) | Kauppila et al. (1981) |
| 100 | 39.7 | — | 34.4 | 32.6 | 30.2 | 29.6 |
| 150 | — | — | 29.5 | 25.9 | 25.9 | 25.2 |
| 200 | 27.0 | 27.3 | 26.6 | 23.3 | 23.2 | 22.3 |
| 300 | 21.1 | 21.4 | 18.4 | 18.6 | 18.6 | 17.9 |
| 400 | 17.6 | 17.8 | 15.0 | 15.4 | 15.8 | 15.3 |
| 500 | 15.2 | 15.3 | 12.7 | 13.4 | 13.8 | 13.4 |
| 750 | — | — | 9.23 | — | — | — |
| 1000 | — | 9.49 | 7.21 | 8.26 | — | — |

**Table 8.5** Total Collision Cross Sections $\sigma_T$ (in $10^{-21}\,m^2$) for Electron-Argon Scattering

| $E$ (eV) | Theory | | Experiment | | |
|---|---|---|---|---|---|
| | SFPEA Khare et al. (1986) | OM Joachain et al. (1977) | de Heer et al. (1979) | Wagenaar and de Heer (1980) | Kauppila et al. (1981) |
| 100 | 95.3 | 93.2 | 82.7 | 81.2 | 76.7 |
| 150 | 76.7 | — | 69.3 | 68.9 | 63.4 |
| 200 | 64.4 | 64.4 | 59.8 | 58.6 | 56.1 |
| 300 | 48.7 | 50.9 | 47.0 | 47.9 | 46.0 |
| 400 | 39.2 | 43.1 | 39.9 | 40.9 | 38.4 |
| 500 | 32.9 | 37.5 | 35.0 | 35.9 | 32.9 |
| 750 | 23.6 | 28.0 | — | — | — |
| 1000 | 18.6 | 24.2 | 22.2 | — | — |

$E \geq 200\,eV$. The total collision cross sections for collisions of positrons with neon and argon atoms are shown in Tables 8.6 and 8.7, respectively. The agreement between the theoretical cross sections and the experimental data is fairly good over the whole energy range (100–1000 eV).

The SFP approximation has been employed by Raj (1981) and Kaushik et al. (1983) to obtain critical points for a number of light atoms. These values are shown in Table 8.8 along with the theoretical values of Fon and Berrington (1981), who employed the $R$-matrix method, and the experimental data of Kollath and Lucas (1979), Register et al. (1980), and Menedez et al. (1980) for the neon atom. The $\theta_c$ obtained by Raj (1981) are slightly lower than those reported by Khare and Raj (1980) in the SFA. However, the $E_c$ of Raj are substantially lower than those given by Khare and Raj. The values of Raj for neon are in fairly good agreement with other theoretical values and the experimental data. Table 8.8 also

**Table 8.6** Total Collision Cross Sections $\sigma_T$ (in $10^{-21}\,m^2$) for Positron-Neon Scattering

| $E$ (eV) | Theory | | | Experiment | | | |
|---|---|---|---|---|---|---|---|
| | SFPEA Lata (1984) | OM Byron and Joachain (1977b) | DWSBA Dewangan and Walters (1977) | Kauppila et al. (1981) | Brenton et al. (1978) | Coleman et al. (1976) | Tsai et al. (1976) |
| 100 | 18.1 | 23.6 | — | 19.1 | 18.0 | 19.8 | 18.3 |
| 150 | 16.3 | — | — | 17.8 | — | 17.1 | 16.4 |
| 200 | 14.5 | 18.9 | 19.3 | 16.7 | 16.2 | 14.9 | 15.7 |
| 300 | 12.7 | 15.9 | 16.4 | 14.1 | — | 10.2 | — |
| 400 | 11.1 | 13.9 | 14.2 | 12.2 | 12.5 | 9.23 | — |
| 500 | 9.76 | 13.2 | 12.5 | 10.8 | — | 5.98 | — |
| 750 | 7.47 | — | — | — | — | — | — |
| 1000 | 5.98 | — | 8.18 | — | 6.59 | — | — |

**Table 8.7** Total Collision Cross Sections $\sigma_T$ (in $10^{-21}\,m^2$) for Positron-Argon Scattering

| $E$ (eV) | Theory | | Experiment | | | |
|---|---|---|---|---|---|---|
| | OM Joachain et al. (1977) | SFPEA Khare et al. (1986) | Tsai et al. (1976) | Brenton et al. (1978) | Griffith-Heyland (1978) | Kauppila et al. (1981) |
| 100 | 63.8 | 70.3 | 61.2 | — | 65.9 | 62.8 |
| 150 | — | 60.8 | 53.7 | — | 54.4 | 53.6 |
| 200 | 49.5 | 53.2 | 46.7 | 50.0 | 42.6 | 48.9 |
| 300 | 41.7 | 41.8 | 38.7 | — | 36.0 | 39.8 |
| 400 | 36.4 | 34.2 | — | 39.7 | 25.5 | 34.5 |
| 500 | 32.4 | 29.0 | — | 31.3 | 24.6 | 29.7 |
| 750 | — | 21.1 | — | — | 16.7 | — |
| 1000 | — | 16.8 | — | 20.6 | — | 21.6 |

shows that the value of the critical energy $E_c$ increases with $Z$. This observation is in accordance with the prediction of Buhring (1968). The number of sets of the critical points $(E_c, \theta_c)$ also increases with $Z$. For example 3, 8, and 12 sets are reported for elastic scattering of electrons by argon $(Z = 18)$, xenon $(Z = 54)$, and mergury $(Z = 80)$, respectively (Walker, 1971; Lucas and Liedike, 1975; Lucas, 1979).

Khare et al. (1983) employed the SFPEA to calculate differential and integrated elastic cross sections for elastic scattering of electrons and positrons by magnesium (a closed-shell atom) in the intermediate energy range (10 to 500 eV). Their DCS are only in moderate agreement with the experimental data of Williams and Trajmar (1978) but agree fairly well with the theoretical values of Gregory and Fink (1974), who solved the Dirac equation for energies varying

**Table 8.8** Critical Energy $(E_c)$ and Critical Angles $(\theta_c)$ for the Elastic Scattering of Electrons by Light Atoms

| Atom | $E_c$ (eV) | $\theta_c$ (deg) | Reference |
|---|---|---|---|
| Be | 5.8 | 94.8 | Kaushik et al. [Theo., 1983] |
| C | 17.49 | 97.63 | Raj [Theo., 1981] |
| N | 26.8 | 97.5 | Raj [Theo., 1981] |
| O | 35.59 | 98.65 | Raj [Theo., 1981] |
| F | 45.2 | 99.73 | Raj [Theo., 1981] |
| Ne | 58.85 | 99.62 | Raj [Theo., 1981] |
| Ne | 64 | 103.6 | Fon and Berrington [Theo., 1981] |
| Ne | $73.7 \pm 1.0$ | $103.0 \pm 0.5$ | Kollath and Lucas [Expt., 1979] |
| Ne | $62.5 \pm 3$ | $101.5 \pm 1.5$ | Register et al. [Expt., 1980] |
| Ne | $64 \pm 1$ | $102 \pm 0.5$ | Menedez et al. [Expt. 1980] |
| Ca | 37.3 | 72.24 | Khare et al. [Theo., 1985] |
| Ca | 39.7 | 141.38 | Khare et al. [Theo., 1985] |
| Ca | 137.8 | 120.94 | Khare et al. [Theo., 1985] |

**Table 8.9** Total Elastic Cross Sections (in $10^{-20} \text{m}^2$) for $e^{\mp}$–Mg Scattering in the SFPE Approximation

| $E$ (eV) | $e^-$–Mg | | | $e^+$–Mg |
|---|---|---|---|---|
| | Theory | | Experiment | Theory |
| | Khare et al. (1983) SFPE | Fabrikant (1980) CCA | Williams and Trajmar (1978) | Khare et al. (1983) SFPE |
| 10 | 55.6 | 25.5 | 29.0 | 9.81 |
| 20 | 29.3 | 12.3 | 16.0 | 5.20 |
| 40 | 9.95 | — | 6.60 | 4.44 |
| 50 | 7.90 | — | — | — |
| 100 | 4.80 | — | — | 3.48 |
| 200 | 3.11 | — | — | 2.45 |
| 300 | 2.38 | — | — | 1.93 |
| 400 | 1.96 | — | — | 1.62 |
| 500 | 1.68 | — | — | 1.40 |

from 100 eV to 2 KeV. The total elastic cross sections obtained by Khare et al. (1983) for electron and positron scattering are shown in Table 8.9 along with the experimental data of Williams and Trajmar (1978) and the theoretical values of Fabrikant (1980), for the electron–magnesium system. For electron scattering, the results of Khare et al. overestimate the cross sections, whereas Fabrikant's results underestimate them. It may be noted that the main contribution to $\sigma_{el}$ comes from the small-angle region, where extrapolation has been employed by Williams and Trajmar to obtain experimental $\sigma_{el}$. As expected, $\sigma_{el}(e^+)$ is smaller than $\sigma_{el}(e^-)$, and with the increase in $E$ the two cross sections come closer to each other.

Khare et al. (1985) investigated elastic scattering of electrons and positrons by calcium, which is also a closed-shell atom but relatively heavy ($Z = 20$). They obtained the DCS, $\sigma_{el}$, and critical points in the SFA and SFPEA. For electrons they also included spin–orbit interactions through (4.2.4) in their optical potential and obtained $|f|$, $|g|$, the relative phase $\phi_{rel}$ between $f$ and $g$, and the parameters $S(\theta)$ and $T(\theta)$ in the energy range 10–500 eV. They obtained three sets of the critical points ($E_c$, $\theta_c$). These are also shown in Table 8.8. In Fig. 8.5 the variation in the parameters $S(\theta)$, $T(\theta)$, $U(\theta)$, and $\phi_{rel}(\theta)$ with the scattering angle $\theta$, as obtained by Khare et al. at the critical energy 39.7 eV, is shown. For this energy the critical angle is 141.4°. It is evident from the figure that all four parameters undergo drastic change near the critical energy. For $\theta$ slightly less than $\theta_c$ the value of $S(\theta)$ is close to $-0.3$, which shows that the scattered electrons are nearly 30% polarized with $m_s = -\frac{1}{2}$. For $\theta$ slightly greater than $\theta_c$ the scattered electrons are about 20% polarized but with $m_s = +\frac{1}{2}$. As discussed in Chapter 4, such behavior accords with expectations.

Figure 8.6 shows the variations of $|f|$ and $|g|$ with $E$ [obtained by Khare et al. (1985)] at critical angles $\theta_c$ for energies close to $E_c$. It may be noted that at $\theta_c = 141.38°$ and $120.94°$, the spin-flip scattering amplitude $|g|$ becomes greater than the corresponding $|f|$ at $E_c$. Since $g$ arises solely from the spin–orbit coupling it may be concluded that the spin–orbit interaction completely dominates the scattering at the critical points, and so cannot be ignored in the region close to them.

In Table 8.10 $\sigma_{el}(e^\pm)$ for the calcium atom, obtained by Khare et al. (1985), are shown. As expected $\sigma_{el}(e^-)$ is greater than $\sigma_{el}(e^+)$. Even at 500 eV, the difference between the two-cross sections is noticeable. This indicates that the energy

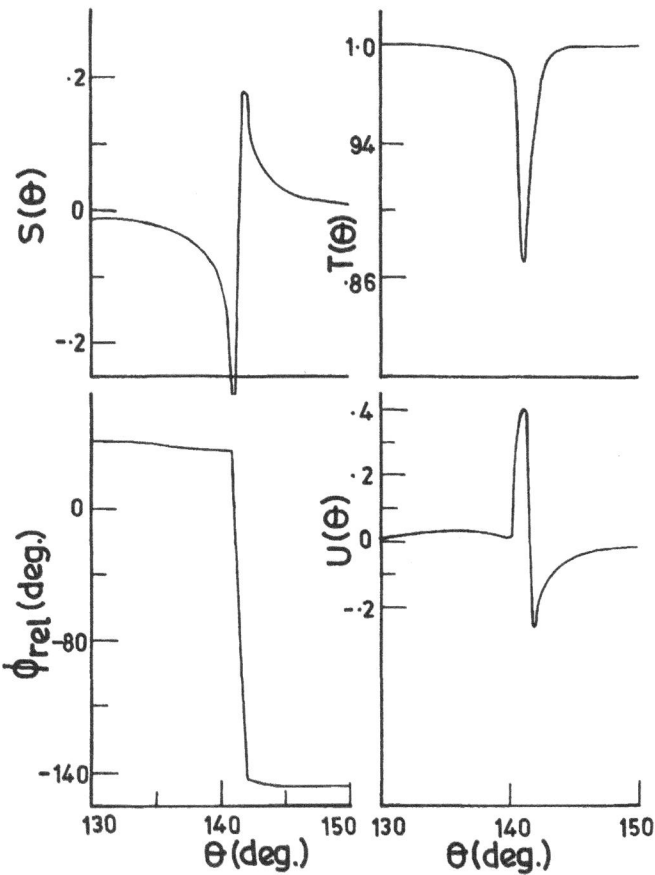

**FIGURE 8.5** Variation of $S(\theta)$, $T(\theta)$, $U(\theta)$, and $\phi_{rel}(\theta)$ with the scattering angle $\theta$ for the elastic scattering of 39.7-eV (critical energy) electrons by calcium atoms near the critical angle ($\theta_c = 141.38°$), as obtained by Khare et al. (1985).

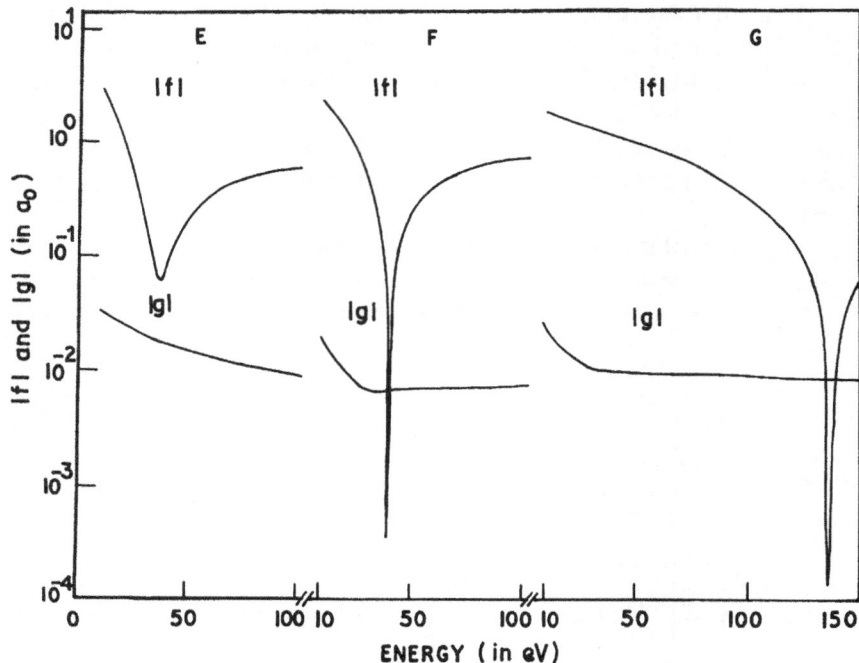

**FIGURE 8.6** Variation of the direct and spin-flip scattering amplitudes $|f|$ and $|g|$ vs. incident electron energy corresponding to the scattering angles 77.24° (E), 141.38° (F), and 120.94° (G). Reproduced from "Elastic scattering of electrons and positrons by Ca atom," S. P. Khare, A. Kumar, and Vijaishri, *J. Phys. B* **18**, 1827, 1985, with permission from the Institute of Physics, Publishing Ltd., UK.

**Table 8.10** Total Elastic Cross Sections (in $10^{-20}\,\mathrm{m}^2$) for $e^{-,+}$–Ca Scattering in the SF and SFPE Approximations (Khare et al. 1985)

| $E$ (eV) | Electron | | Positron | |
|---|---|---|---|---|
| | SF | SFPE | SF | SFPE |
| 10 | 25.6 | 84.9 | 16.3 | 39.56 |
| 20 | 19.7 | 38.9 | 13.3 | 11.94 |
| 30 | 15.1 | 22.5 | 11.5 | 8.17 |
| 40 | 12.9 | 16.8 | 10.3 | 7.81 |
| 50 | 11.5 | 14.0 | 9.38 | 7.67 |
| 75 | 9.41 | 10.6 | 7.85 | 7.04 |
| 100 | 8.15 | 8.90 | 6.86 | 6.39 |
| 150 | 6.58 | 6.96 | — | — |
| 200 | 5.62 | 5.85 | 4.88 | 4.79 |
| 250 | 4.97 | 5.13 | — | — |
| 300 | 4.49 | 4.61 | 3.96 | 3.93 |
| 400 | 3.82 | 3.89 | — | 3.39 |
| 500 | 3.36 | 3.41 | — | 3.01 |

region in which $\sigma_{el}(e^-)$ is not equal to $\sigma_{el}(e^+)$ increases with $Z$. It also shows that the lowest energy at which the FBA becomes applicable increases with $Z$.

Using one-channel approximation, a good number of other investigations for the scattering of electrons by different atoms have been carried out. Real as well as complex optical potentials are employed and the elastic DCS, $\sigma_{el}$, $\sigma_m$, $\sigma_T$, and the polarization parameters $S$, $T$, and $U$ have been evaluated (see McEachran and Stauffer, 1986; Nahar and Wadehra 1991; Yuan and Zhang, 1990a,b, 1991, 1992; Szmytkowski, 1991; Szmytkowski and Sienkiewicz, 1993, 1994; Kumar et al., 1994, 1995; Sienkiewicz and Baylis, 1997; Jain and Tripathi, 1997, etc.).

Recently, Dorn et al. (1998) have used the relativistic Schrödinger equation with a complex potential to calculate the spin polarization for the xenon atom. Neerja et al. (2000) have also employed a complex optical potential with the relativistic Dirac equation to calculate the elastic DCS, $\sigma_{el}$, $\sigma_m$, $\sigma_T$, and the polarization parameters $S$, $T$, and $U$ for the scattering of electrons by ytterbium, radon, and radium atoms in the energy range 2.0–500 eV. Both these calculations show that the absorption potential must be included in the relativistic description for accurate prediction of the $S$, $T$, and $U$ parameters. Further, the $\sigma_{el}$ of Neerja et al. are always found to be smaller than $\sigma'_{el}$, the elastic integrated cross section with $V_{ab} = 0$. But $\sigma'_{el}$ is smaller than the sum $\sigma_{el} + \sigma_{ab}$. These observations are in agreement with those of Lata (1984) and Khare et al. (1986) for neon and argon atoms, respectively.

The one-channel approximation with a complex optical potential described above is found to be reasonable at intermediate and high energies, but becomes unsatisfactory at low $E$. Further, it cannot yield individual excitation cross sections.

## 8.8 The Distorted Wave Born Approximation

The DWBA discussed in Sec. 3.12 for potential scattering has been extended to electron–atom collisions. Using (3.3.13) and (3.3.14) in (3.12.2) and extending the resulting equation to the electron–atom collision, we find that the transition matrix element $T_{ji}$ for the excitation of an atom from $v_i(X)$ to $v_j(X)$ due to electron impact in the DWBA is given by

$$T_{ji} = \langle \chi_j^-(r,X)|V_{1i}|v_i(X)\phi_{k_i}(r)\rangle + \langle \chi_j^-(r,X)|V_{2j}|A\chi_i^+(r,X)\rangle \qquad (8.8.1)$$

where $A$ is the antisymmetrization operator to account for the electron exchange. $\chi_i^+$ is the wave function of the system (electron + atom) distorted by the potential $V_{1i}$ in the initial channel. It is the solution of the differential equation

$$(H_A + H_e + V_{1i} - E_{k_i,i})\chi_i^+(r,X) = 0 \qquad (8.8.2)$$

and satisfies outgoing boundary conditions. The distorted wave $\chi_j^-$ in the final channel also satisfies (8.8.2) but it obeys incoming boundary conditions, and the distortion potential is taken to be $V_{1f}$. The total interaction potential $V = V_{1i} + V_{2i} = V_{1f} + V_{2f}$ is given by (7.2.6). The other symbols occurring in (8.8.1) and (8.8.2) are the same as in (7.2.2) to (7.2.5). The choice of $V_{1i}(V_{1f})$ is not unique, $V_{1i}$ may or may not be equal to $V_{1f}$, but they are taken to be functions only of $r$. Since the wave function $\chi_i^-$ for excitation ($j \neq i$) is orthogonal to $\nu_i(X)$, the matrix element $T_{jM,i}$ for a final magnetic substate $jM$ is

$$T_{jM,i} = \langle \chi_{jM}^-(r, X) | V - V_{1j} | A\chi_i^+(r, X) \rangle \qquad (8.8.3)$$

Now, for an atom having Z electrons,

$$\chi_{i(jM)}^{+(-)} = F^{+(-)}(k_{i(j)}, r)\nu_{i(jM)}(X)S_{i(j)}(1, 2, \ldots, Z) \qquad (8.8.4)$$

where $S_{i(j)}$ is the initial (final) state spin wave function of the system and $F^{+(-)}$ satisfies

$$\left(\nabla_r^2 + k_{i(j)}^2 - U_{i(j)}\right)F^{+(-)}(r) = 0 \qquad (8.8.5)$$

The reduced distortion interaction energy $U_{i(j)}$ is taken to be spherically symmetrical and the method of partial waves is employed to obtain $F^{+(-)}$. The atomic orbitals $\nu_i(X)$ and $\nu_{jM}(X)$ are represented by Hartree–Fock wave functions. Thus $\chi_{i(jM)}^{+(-)}$ and $T_{jMi}^S$ for the spin state $S$ of the system are determined with the help of (8.8.4) and (8.8.3), respectively. Extending (3.3.13) to the atomic excitation, we get

$$f_{jMi}^S = -\sqrt{\frac{k_j}{k_i}}\left(\frac{4\pi^2 m}{\hbar^2}\right)T_{jMi}^S \qquad (8.8.6)$$

To obtain the differential cross section $I_{jMi}$, we sum $|f_{jMi}^S|^2$ over all the final spin states and average them over the initial spin states of the system. This gives

$$I_{jMi} = \frac{1}{2(2S_i + 1)}\sum_S (2S + 1)|f_{jMi}|^2 \qquad (8.8.7)$$

If we do not consider the different final magnetic states separately the differential cross $I_{ji}$ is

$$I_{ji} = \sum_M I_{j_M i} \tag{8.8.8}$$

The general form of DWBA theory has been described by several investigators (see Lahmam-Bennani 1991 and Whelan et al. 1993). This approximation has been extensively utilized to calculate individual excitation cross sections of atoms due to electron and positron impacts and to study the collision dynamics (see Madison, 1979; Beijers et al., 1987; Bartschet and Madison, 1988; Purohit and Mathur, 1993; and Srivastava, 1998, and the references given therein). The usefulness of the DWBA can be tested by comparing the theoretical $I_{jMi}$ and $I_{ji}$ with the experimental data. It is found that the DWBA gives reasonable cross sections at intermediate and high impact energies but is not satisfactory at low values of $E$. For such investigations one is required to solve coupled integro-differential equations, which is discussed in the next section.

## 8.9  The Close-Coupling Approximation

The *close-coupling approximation* (*CCA*) is particularly suited to low-energy electron–atom collisions. In this approximation the ground state and a few lower excited states are explicitly included and the coupled integro-differential equations given by (8.6.2) are solved to obtain elastic and inelastic scattering cross sections. For example, for $e$–H collisions one may include five atomic states $1s$, $2s$, $2p_m$ ($m = 0, \pm1$). To improve the accuracy of the calculation the number of states may be increased. A 14-state close-coupling calculation includes $1s$, $2s$, $2p_m$ ($m = 0, \pm1$), $3s$, $3p_m$ ($m = 0, \pm1$), and $3d_m$ ($m = 0, \pm1, \pm2$) eigenstates. Resonances, discussed in Sec. 8.4.1, are obtained in the CCA. However, the convergence of the results with an increase in the number of atomic states is slow. Hence, to include the effect of the higher excited states of the target, pseudostates are employed. These pseudostates are supposed to mimic the effect of the higher excited target states and the continuum states, which are not explicitly included. Three pseudostates with $l = 0, 1$, and 2 were used by Burke et al. (1969). Two pseudostates $\bar{\nu}_{3p}$ and $\bar{\nu}_{3d}$ for the hydrogen atom were given by Damburg and Karule (1967). Similarly, Matese and Oberoi (1971) proposed three pseudostates $\bar{\nu}_{3s}$, $\bar{\nu}_{3p}$, and $\bar{\nu}_{3d}$. These two sets of normalized pseudostates are given by

$$\bar{\nu}_{3s}(r) = N_{3s}[\nu_s'(r) - \lambda_{1s}\nu_{1s}(r) - \lambda_{2s}\nu_{2s}(r)] \tag{8.9.1}$$

$$\bar{\nu}_{3pm}(r) = N_{3p}[\nu_{pm}'(r) - \lambda_{2p}\nu_{2pm}(r)] \tag{8.9.2}$$

$$\bar{\nu}_{3dm}(r) = N_{3d}(1 + A_d r/a_0)r^2 \exp(-a_d r/a_0)Y_{2m}(\hat{r}) \tag{8.9.3}$$

where the normalized $\nu_s'(r)$ and $\nu_p'(r)$ orbitals are

$$\nu_s'(r) = N_s \exp(-\alpha_s r/a_0) Y_{00}(\hat{r}) \qquad (8.9.4)$$

and

$$\nu_{pm}'(r) = N_p(1 + A_p r/a_0) r \exp(-\alpha_p r/a_0) Y_{1m}(\hat{r}) \qquad (8.9.5)$$

For $p$ orbitals $m = 0, \pm 1$ and for $d$ orbitals $m = 0, \pm 1, \pm 2$. The $\lambda_{2p}$ makes $\bar{\nu}_{3pm}(r)$ orthogonal to $\nu_{2pm}(r)$. Similarly, $\lambda_{1s}$ and $\lambda_{2s}$ ensure that $\bar{\nu}_{3s}(r)$ is orthogonal to $\nu_{1s}(r)$ and $\nu_{2s}(r)$. According to Matese and Oberoi (1971) $\alpha_s = 0.802$, $\alpha_p = 1.450$, $A_p = 0.356$, and $\alpha_d = 1.803$. Burke et al. used only $3\bar{p}$ and $3\bar{d}$ pseudostates with $\alpha_p = \alpha_d = 1$, $A_p = \frac{1}{2}$, and $A_d = \frac{2}{3}$. If the pseudostate channels are open they produce their own resonances, which are not real.

A more general form of the expansion of the antisymmetrized wave function of the system contains: (1) a limited number of target eigenstates, (2) a number of pseudostates, and (3) a set of quadratically integrable functions (Burke, 1985; Joachain, 1990). If we neglect (2) and (3) we recover the close-coupling approximation. The coupled integro-differential equations obtained from the above wave function of the system have given rise to the $R$-matrix method (Burke et al., 1971; Burke, 1987), linear algebraic equations (Seaton, 1974), the noniterative integral equations method (Smith and Henry, 1973a,b), and the matrix variational method (Nesbet, 1980). Burke and Eissener (1983), have reviewed these methods. Recent progress in close-coupling calculations has been discussed by Bartschet (1993). Ghosh et al. (1990) and Walters et al. (1995) describe theoretical calculations of positron collisions with atoms. All these methods are quite useful at low impact energies and have been utilized to obtain individual excitation cross sections and to study collision dynamics.

## 8.10 Electron Impact Excitation of Atoms: The Electron–Photon Delayed Coincidence Technique

We have already seen in Sec. 4.5 that for the scattering of the unpolarized electrons by a potential field a measurement of $I(\theta)$ gives the sum $|f|^2 + |g|^2$ and separate values of $f$ and $g$ cannot be obtained. Furthermore, usually $|f|^2$ is quite large in comparison of $|g|^2$. Thus a measurement of $I(\theta)$ does not yield information about the weak spin–orbit interaction. Hence, polarized electrons are employed and observable $I(\theta)$, $S(\theta)$, $T(\theta)$, and $U(\theta)$ are measured to obtain the values of $|f|$, $|g|$, and $\phi_{\text{rel}}$. Similarly, in the Sec. 7.4, it has been shown that for the elastic scattering of electrons by atoms four different observables are measured to bring out the importance of the exchange interaction. A comparison of the values of the observables with the calculated values provides a more sensitive test of any proposed theoretical model.

Due to electron impact an atom $B$ can be excited to $B^*$. After a time $\tau$ (lifetime of $B^*$), the atom is de-excited. In the radiative decay, in which $B^*$ returns to $B$, the energy $h\nu$ of the produced photon is equal to the excitation energy $\varepsilon_{ex}(B^*)$:

$$e + B \rightarrow e(\theta, \phi) + B^* \searrow B + h\nu(\theta_p, \phi_p) \tag{8.10.1}$$

where $(\theta_p, \phi_p)$ is the direction of the emitted photon, and after the collision the electron is scattered in the direction $(\theta, \phi)$. A measurement of the frequency $\nu$ helps in the identification of the excited state $B^*$. Since a typical collision time is of the order of $10^{-14}$ sec, and the life time of $B^*$ for dipole transition decay is of order of $10^{-9}$ sec, the excitation and the decay can be treated as independent processes.

Up to about 1970 experimentalists were observing either the scattered electrons or the light emitted by excited atoms. The measurements with the scattered electrons yield $I(\theta)$ and $\sigma_{ex}$. However, to obtain absolute cross sections a non-trivial normalization procedure is required. We can measure $In(\theta_M)$, the intensity of light emitted by the atom in the direction of the magic angle $\theta_M$ [see Eq. (8.10.2) below]. Alternatively $In(\|)$ and $In(\perp)$, the intensities of the emitted light with polarization vectors parallel and perpendicular to the direction of the incident electrons, respectively, are measured.

The above measurements are not enough to throw sufficient light on the collision dynamics. In inelastic collisions there are a number of final channels. The energy differences between some of the exited states are so small that it is very difficult to measure $I(\theta)$ for each channel separately. For an excited state of orbital angular momentum $L$ there are $(2L + 1)$ magnetic sublevels. In the absence of any external perturbation these sublevels are degenerate. Hence, the scattered electrons, which have excited different magnetic sublevels, cannot be separated from each other by an electron energy spectrometer. Thus a measurement of $I(\theta)$ gives only an average value of the excitation cross sections over the different magnetic sublevels.

On the other hand, from the measurements of $In(\|)$ and $In(\perp)$ we obtain for the electric dipole radiation

$$In(\theta_P) = \frac{3}{4\pi} \sigma_{ex} \frac{1 - P\cos^2\theta_P}{3 - P} \tag{8.10.2}$$

where $In(\theta_P)$ is the intensity of the light emitted by the excited atoms in the direction $\theta_P$ and the degree of polarization of the emitted light is

$$P = \frac{In(\parallel) - In(\perp)}{In(\parallel) + In(\perp)} \tag{8.10.3}$$

It is evident from (8.10.2) that for $\cos^2 \theta_P = \frac{1}{3}$, $In(\theta_P)$ is independent of the degree of polarization $P$. The angles that satisfy the above condition, known as the magic angles, are 54.7° and 125.3°. Hence, a measurement of $In(\theta_P)$ at a magic angle yields the value of $\sigma_{ex}$. This is to be corrected if the excited level $B^*$ is populated by other excited states, (cascade effect). Furthermore, due to the anisotropic nature of the excitation process the different magnetic sublevels are unequally populated. For example, if we consider the excitation of ground state helium $1^1S$ to the $2^1P$ state then due to unequal population $\sigma_0$, the total cross section for $M = 0$, is not equal to $\sigma_1$, the total cross section for $M = 1$. However, due to symmetry $\sigma_1 = \sigma_{-1}$. Hence, the degree of polarization $P$ of the light traveling perpendicular to the incident beam direction is also given by

$$P = \frac{\sigma_0 - (\sigma_1 + \sigma_{-1})}{\sigma_0 + (\sigma_1 + \sigma_{-1})} = \frac{\sigma_0 - 2\sigma_1}{\sigma_0 + 2\sigma_1} \tag{8.10.4}$$

Thus a measurement of $P$ does provide information about the difference between $\sigma_0$ and $\sigma_1$. However, this information is an average over all the directions of the scattered electrons.

From the above brief discussion it is evident that more observables are needed for a better understanding of the collision dynamics and for testing the proposed theoretical models at a more fundamental level.

A new technique to investigate excitation of helium atoms to the $2^1P$ state was employed by Eminyan et al. (1973, 1974). In this technique the electrons scattered in the direction $(\theta, \phi)$ and the photons emitted by the excited atoms in the direction $(\theta_P, \phi_P)$ are observed in delayed coincidence. The measured coincidence signal corresponds to the electron–photon pairs arising from a single collision event. Hence, the excited atoms $B^*$, which produce photons in the direction of $(\theta_P, \phi_P)$ on de-excitation, having been excited by those electrons that were scattered in the direction of $(\theta, \phi)$ after the collision, form a select ensemble of atoms. An analysis, discussed below, of the electrons and the photons, detected in the delayed coincidence, provides valuable information about the charge cloud of the excited atom, its direction of alignment in space, the angular momentum transferred from the incident electron to the atom, and the coherence of the excitation during the collision. A comparison of the experimental data produced by this technique with the theoretical results also tests any theoretical model at the most fundamental level. Thus the electron–photon coincidence technique gave birth to a new era in atomic collision physics.

Let us consider the excitation of atoms from the $S$ $(L = 0, M = 0)$ state to $P$ $(L = 1, M = 0, \pm 1)$ states. We assume that the spin–orbit and spin–spin inter-actions are negligible and that the scattering process can be described in the $LS$ coupling scheme. With the above assumptions and noting that the excitation and the decay of the excited atoms are independent processes, we represent each excited atom by a single vector $|\psi\rangle$. We expand $|\psi\rangle$ in terms of the basis vectors $|LM\rangle$ with $L = 1$ and $M = 0, \pm 1$. This gives

$$|\psi\rangle = \sum_{M=-1}^{+1} f_M |1M\rangle \qquad (8.10.5)$$

where the expansion coefficients $f_M$ characterize the excitation of the magnetic sublevel $|1M\rangle$. The density matrix $\rho$, which completely describes the excited state $|\psi\rangle$, is given by

$$\rho = \begin{pmatrix} |f_1|^2 & f_1 f_0^* & f_1 f_{-1}^* \\ f_1 f_1^* & |f_0|^2 & f_0 f_{-1}^* \\ f_{-1} f_1^* & f_{-1} f_0^* & |f_{-1}|^2 \end{pmatrix} \qquad (8.10.6)$$

The $|\psi\rangle$ is normalized in such a manner that the relations

$$|f_M|^2 = I_M \qquad (8.10.7)$$

are satisfied, where $I_M$ is the differential cross section for the magnetic sublevel $|1M\rangle$. From (8.10.4) to (8.10.6), we get

$$\langle \psi | \psi \rangle = \text{tr}\, \rho = \sum_M I_M = I \qquad (8.10.8)$$

where $I$ is the differential cross section summed over the three values of $M$. The matrix $\rho$ has nine terms. However, as it is a Hermitian matrix, the number of independent parameters reduces from nine to six. These six parameters can be obtained by determining $f_M$. Since $f_M$ are complex we have three $|f_M|$ and three phases $\chi_M$. We also assume that the spin does not play any role in the collisions. Hence, the excited states possess positive reflection symmetry about the scatter-ing plane. Thus

$$f_1 = -f_{-1} \tag{8.10.9}$$

The above condition further reduces the number of the independent parameters from six to four, namely $|f_0|$, $|f_1|$, $|\chi_0|$, and $|\chi_1|$. The quantum mechanics gives the wave function $\psi$ with an uncertain phase. Hence, if we take $f_0$ to be real, then $\chi_1$ (which we shall now denote by $\chi$) is the phase difference between $f_0$ and $f_1$. Thus finally only three independent parameters are required to determine all the terms of the matrix $\rho$ and the characteristics of the excited state.

Eminyan et al. (1974) chose the three parameters to be $I$, $\lambda$, and $\chi$. The first two have already been defined. The third parameter $\lambda$ is defined by

$$\lambda = I_0 / I \tag{8.10.10}$$

where $I_0$ is given by (8.10.7) with $M = 0$. Using the definitions of I, $\chi$, and $\lambda$, it is easy to show that

$$\cos \chi = \frac{1}{I} \frac{\text{Re}(f_0 f_1^*)}{[0.5\lambda(1-\lambda)]^{1/2}} \tag{8.10.11}$$

Out of the three parameters $I$ can be determined by observing the scattered electrons alone. On the other hand, the two remaining parameters are connected with the interference of $f_0$ and $f_1$, and thus with the off-diagonal elements of $\rho$. The determination of $\lambda$ and $\chi$ has been the primary aim of electron–photon coincidence experiments (Blum and Kleinpoppen, 1979). If we measure only the photons without observing the scattered electrons, the direction of the incident electrons becomes an axis of symmetry and all the off-diagonal elements of $\rho$ become identically zero. In such a case the excitation is incoherent. Hence, for coherent excitation of the magnetic sublevels the excitations process must be not axially symmetric.

To determine $\lambda$ and $\chi$, two types of experimental studies have been done: (a) angular correlation measurements and (b) polarization correlation measurements. In both the photons produced by the decay of the excited atom, traveling in the direction ($\theta_P$, $\phi_P$), are detected in coincidence with the electrons scattered in the direction ($\theta$, $\phi$). A photon detector, which is sensitive only to the polarization $\hat{\varepsilon}$, measures the intensity $In(\hat{\varepsilon}, \theta_P, \phi_P)$ of the radiation. Theoretically $In(\hat{\varepsilon}, \theta_P, \phi_P)$ is calculated through the dipole matrix element $\langle \psi | \hat{\varepsilon} \cdot r | 0 \rangle$, where $|0\rangle$ represents the initial state of the atom. For the decay of the $n^1P$ excited atom, $In(\hat{\varepsilon}, \theta_P, \phi_P)$ emitted into a solid angle $d\Omega$ is given by (Slevin, 1984):

$$In = C\{\tfrac{1}{2}(1-\lambda)(1+\cos^2\theta_P - \sin^2\theta_P \cos 2\phi_P)$$
$$+ \lambda \sin^2\theta_P + [\lambda(1-\lambda)]^{1/2} \cos\chi \sin 2\theta_P \cos\phi_P\} \tag{8.10.12}$$

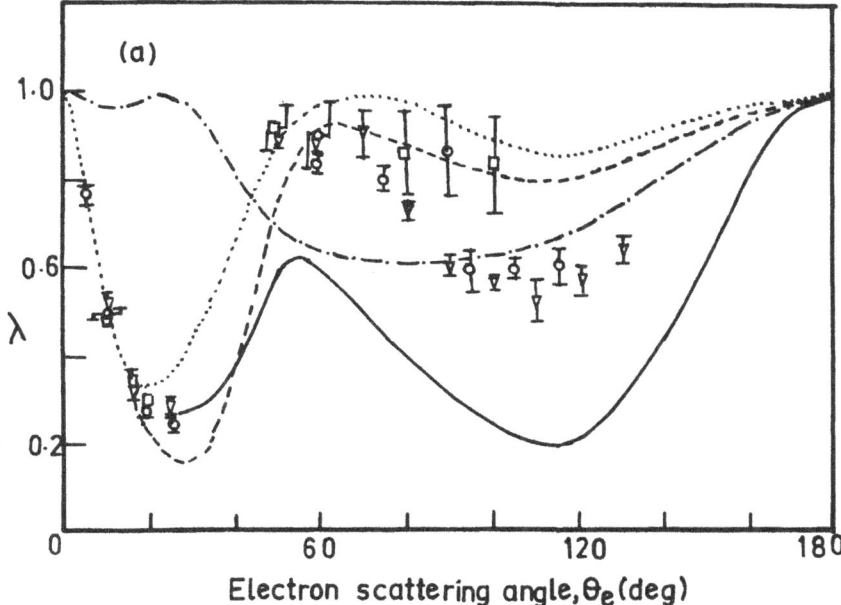

**FIGURE 8.7** Variation of (a) $\lambda$ and (b) $|\chi|$ with electron scattering angle for the excitation of helium atoms from the $1^1S_0$ to the $2^1P$ state by 80-eV electrons. Experiment: $\bigcirc$, Slevin et al. (1980); $\nabla$, Hollywood et al. (1979); $\square$, Steph and Golden (1980). Theory: . . . Madison 1979 (from Sutcliffe et al., 1978); ___, Thomas et al. (1977); _._, Catalan and Roberts (1979); ____, Scott and McDowell (1976). Reproduced from "Coherence in inelastic low-energy electron scattering," J. Slevin, *Rep. Prog. Phys.* **47**, 461, 1984, with permission from the Institute of Physics Publishing Ltd., UK.

where,

$$C = \frac{e^2 \omega^4 d\Omega}{2\pi c^3 \hbar} |\langle 0 \| r \| 1 \rangle|^2 \frac{I(\theta, \phi)}{3K} \qquad (8.10.13)$$

with $K$ as the decay constant. For a given $E$, $\theta$, and $\phi$ the parameters $\lambda$ and $\chi$ are fixed. Hence, in the angular correlation measurement experiment the values of $In(\theta_P, \phi_P)$ for the various values of $(\theta_P, \phi_P)$ but for a fixed value of $(\theta, \phi)$ are measured. The experimental data are fitted to (8.10.12) and the values of $\lambda$ and $\cos \chi$ are obtained. The first successful experiment of this type was performed by Eminyan et al. (1974) for the excitation of helium from the $1^1S_0$ to the $2^1P$ state due to electron impact. In Fig. 8.7 the experimental values of $\lambda$ and $|\chi|$, obtained by Eminyan et al. and other experimental groups, at $E = 80\,\text{eV}$ are shown as functions of the scattering angle $\theta$. These values are compared with various

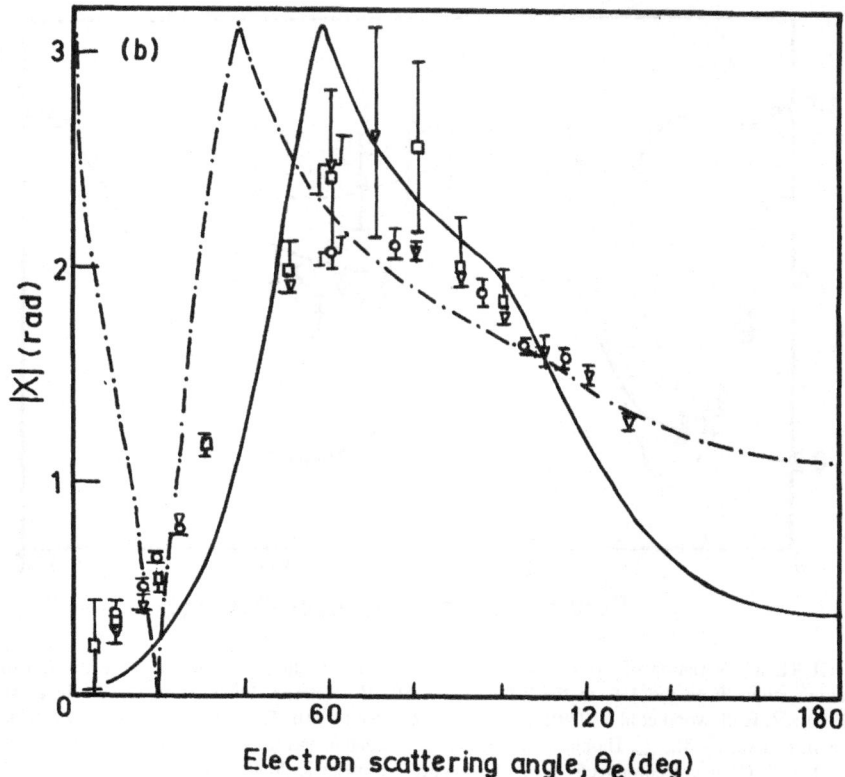

**FIGURE 8.7** (*continued*)

theoretical results. None of the theoretical models is found to be satisfactory for all three parameters $I(\theta)$, $\lambda$, and $|\chi|$. However, at 24 eV the experimental data of Crowe et al. (1983) for $\lambda$ and $|\chi|$ are in very good agreement with the five-state $R$-matrix calculation of Burke and Williams (1977). Since these experiments yield $\cos\chi$, the sign of $\chi$ cannot be determined.

In the polarization correlation measurement experiments the values of the Stokes parameters $\eta_1$, $\eta_2$, and $\eta_3$, defined by (6.8.18), are also determined. The coincidence experiment is carried out in the scattering frame in which the direction of the incident electron is the $z$-axis and the electron detector is placed in the $x$–$z$ plane. The photon detector is usually fixed at $(\theta_P, \phi_P)$.

For such an experimental arrangement we have

$$In(\theta_P,\phi_P)\eta_1 = C\left\{-\lambda(1-\lambda)\cos\theta_P\sin 2\phi_P - 2[\lambda(1-\lambda)]^{1/2}\cos\chi\sin\theta_P\sin\phi_P\right\}$$

$$(8.10.14)$$

$$In(\theta_P,\phi_P)\eta_2 = C\left\{2[\lambda(1-\lambda)]^{1/2}\sin\chi\sin\theta_P\sin\phi_P\right\} \qquad (8.10.15)$$

and

$$In(\theta_P,\phi_P\}\eta_3 = -C\left\{\tfrac{1}{2}(1-\lambda)[\sin^2\theta_P - (1+\cos^2\theta_P)\cos 2\phi_P]-\lambda\sin^2\theta_P\right.$$
$$\left. -[\lambda(1-\lambda)]^{1/2}\cos\chi\sin 2\theta_P\cos\phi_P\right\} \qquad (8.10.16)$$

Hence, for $\theta_P = \phi_P = \pi/2$, from (8.10.12) we have $In = C$ and from (8.10.14) to (8.10.16) we obtain

$$\eta_1 = -2[\lambda(1-\lambda)]^{1/2}\cos\chi \qquad (8.10.17)$$

$$\eta_2 = 2[\lambda(1-\lambda)]^{1/2}\sin\chi \qquad (8.10.18)$$

and

$$\eta_3 = 1-2\lambda \qquad (8.10.19)$$

Thus measurement of the Stokes parameters not only yields the value of $\lambda$ but also gives unambiguous values of $\chi$ through $\cos\chi$ and $\sin\chi$. For these measurements the photon detector is placed on the $y$-axis and a Nicol prism is kept between the excited atom and the detector in such a way that the complete transmission axis of the prism makes an angle $\alpha$ with the $x$-axis, as shown in Fig. 8.8. The transmitted photons are detected in coincidence with the scattered electrons. The intensity of the detected signal $In(\alpha)$ for $\alpha = 0$ and $90°$ gives $\eta_3$.

Similarly with $\alpha = 45°$ and $135°$ we get $\eta_1$. To obtain $\eta_2$ the Nicol prism is replaced by a filter that fully transmits photons of helicity $+1$ $(-1)$. Now measured $In(+)$ and $In(-)$ yield $\eta_2$ through (6.8.18b). It is easy to verify from (8.10.17) to (8.10.19) that

$$P = (\eta_1^2 + \eta_2^2 + \eta_3^2)^{1/2} = 1 \qquad (8.10.20)$$

i.e., the radiation is completely coherent (fully polarized), in agreement with the assumption made in the beginning.

The polarization correlation measurement experiment is difficult for the $2^1P$ state because the emitted radiation lies in the ultraviolet. However Tan et al. (1977) succeeded in performing such an experiment. Their results are in

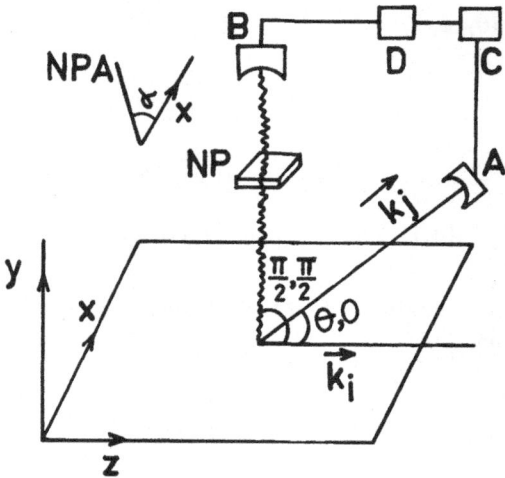

**FIGURE 8.8** Schematic diagram for the measurement of the Stokes parameters to obtain $\lambda$ and $\chi$ parameters for the inelastic scattering of electrons by atoms. The electron detector $A$ is in the scattering ($x$–$z$) plane and the photon detector $B$ is perpendicular to the $x$–$z$ plane. The scattered electrons in the direction of ($\theta$, $\phi$) are detected in coincidence with photons emitted in the direction of ($\pi/2$, $\pi/2$) by the network $C$. The axis of the complete transmission of the Nicol prism (NPA) makes an angle $\alpha$ with the $x$-axis.

agreement with those obtained by the angular correlation experiment. This shows that the two methods are equivalent. But the angular correlation gives $|\chi|$, whereas the polarization correlation gives $\chi$ itself.

In the collision frame the population of the excited atom can be described by an alignment tensor $A$ and an orientation vector $O$. The nonzero components of $A$ and $O$ for an $S$ to $P$ transition are given by (Morgan and McDowell, 1975, 1977; Mathur, 1998)

$$A_0 = \langle 3L_z^2 - L^2 \rangle / 2\hbar^2 = \tfrac{1}{2}(1 - 3\lambda) \tag{8.10.21}$$

$$A_{1+} = \langle L_x L_z + L_z L_x \rangle / 2\hbar^2 = [\lambda(1-\lambda)]^{1/2} \cos\chi \tag{8.10.22}$$

$$A_{2+} = \langle L_x^2 - L_z^2 \rangle / 2\hbar^2 = (\lambda - 1)/2 \tag{8.10.23}$$

and

$$O_{1-} = \langle L_y \rangle / 2\hbar^2 = [\lambda(1-\lambda)]^{1/2} \sin\chi \tag{8.10.24}$$

Hence

$$\langle L_y \rangle = 2[\lambda(1-\lambda)]^{1/2} \hbar \sin \chi \qquad (8.10.25)$$

The above equation shows that for a given value of $\lambda$, the extent to which the final atomic excited $P$ state is oriented depends upon the value of $\chi$. Thus $O_{1-}$ is directly related to the dynamics of the collision process.

Since the pioneer work of Eminyan et al., the coincidence technique has been applied to a large number of atoms and their alignment and orientation parameters have been determined. Furthermore, most of the methods discussed in Chapter 7 and in this chapter have been applied to determine these parameters theoretically. A comparison of the experimental data with the theoretical results tests the various theoretical models at the most fundamental level. Anderson et al. (1988), Slevin and Chwirot (1990), and Becker et al. (1992) have reviewed this field extensively. Recently Verma and Srivastava (1998) and Mathur (1998) have successfully employed the DWBA to obtain alignment and orientation parameters for the various transitions occurring in a number of atoms. These two papers may be consulted for references to other recent investigations.

Hertel and Stoll (1974, 1978) developed an experimental technique in which reaction (8.10.1) proceeds in the opposite direction. In this process the atoms $B$ are excited to a selective excited state $B^*$ by a suitable laser. These excited atoms collide with a monoenergetic beam of electrons of energy $E$. Due to superelastic collisions the atoms $B^*$ are de-excited to $B$ and the scattered electrons acquire kinetic energy equal to $E + \varepsilon_{ex}$. The first experiment using polarized electrons and atoms for superelastic scattering of electrons from laser excited atoms was performed by McClelland et al. (1986). A suitable theoretical method for the analysis of such an experiment has been developed by Hertel et al. (1987).

## Questions and Problems

8.1  Why is it not possible to calculate the exact scattering amplitude even for the elastic scattering of electrons by ground state hydrogen atom? 13-eV electrons are scattered by ground state hydrogen atoms. How many channels, including the degenerate channels, are open and how many are closed? Calculate the minimum energy of the scattered electrons.

8.2  What is the static field approximation? Justify its name and derive an expression for the static field for the elastic scattering of electrons by the ground state helium atom. Show that it is of short range. Represent the helium atom by the wave function given in the problem 7.1.

8.3 Use the expression of the static field derived in the previous problem and obtain $f_{B1}(K)$ for the scattering of electrons by this field in the FBA. Verify that the expression $f_{B1}(K)$ so obtained is identical to that obtained from (7.3.6) for the helium atom.

8.4 Justify the following statement: "The polarization potential in electron–atom scattering is due to virtual transitions whereas the absorption potential arises due to real transitions."

8.5 Discuss the phenomenon of resonance in electron–atom collisions. Differentiate between the Feshbach and the shape resonance.

8.6 The following table gives the phase shifts in the static field (SF) and the static plus dipole polarization (SFP) approximations for the scattering of 100-eV electrons by helium atoms. Calculate the differential cross sections $I_{SF}(K)$ and $I_{SFP}(K)$ for the forward direction and the integrated elastic cross section $\sigma_{el}$. How does the difference $[I_{SEP}(K) - I_{SF}(K)]$ change with an increase in $K$? Explain your answer.

|     | Phase shifts $\eta_l$ (in radians) | |
| --- | --- | --- |
| $l$ | SF | SFP |
| 0 | 0.941 | 1.022 |
| 1 | 0.256 | 0.347 |
| 2 | 0.080 | 0.146 |
| 3 | 0.027 | 0.072 |
| 4 | 0.010 | 0.039 |

8.7 Use (8.6.18) to calculate $U_{ex}^{\pm}(r)$ (in $a_0^{-2}$) for the elastic scattering of 200-eV electrons by ground state hydrogen atoms. Plot them as functions of $r/a_0$ and compare them with $U_{SF}(r)$.

8.8 Using (8.9.1), (8.9.4), and $\alpha_S = 0.802$ calculate the values of $N_S$, $\lambda_{1s}$, $\lambda_{2s}$, and $N_{3S}$.

8.9 Discuss two theoretical methods that employ partial waves to calculate the excitation cross sections of the individual excited channels.

8.10 In a polarization correlation measurement experiment using an electron–photon delayed coincidence technique, helium atoms were excited by 40-eV electrons to the $2^1P$ state. The experiment was carried out in a scattering frame in which the incident electrons were moving along the $z$-axis and the scattered electrons were detected in the $x$–$z$ plane. The photon detector was on the $y$-axis. For a 16° scattering angle the values of the $\lambda$ and $|\chi|$ parameters were 0.70 and 0.53 rad, respectively. Calculate the values of the Stokes parameters $\eta_1$, $\eta_2$, and $\eta_3$ and show that the emitted radiation was completely coherent.

# *Collision of Electrons with Molecules*

## *9.1  Introduction*

Collision of electrons and molecules represents a more complex problem than electron–atom collisions. The reason for this complexity is easy to understand. A molecule is usually defined as a group of atoms held together by valence forces. We may regard even the unstable systems of atomic nuclei and electrons as molecules. Thus, whereas atoms have only one center (nucleus), molecules are essentially multicenter objects. Due to the motion of the nuclei, molecules possess additional degrees of freedom. The motion of the electrons gives rise to quantized electronic states and the motion of the nuclei produces vibrational and rotational states. These two motions are not independent of each other but to a good approximation they can be separated from one another (see Sec. 9.2). Each electronic state has a number of vibrational states and each vibrational state contains a number of rotational states. Thus the energy spectra of molecules are much more complex than the atomic spectra.

Heteronuclear diatomic molecules and a large number of polyatomic molecules do not have a center of symmetry. This gives rise to a noncentral interaction between the molecular target and the incident electron. Hence, drastic assumptions are required if we wish to use the method of partial waves, discussed in the previous chapter, to analyze electron–molecule collisions. Further, in such collisions, dissociative channels also exist. Due to these complexities it is not surprising that there are fewer theoretical studies of electron–molecule collisions than of the electron–atom scattering.

In this chapter we shall briefly discuss the collision of electrons with a few diatomic and polyatomic molecules mainly at intermediate and high impact energies. For studies of low-energy collisions employing more sophisticated approximations, readers are referred to reviews by Lane (1980), Shimamura and Takayanagi (1984), Gianturco and Jain (1986), Morrison (1988), Burke (1993), and Gianturco (1995). Since the hydrogen molecule is the simplest neutral molecule, quite often we shall use it to develop various approximate methods for electron–molecule collisions.

## 9.2 The Born–Oppenheimer Approximation and the Franck–Condon Principle

To evaluate the scattering amplitude for electron–molecule collisions we need the molecular wave functions. An *ab-initio* evaluation of a molecular wave function is an impossible task. However, great simplification is introduced if we note that the nuclei are much heavier than the electrons. Hence, in a molecule, the motion of the electrons is much faster than that of the nuclei. Thus to a first approximation we may consider the motion of the electrons in a space having fixed-nuclei. This fixed-nuclei approximation leads to the *Born–Oppenheimer approximation* (BOA) (not to be confused with the exchange Born-Oppenheimer approximation discussed in Sec.7.5). According to the BOA, the quantum mechanical wave function of the molecule, $\psi$, is equal to the product of $\Phi_n$ and $v_i$, where $\Phi_n$ depends only upon the coordinates of the nuclei and the electronic wave function $v_i$ satisfies

$$\left[ -\sum_i \frac{\hbar^2}{2m} \nabla_i^2 + V_{ne} + V_{ee} \right] v_i = \varepsilon_i(R) v_i \qquad (9.2.1)$$

The first term of the above equation is the kinetic energy operator of the molecular electrons, $V_{ne}$ is the interaction energy between the molecular nuclei and the electrons and $V_{ee}$ is the electron–electron interaction energy. This equation is solved for fixed nuclei, so that in the expression for the eigenenergy $\varepsilon_i(R)$, the coordinates of the nuclei enter as parameters. For a diatomic molecule, $\varepsilon_i(R)$ depends upon the internuclear distance $R$. A typical variation of $\varepsilon_i(R)$ with $R$ for a stable state $G$ of a molecule is shown in Fig. 9.1. The binding energy of the molecule is approximately equal to $\varepsilon_i$ at $R = R_0$, where $R_0$ is the equilibrium internuclear distance. The variation of $\varepsilon_i(R)$ with $R$ for another electronic state $E$ is shown in the same figure. Since the curve for this state has no minimum it is an unstable state. A molecule in this state automatically dissociates into its constituent atoms. The curves $G$, $E$, etc. are known as the potential energy functions. To obtain the wave function of the nuclei, their motions are considered in the potential energy $\varepsilon_i(R)$ This approximation is known as the adiabatic nuclei approximation, and the Schrödinger equation for the motion of the nuclei is given by

$$\left[-\sum_p \frac{\hbar^2}{2M_p} \nabla_p^2 + V_{nn} + \varepsilon_{el}(R)\right]\Phi_n = E_T\Phi_n \tag{9.2.2}$$

where $M_p$ is the mass of the $p$th nucleus, $V_{nn}$ is the nuclear–nuclear interaction energy, and $E_T$ is the total energy of the molecule.

For a diatomic molecule Eq. (9.2.2) is for two particles. This may again be reduced to two one-body equations. For spherically symmetric $V_{nn}$ we have

$$\Phi_n(R) = \varphi_v(R)\,\chi_{JM}(\theta, \varphi) \tag{9.2.3}$$

where $(\theta, \varphi)$ are the polar coordinates of the internuclear axis relative to a space-fixed axis. Hence, in the BOA, the molecular wave function is the product of the electronic, vibrational, and rotational wave functions. In the electronic wave function the intermolecular distance $R$ enters as a parameter.

When a molecule makes a transition from one electronic state to another the vibrational and rotational states also change. Again due to the large mass of a nucleus as compared to that of electron, during the transition the nuclei hardly move. Hence, the internuclear distances in the initial and final states immediately after the transition do not differ from one another by any significant amount. The Franck–Condon principle assumes that in the transition from one electronic state, say, $G$, to another, say, $E$, the internuclear distance does not change, i.e., the transition is vertical with $R_G = R_E$. This principle plays a useful role in analysis of molecular spectra.

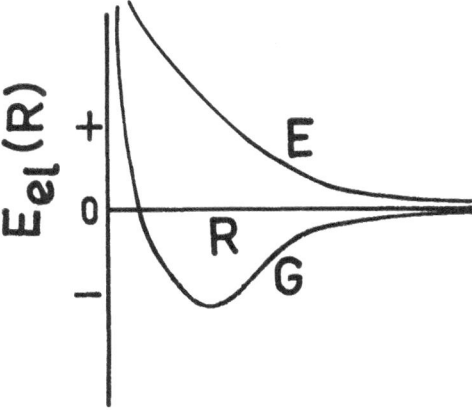

**FIGURE. 9.1** Variation of the electronic energy $E_{el}(R)$ with the internuclear distance R for a diatomic molecule. $G$ and $E$ are stable and unstable states of the molecule, respectively.

## 9.3    Electronic Excitation of Molecules

Let us consider the excitation of a diatomic molecule from its initial state $|iv\,JM\rangle$ to its final state $|jv'J'M'\rangle$ due to electron impact. We represent the initial and final electronic states by $|i\rangle$ and $|j\rangle$, respectively. From (7.2.11) the transition matrix element for the above excitation is given by

$$T_{ivJM}^{jv'J'M'} = \langle k_j\, jv'J'M'|T|k_i ivJM\rangle \tag{9.3.1}$$

The value of $k_j$ will be different for different final rotational states. However, the energy differences among various rotational states are usually quite small in comparison to $\hbar^2 k_j^2/2m$. Hence, we neglect the variation of $k_j$ with $J'M'$ and sum $\left|T_{ivJM}^{jv'J'M'}\right|^2$ over all the final rotational states. We also average the sum over the initial rotational states $|JM\rangle$. Using the closure relationship

$$\sum_{J'M'} \chi_{J'M'}^{*}(\hat{R})\, \chi_{J'M'}(\hat{R}') = \delta(\hat{R} - \hat{R}') \tag{9.3.2}$$

we get

$$\sum_{JM} \chi_{JM}^{*}(\hat{R})\, \chi_{JM}(\hat{R}') \sum_{J'M'} \chi_{J'M'}(\hat{R}) \chi_{J'M'}^{*}(\hat{R}')$$

$$= \sum_{JM} \chi_{JM}^{*}(\hat{R})\, \chi_{JM}(\hat{R}')\delta(\hat{R} - \hat{R}') = \frac{1}{4\pi}\sum_{j}(2J+1)$$

where the rotational states $\chi$ are spherical harmonics. Since the number of rotational states for a given $J$ is $(2J+1)$, the average value is given by

$$\left\langle \sum_{J'M'} \left|T_{ivJM}^{jv'J'M'}\right|^2 \right\rangle_{av} = \frac{1}{4\pi}\int\left(\int\left|T_{iv}^{jv'}\right|^2 R^2 dR\right)\sin\theta_R d\theta_R\, d\phi_R \tag{9.3.3}$$

where $(\theta_R, \varphi_R)$ are the polar angles of the internuclear axis with respect to an axis fixed in space. The electronic transition matrix element is

$$T_{iv}^{jv'} = \langle v'|T_{ji}|v\rangle \tag{9.3.4a}$$

and

$$T_{ji} = \langle k_j\, j|T|k_i i\rangle \tag{9.3.4b}$$

where $T_{ji}$ is a function of the vector $R$ but the vibrational wave function $\varphi_v$ depends only upon the scalar $R$. The electronic wave functions vary slowly with $R$ but $|\varphi_v(R)|^2$ is a maximum at $R \approx R_0$, the initial equilibrium nuclear separation. Hence, the transition matrix element $T_{ji}$ may be evaluated only at $R = R_0$. This yields

$$\left\langle \sum_{J'M'} |T_{ivJM}^{jv'J'M'}|^2 \right\rangle \mathrm{av} = \frac{1}{4\pi} \int |T_{ji}(R_0)|^2 \sin\theta \, d\theta \, d\varphi \, P_{v'v} \qquad (9.3.5)$$

where now $(\theta, \varphi)$ are the polar coordinates of $R_0$. The Franck–Condon factor $P_{v'v}$ is defined by

$$P_{v'v} = \left| \int \varphi_{v'}^*(R)\varphi_v(R)R^2 dR \right|^2 \qquad (9.3.6)$$

Under the above assumptions, using (7.2.10) we get

$$\left\langle \left| \sum_{J'M'} f_{ivJM}^{jv'J'M'} \right|^2 \right\rangle_{\mathrm{av}} = \frac{1}{4\pi} \int |f_{ji}(R_0)|^2 d\Omega(\hat{R}_0) P_{v'v} \qquad (9.3.7)$$

where

$$f_{ji}(R_0) = -2\pi^2 \langle k_j j | U | \psi_{k_i}^+, i \rangle \qquad (9.3.8)$$

The target wave functions depend upon $|R_0|$ and the reduced interaction energy is a function of $R_0$.

If we further assume that $k_j$ is independent of the final vibrational states, the differential cross sections summed over all the final rotational and vibrational states and averaged over the initial rotational states are given by

$$\bar{I}_{ji} = \frac{k_j}{k_i} \frac{1}{4\pi} \int |f_{ji}|^2 d\Omega(\hat{R}_0) \qquad (9.3.9)$$

because

$$\sum_v P_{v'v} = 1. \qquad (9.3.10)$$

## 9.4   Electronic Wave Functions and States of Diatomic Molecules

To evaluate $T_{ji}$ we require the electronic wave functions of the molecule. A multielectron molecular wave function is constructed from a set of molecular orbitals. A molecular orbital is a function of the coordinates of a single electron.

Thus a molecular orbital multiplied by a spin wave function ($\alpha$ or $\beta$) is known as the molecular spinor and, according to the Pauli exclusion principle, a molecular spinor cannot have more than one electron. However, it can extend to any number of atoms of the molecule and reflects the molecule's basic symmetry. The electronic states of a diatomic molecule are represented by the symbol $^{2S+1}\Lambda$, where $S$ denotes the spin of the molecule and $\Lambda = 0, \pm1, \pm2$, etc., correspond to $\Sigma, \Pi, \Delta$, etc., electronic states of the molecule. Further we have $\Sigma^+$ and $\Sigma^-$ states. The wave function of $\Sigma^+$ state does not change when the wave function is reflected in a plane containing an internuclear axis whereas the wave function of the $\Sigma^-$ state changes its sign in the above operation. In addition, homonuclear diatomic molecules have a center of symmetry (middle point of the internuclear axis). When the electronic wave function of such a molecule is inverted about the center of symmetry then it either does or does not change its sign. The states that remain unchanged are said to be gerade states and are designated by the symbol $g$. For example, we may have $\Sigma_g^+, \Sigma_g^-, \Pi_g, \Delta_g$, etc., states. The states that change their sign in the above operation are known as ungerade states and are associated with the symbol $u$. We have $\Sigma_u^+, \Sigma_u^-, \Pi_u, \Delta_u$, etc., states. This type of symmetry does not exist for heteronuclear molecules.

Let us consider the following two electronic space wave functions for the $H_2$ molecule

$$v_\pm(r_1, r_2) = \frac{N_\pm}{\sqrt{2}}[\varphi_A(1)\,\varphi_B(2) \pm \varphi_A(2)\,\varphi_B(1)] \tag{9.4.1}$$

where $N_\pm$ are the normalization constants and $A$ and $B$ represent two nuclei of the molecule. The atomic orbitals are given by (6.10.7). It is evident from (9.4.1) that if we exchange 1 and 2, $v_+$ does not change while $v_-$ does change its sign. Hence, the $v_\pm$ are associated with the antisymmetric (singlet) and symmetric (triplet) spin wave functions, respectively. Further, the orbitals represented by $\phi_A$ are $\phi_B$ spherically symmetric, so the $v_\pm$ are $\Sigma(\Lambda = 0)$ states.

Let us now reflect electrons 1 and 2 in a plane containing the internuclear axis $AB$. It is evident from Fig. 9.2 that $r_{1A} = r_{1'A}$ and $r_{1B} = r_{1'B}$. Similar relationships are true for the second electron. Hence, both the $v_\pm$ remain unchanged in the above operation, and both have $+$ symmetry. Now we invert 1 and 2 about the center of symmetry 0. Considering the equilateral triangles $\Delta A10$ and $\Delta B1'0$ (in Fig. 9.3) we find that $r_{1A} = r_{1'B}$. A similar relationship also holds for the other coordinates. Hence, on inversion,

$$v'_\pm = \frac{N_\pm}{\sqrt{2}}[\varphi_B(1)\,\varphi_A(2) \pm \varphi_B(2)\,\varphi_A(1)] \tag{9.4.2}$$

Thus $v'_+ = v_+$ but $v'_- = -v_-$. All the above considerations show that the electronic states $v_\pm$ are $^1\Sigma_g^+$ and $^3\Sigma_u^+$, respectively.

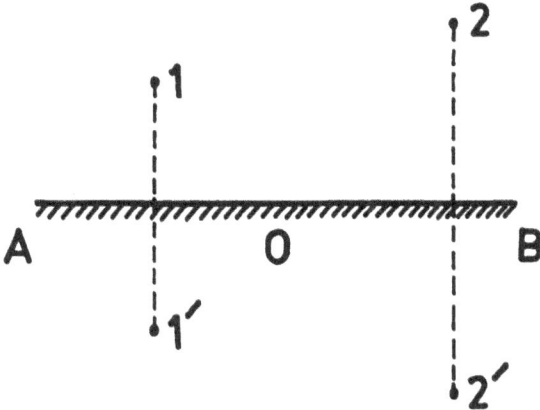

**FIGURE 9.2** Reflection of the coordinates of electrons of a diatomic molecule about a plane containing an internuclear axis AB.

We now proceed to evaluate the normalization constants $N\pm$. From (9.4.1) and (9.4.2), we obtain

$$\langle v_\pm | v_\pm \rangle = |N_\pm|^2 (1 \pm S^2) \qquad (9.4.3)$$

where the overlap integral

$$S = \langle \varphi_A(i) | \varphi_B(i) \rangle \qquad (9.4.4)$$

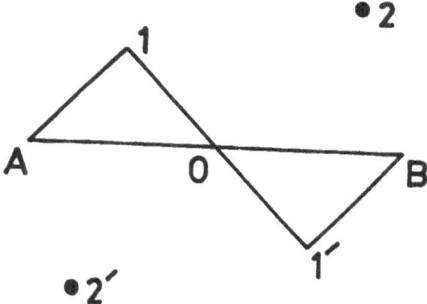

**FIGURE 9.3** Inversion of the coordinates of electrons of a diatomic molecule about the center of symmetry 0.

with $i = 1$ or 2. Using (6.10.7) and dropping the $i$, we get

$$S = \frac{Z^3}{\pi a_0^3} \int e^{-Z(r_A + r_B)/a_0} d\mathbf{r} \tag{9.4.5}$$

To evaluate the above integral we employ spheroidal coordinates $(\lambda, \mu, \varphi)$; $\lambda$ and $\mu$ are given by

$$\lambda = (r_A + r_B)/R \qquad \text{and} \qquad \mu = (r_A - r_B)/R. \tag{9.4.6}$$

The polar angle $\varphi$ is the same as defined in the spherical polar coordinate system. It is easy to see that in the space $\lambda$ and $\mu$ vary from 1 to $\infty$ and from $-1$ to $+1$, respectively. The volume element in this system is $(R^3/8)(\lambda^2 - \mu^2)d\lambda\, d\mu\, d\varphi$. Hence, from (9.4.6),

$$S = (ZR/a_0)^3 \frac{1}{8\pi} \int_1^\infty \int_{-1}^{+1} \int_0^{2\pi} e^{-Z\lambda R/a_0} (\lambda^2 - \mu^2) d\lambda\, d\mu\, d\varphi \tag{9.4.7}$$

Integration over $\mu$ and $\varphi$ gives

$$S = \frac{1}{2}(ZR/a_0)^3 \int_1^\infty e^{-Z\lambda R/a_0}(\lambda^2 - \tfrac{1}{3})d\lambda \tag{9.4.8}$$

Using (8.3.9), we get

$$S = \exp(-ZR/a_0)\left[1 + ZR/a_0 + \tfrac{1}{3}(ZR/a_0)^2\right] \tag{9.4.9}$$

Hence, from (9.4.3),

$$N_\pm = [1 \pm \exp(-2\,ZR/a_0)]\left\{\left[1 + ZR/a_0 + \tfrac{1}{3}(ZR/a_0)^2\right]^2\right\}^{-1/2} \tag{9.4.10}$$

As $R \to 0$ we get a single-center object having two protons and two electrons, which is a helium atom that does not have any neutrons and so is unstable. However, helium is referred to as the united atom limit (UAL) of $H_2$. In this limit $\varphi_A(i) = \varphi_B(i)$. Hence $v_-$ vanishes identically, and for $v_+$, $S = 1$ and $N_+ = 1/\sqrt{2}$. On the other hand, as $R \to \infty$ we get $S = 0$ and $N_\pm = 1$ and both states, given by (9.4.1), dissociate into two ground state hydrogen atoms, which thus represent the separated atom limit (SAL) of $H_2$.

It is evident from (9.4.1) that $v_\pm (r_1, r_2)$ are two-center wave functions. In collision investigations sometimes it is more convenient to employ single-center wave functions. Here the middle point of the internuclear axis is taken as the origin and the polar coordinates of the electrons and the nuclei are taken with respect to this origin. A set of simple one-center wave functions for the ground and excited states of the $H_2$ molecule are given by Huzinaga (1957). These wave functions were successfully utilized by Khare (1966a,b, 1967) to investigate the excitation of the hydrogen molecule to the singlet and triplet excited states due to electron impact. Huzinaga's ground state one-center wave function is given by

$$v_0(X, 1s\sigma^1\Sigma_g^+) = \frac{C_1}{\sqrt{2}} [\varphi_{in}^0(r_1)\varphi_{out}^0(r_2) + \varphi_{out}^0(r_1)\varphi_{in}^0(r_2)$$
$$+ C_2\varphi_{1r}^0(r_1)\varphi_{1r}^0(r_2)] \qquad (9.4.11)$$

where, in atomic units,

$$\varphi_{in}^0(r) = C_3[N(1, \xi_1)e^{-\xi_1 r}Y_{00} + \delta N(4, \xi_2)r^3e^{-\xi_2 r}Y_{00}$$
$$+ \lambda N(4, \xi_2)r^3e^{-\xi_2 r}Y_{20}] \qquad (9.4.12)$$

$$\varphi_{1r}^0(r) = N(2, Z)re^{-Zr}Y_{10} \qquad (9.4.13)$$

$$\varphi_{out}^0(r) = N(1, \eta_1)e^{-\eta_1 r}Y_{00} \qquad (9.4.14)$$

with

$$N(n, x) = \frac{(2x)^{n+1/2}}{[(2n)!]^{1/2}} \qquad (9.4.15)$$

In atomic units the values of the various parameters are $C_1 = 0.99560365$, $C_2 = -0.09366858$, $C_3 = 0.489949475$, $\xi_1 = 1.1$, $\delta = 0.524208$, $\xi_2 = 4.3$, $\lambda = 0.273048$, $Z = 1.6$, and $\eta_1 = 0.8$. The energy of the ground state $H_2$ molecute obtained from the above wave function is in close agreement with that obtained from the two-center wave function of Kolos and Roothan (1960). Huzinaga represented the singlet and triplet excited states of $H_2$ by

$$v_n^{\pm}(r_1, r_2, R_0) = \frac{1}{\sqrt{2}} [\varphi_{in}^n(r_1)\varphi_{out}^n(r_2) \pm \varphi_{out}^n(r_1)\varphi_{in}^n(r_2)] \qquad (9.4.16)$$

The plus sign in the above equation corresponds to the singlet states while the minus sign is for the triplet states. The excited states considered by Khare (1966a,b, 1967) are $B(2p\sigma^1\Sigma_u^+)$, $C(2p\pi^1\Pi_u)$, $D(3p\pi^1\Pi_u)$, $a(2s\sigma^3\Sigma_g^+)$, $b(2p\sigma^3\Sigma_u^+)$ and $c(2p\pi^3\Pi_u)$. For these states

$$\varphi_{in}^{n}(r) = p_1 N(1, \alpha) \exp(-\alpha r) Y_{00} + p_2 N(4, \beta) r^3 \exp(-\beta r) Y_{00}$$
$$+ p_3 N(4, \beta) r^3 \exp(-\beta r) Y_{20} \qquad (9.4.17)$$

with $p_1 = 0.761157$, $p_2 = 0.253422$, $p_3 = 0.110674$, $\alpha = 1.1$, and $\beta = 4.3$ (all in a.u.). $\varphi_{out}^{n}(r)$ for different states are:

$$\varphi_{out}^{a}(r) = N(2, \eta_2^{a}) r \exp(-\eta_2^{a} r) Y_{10} \qquad (9.4.18)$$

$$\varphi_{out}^{B,b}(r) = N(2, \eta_2^{B,b}) r \exp(-\eta_2^{B,b} r) Y_{10} \qquad (9.4.19)$$

$$\varphi_{out}^{C,c}(r) = N(2, \eta_2^{C,c}) r \exp(-\eta_2^{C,c} r) \frac{1}{\sqrt{2}} (Y_{11} - Y_{1\bar{1}}) \qquad (9.4.20)$$

and

$$\varphi_{out}^{D}(r) = N(2, \eta_2^{D}) r \exp(-\eta_2^{D} r) \frac{1}{\sqrt{2}} (Y_{11} - Y_{1\bar{1}}) \qquad (9.4.21)$$

The variational values of $\eta_2$ at $R_0 (=1.4a_0)$ in a.u. are $\eta_2^{B} = 0.520$, $\eta_2^{c} = 0.436$, $\eta_2^{D} = 0.338$, $\eta_2^{b} = 0.886$, and $\eta_2^{c} = 0.566$ (Khare, 1967). Several other single-center wave functions for $H_2$ are available (see, e.g., Carter et al., 1958; Joy and Parr, 1958; Hayes, 1967).

## 9.5   Elastic Collision of Electrons with the Ground State Hydrogen Molecule

Let us consider the integral approach and start with the simplest approximation, namely the FBA. In this approximation (9.3.8) for elastic scattering reduces to

$$f_{ii}^{B1}(R_0) = -4m \left( \frac{\pi e}{h} \right)^2 \left\langle k_j, i \left| -\frac{1}{|R_A - r|} - \frac{1}{|R_B - r|} \right. \right.$$
$$\left. \left. + \frac{1}{|r_1 - r|} + \frac{1}{|r_2 - r|} \right| k_i, i \right\rangle \qquad (9.5.1)$$

where $r$, $R_A$, $R_B$, $r_1$, and $r_2$ are the coordinates of the incident electron, nuclei $A$ and $B$, and the molecular electrons, respectively, with respect to 0, the mid point of the internuclear axis $AB$ (see Fig. 9.1). Representing the incident and scattered electrons by the normalized plane wave and integrating over $r$, we get

$$f_{ii}^{B1}(\mathbf{R}_0) = \frac{2}{K^2 a_0}(e^{i\mathbf{K}\cdot\mathbf{R}_A} + e^{i\mathbf{K}\cdot\mathbf{R}_B} - \langle i|e^{i\mathbf{K}\cdot\mathbf{r}_1} + e^{i\mathbf{K}\cdot\mathbf{r}_2}|i\rangle) \qquad (9.5.2)$$

Since $\langle i|i\rangle = 1$. Now taking

$$\mathbf{R}_A = -\mathbf{R}_B = \frac{\mathbf{R}_0}{2}$$

in the above equation we get

$$f_{ii}^{B1}(\mathbf{R}) = \frac{4}{K^2 a_0}[\cos(\mathbf{K}\cdot\mathbf{R}/2) - \langle i|e^{i\mathbf{K}\cdot\mathbf{r}_1}|i\rangle] \qquad (9.5.3)$$

To simplify the notation the subscript of $R_0$ has been dropped and the equilibrium internuclear distance is represented by $R$.

Use of (9.4.1) and (9.4.5) in (9.5.3) and integration over $\mathbf{r}_2$ gives

$$f_{ii}^{B1}(\mathbf{R}) = \frac{4}{K^2 a_0}\left\{\cos(\mathbf{K}\cdot\mathbf{R}/2) - N_+^2/2 \int e^{i\mathbf{K}\cdot\mathbf{r}_1} d\mathbf{r}_1 \right.$$
$$\left. \times \left[|\varphi_A(1)|^2 + |\varphi_B(1)|^2 + 2S\varphi_A(1)\varphi_B(1)\right]\right\} \qquad (9.5.4)$$

We use (6.11.1) for $\varphi_A$ and $\varphi_B$ and integrate over $\mathbf{r}_1$ This gives

$$f_{ii}^{B1}(\mathbf{R}) = \frac{4}{K^2 a_0}\left[\cos(\mathbf{K}\cdot\mathbf{R}/2) - N_+^2\cos(\mathbf{K}\cdot\mathbf{R}/2)\frac{16Z^4}{(4Z^2 + K^2 a_0^2)^2} - N_+^2 S I\right] \qquad (9.5.5)$$

where the variational parameter $Z_S$ is replaced by $Z$ and the overlap integral $I$ is given by

$$I = \int e^{i\mathbf{K}\cdot\mathbf{r}_1}\varphi_A(r_1)\,\varphi_B(r_1)dr_1 \qquad (9.5.6)$$

The evaluation of $I$ is rather difficult. Let us first consider (9.5.5) in the separated atom limit of $H_2$, i.e., $S = 0$ and $N_+ = 1$. Then

$$f_{SAL}^{B1} = \frac{4(8Z^2 + K^2 a_0^2)}{(4Z^2 + K^2 a_0^2)^2} a_0 \cos(\mathbf{K}\cdot\mathbf{R}/2) \qquad (9.5.7)$$

$$= 2f_H(Z)\cos(\mathbf{K}\cdot\mathbf{R}/2) \qquad (9.5.8)$$

where $f_H(Z)$ is the scattering amplitude in the FBA for the elastic scattering of electrons by a hydrogenic atom having a charge $Z$[Sec. (7.3.1)]. Since the molecule rotates in space, the polar angle $\theta$ of the internuclear axis $\boldsymbol{R}$, with respect to a fixed $\boldsymbol{K}$, varies from 0 to $\pi$. Hence, the DCS is averaged over $\theta$, which gives

$$\bar{I}_{H_2}(R) = 2I_H(Z)\int_{-1}^{+1}\cos^2(KR\mu/2)d\mu \tag{9.5.9}$$

where $\mu = \cos\theta$ and $I_H(Z)$ is given by (7.3.26). Evaluation of the above integral gives

$$\bar{I}_{H_2}(R) = 2I_H(Z)\left[1 + \frac{\sin(KR)}{KR}\right] \tag{9.5.10}$$

It may again be noted that the above equation gives the DCS averaged over all the initial rotational states and summed over all the final rotational and vibrational states in the separated atom limit for the elastic scattering of electrons by the ground state of the hydrogen atom.

It is evident that $\bar{I}_{H_2}$ oscillates with $K$. The amplitude of the oscillation decreases with $K$. In the forward direction $\bar{I}_{H_2}$ is $4I_H(Z)$, but at large $K$ it is equal to $2I_H(Z)$. Thus in the forward direction the scattering amplitudes $f_H(Z)$ from each atom are added (coherent addition), whereas for large $K$ the differential cross sections $I_H(Z)$ are added (incoherent addition) to yield $\bar{I}_{H_2}$. With the help of (7.3.26) and (7.3.30), we also get from (9.5.10)

$$\bar{I}_{H_2}(R) = \tfrac{1}{2}I_{He}(Z)\left(1 + \frac{\sin(KR)}{KR}\right) \tag{9.5.11}$$

where $I_{He}(Z)$ is the differential cross section of the helium atom having the same $Z$. Epuation (9.5.10) clearly shows the diffraction of the de Broglie wave of the incident electrons by the two-center hydrogen molecule. Variational calculation for the $v_+(1^1\Sigma_g^+)$ state has given $Z = 1.166$ and $R = 1.404a_0$.

Khare and Moisewitsch (1965) examined the effect of the overlap term on $\bar{I}_{H_2}$ They represented the molecule in the separable form

$$v_+(\boldsymbol{r}_1, \boldsymbol{r}_2) = u(\boldsymbol{r}_1)u(\boldsymbol{r}_2) \tag{9.5.12}$$

with

$$u(r) = N[\varphi(r_A) + \varphi(r_B)] \tag{9.5.13}$$

and

$$\varphi(r) = e^{-Zr/a_0}/a_0^{3/2} \tag{9.5.14}$$

Thus we get

$$N_+^2 = \frac{Z^3}{2\pi(1+S)} \qquad (9.5.15)$$

where $S$ is given by (9.4.9). To evaluate the overlap integral $I$, given by (9.5.6), Khare and Moiseiwitsch expanded the plane wave in the spheroidal coordinates and finally obtained

$$\begin{aligned}
\bar{I}_{H_2} &= \frac{8}{K^4 a_0^2}\left[\left(1+\frac{\sin KR}{KR}\right)\left\{1-\frac{32\pi N^2 Z}{(K^2 a_0^2+4Z^2)^2}\right\}^2\right. \\
&\quad +4\pi^2 N^4\left(\frac{R}{a_0}\right)^6 \sum_{l(\text{even})}\frac{(-1)^l}{\Lambda_{0l}(C)}[M_{0l}(C)]^2 \\
&\quad \left.-8\pi N^2\left(\frac{R}{a_0}\right)^3\left[1-\frac{32\pi N^2 Z}{(K^2 a_0^2+4Z^2)^2}\right]\sum_{l(\text{even})}\frac{(-1)^l}{\Lambda_{0l}(C)}M_{0l}(C)je_{0l}(C,1)\right] \qquad (9.5.16)
\end{aligned}$$

where

$$M_{0l}(C) = \int_1^\infty je_{01}(C,\lambda)[(\lambda^2-\tfrac{1}{3})d_0(C/0l)-\tfrac{2}{15}d_2(C/0l)]e^{-\lambda RZ/a_0}d\lambda \qquad (9.5.17)$$

and

$$\Lambda_{0l}(C) = \sum_n [d_n(C/0l)]^2\frac{2}{2n+1} \qquad (9.5.18)$$

where $n$ is an even integer, $C = \frac{1}{2}KR$, and $je_{0l}$ are the spheroidal functions defined by Morse and Feshbach (1953). The differential cross section $\bar{I}_{H_2}$ has been evaluated using the tables of the spheroidal wave functions compiled by Stratton et al. (1956). In the separated atom limit the first term of (9.5.16) goes to (9.5.10) (which we now denote by $\bar{I}_S$). The second and third terms of (9.5.16) arise due to overlap. According to Khare and Moiseiwitsch, the ratio $\bar{I}_{H_2} / \bar{I}_S$ as a function of $C$ lies between $1 \pm 0.04$. Hence, the separated atom is a satisfactory approximation for $\bar{I}_{H_2}$. Liu and Smith (1973) and Ford and Browne (1973) employed accurate two-center wave functions to investigate elastic $e - H_2$ scattering in FBA. Their results are not too much different from those obtained by Khare and Moiseiwitsch.

It may be noted that in (9.5.14) the value of $Z$ is 1.193 and not 1, which accounts for the valence bond effect. In the case of a diatomic molecule, the formation of a bond between two atoms tends to localize the electron densities between the atoms and changes the effective size of the molecule compared to the collection of noninteracting atoms arranged in the same geometrical configuration. The result is usually a decrease in the cross section below that expected for the free (independent) atom model. The differential cross section for $H_2$ in the independent atom model (IAM) is given by (9.5.10) with $Z = 1$ At high $E$ the difference between $\bar{I}_{H_2}^{Bl}$ and $\bar{I}_{H_2}^{IAM}$ arises due to molecular binding effects. $\Delta N_{el}$, defined by

$$\Delta N_{el} = \left(\frac{K^2}{4}\right)[\bar{I}_{H_2}^{Bl} - \bar{I}_{H_2}^{IAM}] \tag{9.5.19}$$

was measured experimentally at $E = 25\,keV$ by Ulsh et al. (1974). Hence, a comparison of the theoretical values of $\Delta N_{el}$ with the experimental data gives an idea of the accuracy of the molecular wave function.

Gupta and Khare (1978) employed the 57-term single-center wave function of Hayes (1967) for the ground state of $H_2$ to calculate $\bar{I}_{H_2}$ in the FBA. The use of spherical harmonics associated with the single-center wave function allows evaluation of all the integrals analytically.

Figure 9.4 compares the experimental data for $\Delta N_{el}$ (Ulsh et al., 1974) with the theoretical results of Liu and Smith (1973) (with a two-center wave function) and Gupta and Khare. Although the experimental accuracy is poor, it can be seen from the figure that the theoretical calculations are in agreement with the experimental data. This also shows the suitability of one-center wave functions for the calculation of the differential and the total cross sections for the scattering of electrons by hydrogen molecules. However, the FBA, which completely neglects the distortion of the projectile and the target wave function and the exchange effect, is not found to be satisfactory for complex molecules even at the intermediate energies.

### 9.5.1 The Plane Wave Approximation

In the *plane wave approximation* (discussed in Sec. 7.7) the summed elastic differential cross section (sum of pure elastic and rotational excitations) is given by

$$\bar{I}_{H_2} = \overline{|f_{Bl} - g|^2} + 2\overline{(f_{B_l} - g)f_{dp}} \tag{9.5.20}$$

where the first Born scattering amplitude $f_{Bl}$ is given by (9.5.8). The exchange scattering amplitude $g$ in the Ochkur approximation is

$$g = -\frac{2}{k_i^2 a_0}\langle i|e^{iK\cdot r_1}|i\rangle \tag{9.5.21}$$

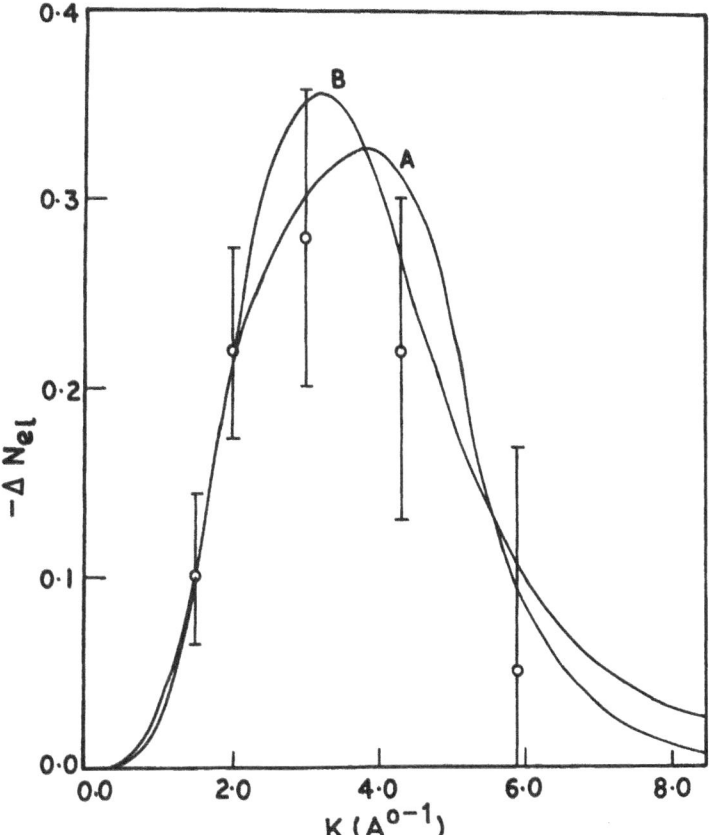

**FIGURE 9.4** Variation of $\Delta N_{el}$ with respect to K (in 1/Å). A—Gupta and Khare (1978) with the Hayes (1967) wave function, B—Liu and Smith (1973) with Davidson and Jones (1962) wave function. O are experimental data of Ulsh et al. (1974). Reproduced from "Elastic scattering of electrons by molecular hydrogen for incident energies 100–2000 eV," P. Gupta and S. P. Khare, *J. Chem. Phys.* **68**, 2193, 1978, with permission from the American Institute of Physics, USA.

and the second-order scattering amplitude $f_{dp}$, due to the dynamic polarization potential $V_{dp}$, is

$$f_{dp} = -\frac{m}{2\pi\hbar^2}\int e^{iK\cdot r}\, V_{dp}(r)dr \qquad (9.5.22)$$

Gupta and Khare (1978) employed the 57-term single-center wave function of Hayes (1967) to evaluate $f_{B1}$ and $g$. They extended $V_{dp}$, given by (7.7.37), to $e$–$H_2$ scattering and took

$$V_{dp}(r) = -\frac{e^2}{2}\left[a_d r^2 (r^2 + d^2)^{-3} + a_d' r^2 (r^2 + d^2)^{-3} P_2(\cos\Theta)\right.$$
$$\left. + a_q r^4 (r^2 + d^2)^{-5}\right] \tag{9.5.23}$$

where $\alpha_d$ is the mean dipole polarizability and $\alpha_d'$ is a measure of the anisotropy of the polarizability,

$$a_d = \tfrac{1}{3}(a_2 + 2a_x) \tag{9.5.24}$$

and

$$a_d' = \tfrac{2}{3}(a_z - a_x) \tag{9.5.25}$$

where $\alpha_z$ and $\alpha_x$ are the dipole polarizabilities of the hydrogen molecule along the internuclear axis and perpendicular to the axis, respectively, and $\Theta$ is the angle between $r$ and $R$. The energy-dependent cut-off parameter is given by (7.7.39). Using the values of Kolos and Wolnicwicz (1965) for $\alpha_z$ and $\alpha_x$, one gets $\alpha_d = 5.18a_0^3$ and $\alpha_d' = 1.20a_0^3$. These values along with $\alpha_q = 17.27a_0^5$ are utilized by Gupta and Khare to evaluate $f_{dp}$ and $\bar{I}_{H_2}$. Their theoretical cross sections in the energy range 100 to 2000 eV are in good agreement with the absolute experimental differential cross sections of van Wingerden et al. (1977). Curve $B$ of Fig. 9.5 shows that the FBA seriously underestimates the cross sections at small scattering angles. The inclusion of the exchange and the polarization gives DCS (curve $A$) that are in good agreement with the experimental data over the whole angular range. A substantial increase in $\bar{I}_{H_2}$ at small scattering angles is due to $f_{dp}$. This is expected because the static field is a short-range potential, whereas $V_{dp}(r)$ is a long-range potential. As already noted, in the plane wave approximation the distortion of the projectile's wave function is completely neglected and the distortion of the target wave function is included only up to the first order. To a certain extent these defects are removed in the modified Glauber approximation.

### 9.5.2   The Modified Glauber Approximation

In the modified glauber approximation (MGA) the scattering amplitude $f_{MG}$ is given by (7.8.5). This may also be written as

$$f_{MG} = f_{B1} + \bar{f}_{B2} + \sum_{n\geq 3}^{\infty} \bar{f}_{Gn} \tag{9.5.26}$$

The evaluation of $\bar{f}_{B2}$ and $\bar{f}_{Gn}$ for e–$H_2$ elastic scattering is an involved task. However, we have seen that in the separated atom limit

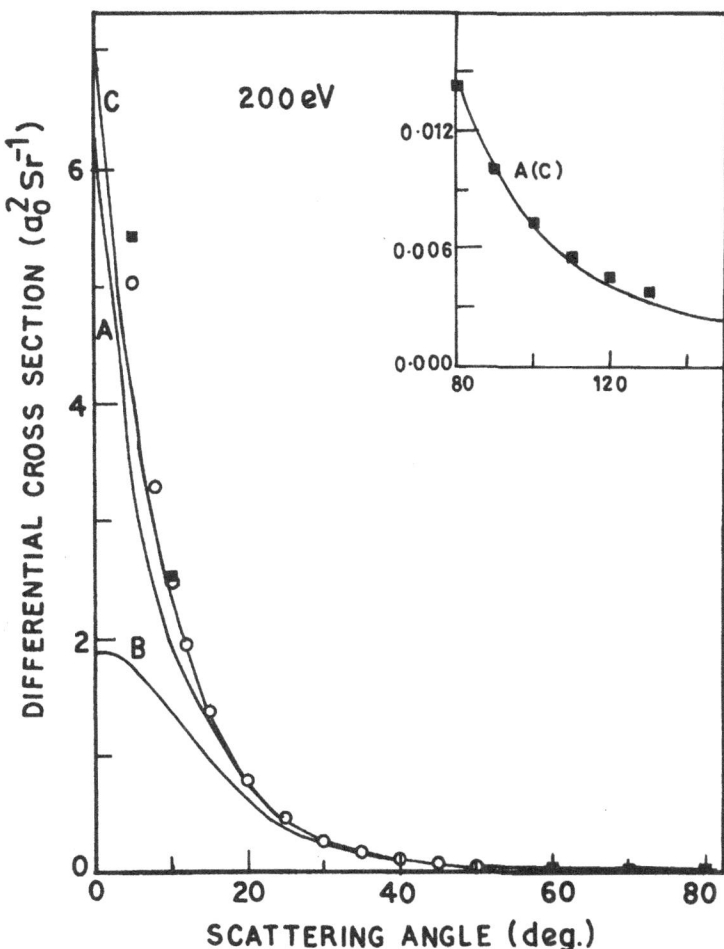

**FIGURE 9.5** Differential scattering cross sections for 200-eV electrons scattered elastically by hydrogen molecules. Curve A, cross sections of Gupta and Khare (1978) in the plane wave approximation with the Hayes wave function. Curve B, cross sections in the FBA with the same wave function. Curve C also includes nonadiabatic correction [see (7.7.34)]. O and ■ are the experimental data of van Wingerden et al. (1977) and Fink et al. (1975) (as renormalized by van Wingerden et al.). Reproduced from the reference given in Fig. 9.4, with permission from the American Institute of Physics, USA.

$$f_{H_2}^{B1}(Z) = 2\cos(K \cdot R/2)f_H^{B1}(Z) = \cos(K \cdot R/2)f_{He}^{B1} \tag{9.5.27}$$

where $f_{He}^{B1}(Z)$ is the scattering amplitude in the FBA for the elastic scattering of electrons by the united atom limit of the hydrogen molecule, i.e., a helium atom having same $Z$ as a hydrogen molecule. To obtain $\bar{f}_{B2}$ Jhanwar et al. (1982b) represented the hydrogen molecule by the following simple but fairly accurate two-center wave function:

$$v(r_1, r_2) = N[u_g(r_1)u_g(r_2) + Cu_u(r_1)u_u(r_2)] \tag{9.5.28}$$

where the gerade and ungerade orbitals are given by

$$u_{g,u}(r_i) = N_{g,u}[\varphi(r_{iA}) \pm \varphi(r_{iB})] \tag{9.5.29}$$

The atomic orbitals $\varphi(r_i)$ are represented by (6.11.1) with $Z_s = Z$ and

$$N_{g,u}^2 = \frac{1}{2(1 \pm S)} \tag{9.5.30}$$

For $R = 1.4a_0$ the variational parameters $Z$ and $C$ are 1.2005 and $-0.5814$, respectively. The normalization constant $N$ is equal to $(1 + C^2)^{1/2}$. In the united atom limit (9.5.28) reduces to

$$v(r_1, r_2) = u_g(r_1)u_g(r_2) \tag{9.5.31}$$

The reduced interaction potential energy $U$ of the system is given by

$$U(r, r_1, r_2) = U_A(r_1)U_B(r_2) \tag{9.5.32}$$

with

$$U_A(r_1) = \frac{2}{a_0}\left(-\frac{1}{|r - R_A|} + \frac{1}{|r - r_1|}\right) \tag{9.5.33}$$

and

$$U_B(r_2) = \frac{2}{a_0}\left(-\frac{1}{|r - R_B|} + \frac{1}{|r - r_2|}\right) \tag{9.5.34}$$

Putting (9.5.33) and (9.5.34) into (7.7.3), we get

$$\bar{f}_{H_2}^{B2} = f_{AA} + f_{BB} + f_{AB} + f_{BA} \tag{9.5.35}$$

where

$$f_{AA} = -2\pi^2 \underset{p}{S} \int \frac{\langle k_j, i|U_A|k_q, p\rangle\langle k_q, p|U_A|k_i, i\rangle}{k_p^2 - k_q^2 + i\varepsilon}\, dk_q \qquad (9.5.36)$$

and

$$f_{AB} = -2\pi^2 \underset{p}{S} \int \frac{\langle k_j, i|U_A|k_q, p\rangle\langle k_q, p|U_B|k_i, i\rangle}{k_i^2 - k_q^2 + i\varepsilon}\, dk_q \qquad (9.5.37)$$

The expressions for $f_{BB}$ and $f_{BA}$ are similar. Both $f_{AA}$ and $f_{AB}$ are double-scattering terms, but $f_{AA}$ represents the scattering process in which the incident electron is scattered twice by the same atom $A$, whereas in $f_{AB}$ the incident electron is first scattered by atom $B$ and then by atom $A$. The evaluation of $f_{AA}, f_{AB}$, etc., is quite difficult even for a simple diatomic molecule such as $H_2$. An approximate evaluation was carried out by Jhanwar et al. (1982b), who obtained

$$\bar{f}_{H_2}^{B2} = \cos(\boldsymbol{K}\cdot\boldsymbol{R}/2)\bar{f}_{He}^{B2}(Z) \qquad (9.5.38)$$

where $\bar{f}_{He}^{B2}(Z)$ is the second Born scattering term for an object represented by (9.5.31). They assumed a similar relationship between $\bar{f}_{H_2}^{Gn}$ and $\bar{f}_{He}^{Gn}(Z)$ for $n \geq 3$ and took

$$f_{H_2}^{MG} = \cos(\boldsymbol{K}\cdot\boldsymbol{R}/2)\left[ f_{He}^{B1}(Z) + \bar{f}_{He}^{B2}(Z) + \sum_{n\geq 3}^{\infty} \bar{f}_{He}^{MG}(Z) \right] \qquad (9.5.39)$$

or

$$f_{H_2}^{MG} = \cos(\boldsymbol{K}\cdot\boldsymbol{R}/2)[ f_{He}^{G}(Z) - \bar{f}_{He}^{G2}(Z) + \bar{f}_{He}^{B2}(Z)] \qquad (9.5.40)$$

The term $\bar{f}_{He}^{B2}(Z)$ was evaluated in the closure approximation with $Z = 1.2005$ and $\Delta = 29.39\,\mathrm{eV}$ by Jhanwar et al. (1982b). To include exchange they employed the *Glauber–Ochkur (GO) exchange approximation* and took

$$g_{H_2}^{GO} = \cos(\boldsymbol{K}\cdot\boldsymbol{R}/2)g_{He}^{GO}(Z) \qquad (9.5.41)$$

The expression for $g_{He}^{GO}(Z)$ was derived by following Dewangan (1976) and Khayrallah (1976). Finally in the separated atom limit we get

$$\bar{I}_{H_2}(K) = \frac{1}{2}\left(1 + \frac{\sin KR}{KR}\right)|f_{He}^{MG}(Z) - g_{He}^{GO}(Z)|^2 \qquad (9.5.42)$$

Furthermore, the imaginary parts of the forward amplitudes of $|f_{He}^{MG}(Z) - g_{H_2}^{MG}(Z)|$ and $|f_{H_2}^{MG}(Z)|$ yield total scattering cross sections through the optical theorem for electron and positron impacts, respectively.

Jhanwar et al. (1982b) calculated $\bar{I}_{H_2}$ ($\theta$) for incident energies varying from 50 to 1000 eV. Their theoretical cross sections are in fair agreement with the experimental data for $E \geq 100$ eV. At 50 eV the theory overestimates the cross sections, which is not unexpected because the Glauber approximation is a high-energy approximation. In Table 9.1, $\bar{I}_{H_2}$ obtained in different approximations at $E = 200$ eV is compared with the experimental data. In the FBA and PWA, the one-center wave functions of Joy and Parr (1958) was employed while calculations in EBS and MGA were performed with the two-center wave function of Weinbaum (1933). The experimental data in the last three rows are as renormalized by van Wingerden et al. As expected, the FBA underestimates the cross sections at low $K$. The agreement between the cross sections obtained in the MGA and the experimental data is good over the whole angular range.

The total collision cross sections $\sigma_T$ are shown in Table 9.2. Theoretical values of Jhanwar et al. (1982b) for electron as well as positron scattering are in good agreement with the experimental data. The table also shows that the ratios $\sigma_T(e+)/\sigma_T(e-)$ are close to unity, theoretically as well as experimentally. The theoretical values are always less than or equal to unity, whereas the experimental

**Table 9.1** Differential Cross Sections (in $10^{-21}$ m$^2$) for the Elastic Scattering of 200-eV Electrons by Hydrogen Molecules

| $\theta$ (deg.) | Theory | | | | Experiment | | | |
|---|---|---|---|---|---|---|---|---|
| | FBA     PWA Khare and Shobha (1972, 1974) | | EBS Khare and Lata (1985) | MGA Jhanwar et al. (1982b) | van Wingerden et al. (1977) | Fink et al. (1975) | Llyod et al. (1974) | Williams (1969) |
| 5 | 5.20 | 11.7 | 13.2 | 12.9 | 14.0 | 15.2 | — | — |
| 10 | 4.11 | 6.69 | 6.66 | 6.66 | 7.02 | 7.08 | — | — |
| 15 | 2.85 | 3.83 | 3.72 | 3.81 | 3.89 | — | — | — |
| 20 | 1.84 | 2.21 | 2.10 | 2.21 | 2.24 | 2.18 | 1.41 | 1.81 |
| 25 | 1.13 | 1.28 | 1.19 | — | 1.28 | — | — | — |
| 30 | 0.680 | 0.753 | 0.680 | 0.725 | 0.772 | — | 0.713 | 0.680 |
| 40 | 0.263 | 0.288 | 0.252 | 0.260 | 0.282 | 0.256 | 0.263 | 0.285 |
| 50 | 0.127 | 0.138 | 0.126 | 0.119 | 0.133 | 0.139 | 0.152 | 0.136 |
| 60 | 0.078 | 0.084 | 0.086 | — | — | 0.083 | 0.081 | 0.083 |
| 70 | 0.055 | 0.059 | 0.069 | 0.071 | — | 0.058 | 0.054 | 0.059 |
| 80 | 0.039 | 0.041 | 0.058 | — | — | 0.040 | 0.040 | 0.042 |
| 90 | 0.028 | 0.029 | — | — | — | 0.029 | 0.028 | 0.036 |
| 100 | 0.020 | 0.021 | 0.039 | — | — | 0.021 | 0.025 | 0.022 |
| 120 | 0.011 | 0.012 | — | — | — | 0.013 | — | — |

**Table 9.2** Total Collisional Cross Sections (in $10^{-20}\,\mathrm{m}^2$) for $e^{\pm}$–$H_2$ Scattering

| $E$ (eV) | Theory | | | Experiment | | | |
|---|---|---|---|---|---|---|---|
| | EBS Khare and Lata (1985) | MGA Jhanwar et al. (1982b) | | Charlton et al. (1980) | van Wingerden et al. (1980) | Hoffman et al. (1982) | |
| | $e^+$ | $e^-$ | $e^+$ | $e^+$ | $e^-$ | $e^-$ | $e^+$ |
| 100 | 3.22 | 2.68 | 2.64 | 2.96 | 2.52 | 2.56 | 2.68 |
| 150 | 2.36 | 2.09 | 2.08 | 2.10 | 2.00 | 1.98 | — |
| 200 | 1.88 | 1.71 | 1.71 | — | 1.61 | 1.69 | 1.71 |
| 300 | 1.35 | 1.27 | 1.26 | 1.29 | 1.22 | 1.27 | 1.27 |
| 400 | 1.06 | 1.01 | 1.01 | 1.06 | 0.965 | 1.04 | 0.999 |
| 500 | 0.880 | 0.845 | 0.845 | — | 0.831 | 0.870 | 0.859 |
| 700 | 0.668 | 0.641 | 0.641 | — | 0.638 | — | — |
| 1000 | 0.487 | 0.476 | 0.476 | — | — | — | — |

values, up to 300 eV, as obtained from the data of Hoffmann et al. (1982), are always greater than or equal to unity.

## 9.6   Excitation of the Hydrogen Molecule

Since the ground state of the hydrogen molecule is a singlet state, its excitation to the higher singlet state proceeds via direct as well as exchange scattering. However, excitations from the singlet to triplet states are possible only through exchange scattering.

For singlet-to-singlet excitations the Ochkur exchange scattering amplitude $g_{ji}$, as for the helium atom, is given by

$$g_{H_2}^{OC} = \frac{K^2}{2k_i^2} f_{H_2}^{B1} \tag{9.6.1}$$

Hence, in the Born–Ochkur approximation (9.3.9) modifies to

$$\bar{I}_{ji} = \frac{k_j}{k_i} \frac{1}{4\pi} \int \left(1 - \frac{K^2}{2k_i^2}\right)^2 |f_{ij}|^2 \, d\Omega(\hat{R}) \tag{9.6.2}$$

From (7.3.4) and (7.3.5) we have

$$f_{ji} = -\frac{2}{K^2 a_0} \langle v_j(r_1, r_2) | e^{iK \cdot r_1} + e^{iK \cdot r_2} | v_i(r_1, r_2) \rangle \tag{9.6.3}$$

Khare (1966a,b) employed single-center wave functions given by (9.4.11) to (9.4.21) to evaluate the differential cross sections, generalized oscillator strengths, and integrated cross sections for the excitation of the hydrogen molecule from the ground state $X(1s\sigma_g^1{}^1\Sigma_g^+)$ to $B(2p\sigma^1\Sigma_u^+)$, $C(2p\pi^1\Pi_u)$, and $D(3p\pi^1\Pi_u)$ states in the first Born–Ochkur approximation. A sum of the excitation cross sections to $B$ and $C$ electronic states obtained by Khare (1966a) is compared with the experimental data of Geiger (1964) at 25 keV in Fig. 9.6. The agreement between theory and experiment is satisfactory. For the $D$ electronic state the optical strength of Khare (1966b) is also in good agreement with the experimental value of Geiger (1964) and the theoretical value of Mulliken and Reike (1941). Khare (1968) employed single-center wave functions to obtain photoionization cross sections of the hydrogen molecule. Here again the agreement between the theoretical optical oscillator strength and the corresponding experimental data is found to be satisfactory. Such agreement indicates the suitability of the single-center wave functions for bound-to-bound and bound-to-free transitions.

For singlet–triplet excitations, the direct scattering amplitude is zero and the exchange scattering amplitude $g_{ji}$ in the Born–Oppenheimer approximation is given by

$$g_{ji} = \sqrt{3}(-2\pi^2)\langle k_j(r_1)v_j(r,r_2)|U(r,r_1,r_2)|k_i(r)v_i(r_1,r_2)\rangle \qquad (9.6.4)$$

where

$$U(r,r_1,r_2) = \frac{2}{a_0}\left(-\frac{1}{|r-R_A|} - \frac{1}{|r-R_B|} + \frac{1}{|r-r_1|} + \frac{1}{|r-r_2|}\right) \qquad (9.6.5)$$

The factor $\sqrt{3}$ is due to the fact that the excited state consists of three degenerate excited states having $M_s = 1$, 0, and $-1$. In the Ochkur approximation, which is correct up to $k_i^{-2}$, the nuclear terms and $1/|r-r_2|$ are dropped because their contributions fall faster than $k_i^{-2}$, and $1/|r-r_1|$ is approximated by $4\pi\delta(r-r_1)/k_i^2$ [see Eq. (7.5.5)]. Hence, in the above approximation, with Dirac delta function normalized plane waves, we get

$$g_{ji} = -\frac{2\sqrt{3}}{k_i^2 a_0}\langle v_j(r_1,r_2)|e^{iK\cdot r_1}|v_i(r_1,r_2)\rangle \qquad (9.6.6)$$

Khare and Moiseiwitsch (1966) have investigated the excitation of the ground state $v(X1s\sigma1\Sigma_g^+)$ of the hydrogen molecule to its lowest triplet state $v(b2p\sigma^3\Sigma_u^+)$ due to electron impact. They took

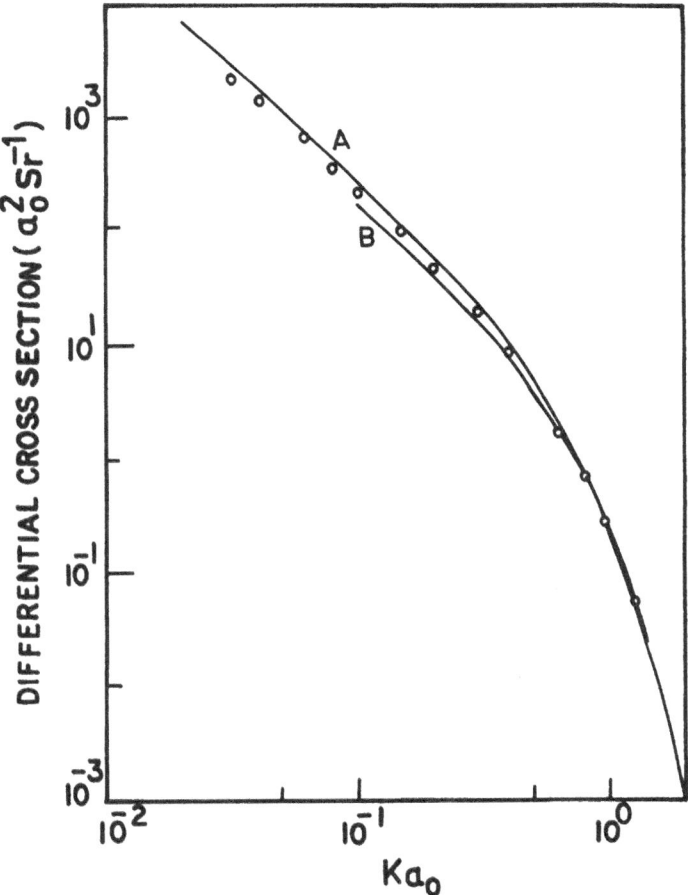

**FIGURE 9.6** Differential cross sections for elastic scattering of 25-keV electrons that excite the ground state hydrogen molecule to B and C electronic states. A, Khare (1966a); B, Roscoe (1941); O, experimental points of Geiger (1964). Reproduced from "Excitation of hydrogen molecule by electron impact," S. P. Khare, *Phys. Rev.* **149**, 33, 1966, with permission from the American Institute of Physics, USA.

$$v_i = \frac{N_i}{\sqrt{2}} [\varphi_i(r_{1A})\varphi_i(r_{2B}) + \varphi_i(r_{2A})\varphi_i(r_{1B})] \qquad (9.6.7)$$

and

$$v_j = \frac{N_j}{\sqrt{2}} [\varphi_j(r_{1A})\varphi_j(r_{2B}) - \varphi_j(r_{2A})\varphi_j(r_{1B})] \qquad (9.6.8)$$

with $\phi_{i,j}$ given by (6.11.1) having nuclear charges $Z_i$ and $Z_j$, respectively. Using (9.6.7) and (9.6.8) in (9.6.6) and proceeding to large values of $R$ (separated atom limit of the molecule), we get

$$g_{ji} = -\frac{\sqrt{3}}{k_i^2 a_0} \langle \varphi_j(r)|\varphi_i(r)\rangle [\langle \varphi_j(r_{1A})|e^{iK\cdot r_1}|\varphi_i(r_{1A})\rangle$$
$$-\langle \varphi_j(r_{1B})|e^{iK\cdot r_1}|\varphi_i(r_{1B})\rangle] \tag{9.6.9}$$

$$= -\frac{2i\sqrt{3}}{k_i^2 a_0} \sin(K\cdot R/2)\langle \varphi_j(r)|\varphi_i(r)\rangle\langle \varphi_j(r)|e^{iK\cdot r}|\varphi_i(r)\rangle \tag{9.6.10}$$

Evaluating the values of $\langle \varphi_j(r)|\varphi_i(r)\rangle$ and $\langle \varphi_j(r)|e^{iK\cdot r}|\varphi_i(r)\rangle$ with the help of (9.6.7) and (9.6.9) we get

$$g_{ji} = \sin(K\cdot R/2)g'_{ji} \tag{9.6.11}$$

where

$$g'_{ji} = -\frac{128\sqrt{3}i}{k_i^2 a_0} \frac{(Z_iZ_j)^3}{\left[(Z_i+Z_j)^2 + K^2 a_0^2\right]^2} \frac{1}{(Z_i+Z_j)^2} \tag{9.6.12}$$

Finally averaging over the orientations of the molecular axis, the average differential excitation cross section for the excitation of the molecule from the ground state to the triplet excited state $b(^3\Sigma_u^+)$ is given by

$$\bar{I}_{H_2}(X^1\Sigma_g^+ \rightarrow b^3\Sigma_u^+) = \frac{1}{2}\left(1 - \frac{\sin KR}{KR}\right)|g'_{ji}|^2 \tag{9.6.13}$$

The integration of (9.6.13) over the scattering angles gives the total excitation cross section. Since $b(^3\Sigma_u^+)$ is a repulsive state, after the excitation the hydrogen molecule dissociates into two ground state hydrogen atoms. Khare (1967) also employed the Ochkur approximation but used the single-center wave functions given by (9.4.11) to (9.4.20) to calculate the total cross section for the excitation of the hydrogen molecule from the ground state $X(^1\Sigma_g^+)$ to the triplet $a(2s\sigma^3\Sigma_g^+)$, $b(2p\sigma^3\Sigma_u^+)$, and $c(2p\pi^3\Pi_u)$ electronic states due to electron impact. All three excited states are unstable and dissociate into two hydrogen atoms. The cross sections obtained by Khare (1967) for the sum of all three states are shown in Fig. 9.7. The cross-section curve shows a sharp maximum close to the threshold of the excitation and then falls off quite rapidly with the increase in impact energy. For high impact energies, the cross sections fall as $E^{-3}$. The

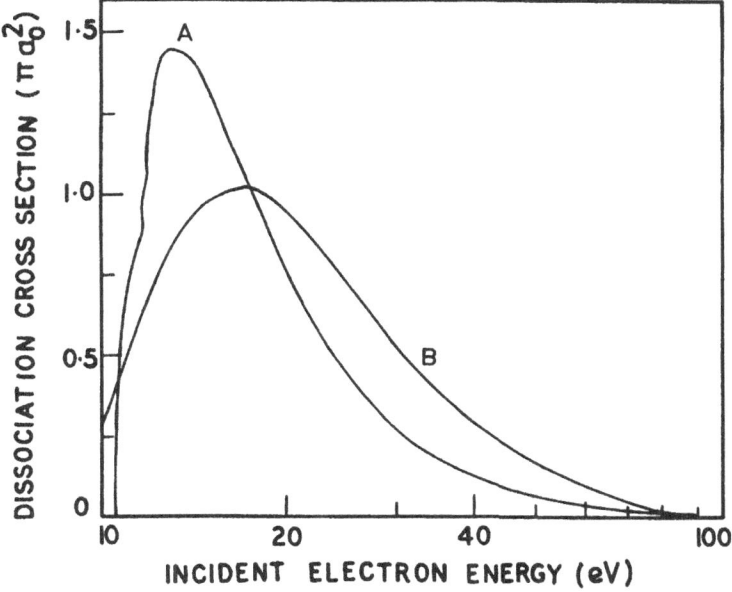

**FIGURE 9.7** Cross sections for the dissociation of the ground state hydrogen molecule due to singlet–triplet excitation produced by electron impact. Curve A gives the theoretical cross sections of Khare (1967) and curve B gives the experimental values of Corrigan (1965). Reproduced from "Excitation of hydrogen molecule by electron impact: III. Singlet-triplet excitation," S. P. Khare, *Phys. Rev.* **157**, 107, 1967, with permission from the American Institute of Physics, USA.

cross section for the excitation to the $b(^3\Sigma_u^+)$ state has the largest value; hence, this excitation dominates in the dissociation of the $H_2$ molecule. It was pointed out by Cartwright and Kupperman (1967) that the cross sections obtained by the Ochkur approximation are sensitive to the wave functions of the $H_2$ molecule. Chung and Lin (1978) also employed the FBA to investigate excitation of $H_2$ molecules.

## 9.7   Ionization of Molecules

In Sec.7.6 we discussed the ionization of atoms and noted the difficulties associated with the calculation of ionization cross sections. Replacement of the atom by a molecule further complicates the investigation. Due to the multicentered nature of the target the differential cross section is eightfold differential (Champion et al., 2001). Even in the FBA, one is obliged to generate continuum generalized oscillator strengths (CGOS), which is a difficult task for molecules. Hence, a number of semiempirical and semiclassical formulas for ionization cross

sections have been proposed. Such methods formulated till the early 1980s
are discussed by Younger and Mark (1985). Recently Deutsch et al. (2000a)
reviewed the methods developed by Khare and his associates, Kim and his
associates, and Deutsch and his associates. In this section we shall describe these
methods briefly.

### 9.7.1   The Kim and Rudd Model

Kim and Rudd (1994) started with the Mott differential cross section
for the collision between two free electrons, one at rest and the other moving
with an energy $E$. Let in the collision energy $\varepsilon$ be transferred from the moving
electron to the other electron. The Mott differential cross section for the
above collision is given by (7.6.56). If we take $t = E/I$, $\omega = \varepsilon/I$, and replace
$d\sigma_i/d\varepsilon$ by the symbol $I_M(t,\omega)$, the above equation reduces to a symmetrized form,
given by

$$I_M(t,\omega) = \frac{S}{I} \sum_{n=1}^{2} F_n(t) \left\{ \frac{1}{(\omega+1)^n} + \frac{1}{(t-\omega)^n} \right\} \tag{9.7.1}$$

where

$$S = 4\pi a_0^2 R^2 \, N/I^2, \quad F_2(t) = 1/t, \quad F_1(t) = -F_2(t)/(t+1) \tag{9.7.2}$$

with $R$ being the Rydberg energy. With exchange, the maximum value of $\omega$ is
$(t-1)/2$. Hence, the Mott ionization cross section is

$$\begin{aligned}
\sigma_M(t) &= S \sum_{n=1}^{2} \int_0^{(t-1)/2} F_n(t) \left\{ \frac{1}{(\omega+1)^n} + \frac{1}{(t-\omega)^n} \right\} d\omega \\
&= S \sum_{n=1}^{2} \int_0^{t-1} F_n(t) \frac{1}{(\omega+1)^n} d\omega \\
&= S[F_1(t)\ln t + F_2(1-1/t)]
\end{aligned} \tag{9.7.3}$$

For soft collisions (Bethe term), Kim and Rudd employed the Born–Bethe
approximation and took

$$\begin{aligned}
I_B(E,W) &= \frac{4\pi a_0^2 R^2}{E} \frac{1}{W} \frac{df(W,0)}{dW} \ln(E/I) \\
&= \frac{S}{NI} \frac{1}{t(\omega+1)} \frac{df(\omega,0)}{d\omega} \ln(t)
\end{aligned} \tag{9.7.5}$$

Hence, on integration over $\omega$ the total Bethe cross section is given by

$$\sigma_B(t) = \frac{S}{t} D(t) \ln t \qquad (9.7.6)$$

where

$$D(t) = \frac{1}{n} \int_0^{(t-1)/2} \frac{1}{\omega+1} \frac{df(\omega,0)}{d\omega} d\omega \qquad (9.7.7)$$

To combine (9.7.3) and (9.7.6), they obtained the total stopping power cross section from (9.7.1) and (9.7.5) and equated it to the Born–Bethe stopping power cross section at large $t$. This gives

$$t(\ln 2)F_1(t) + F_2(t)\ln t + \frac{N_i}{N} \frac{\ln t}{t} = 2\frac{\ln t}{t} \qquad (9.7.8)$$

where

$$N_i = \int_0^\infty \frac{df(\omega,0)}{d\omega} d\omega \qquad (9.7.9)$$

They then took $F_2 = a/t$ and compared the coefficient of $\ln t/t$ on the two sides of (9.7.8), obtaining

$$a = 2 - N_i/N \qquad (9.7.10)$$

Thus, finally

$$\sigma_i = \sigma_M + \sigma_B$$
$$= \frac{S}{t}\{D(t)\ln t + (2 - N_i/N)[1 - 1/t - \ln t/(t+1)]\} \qquad (9.7.11)$$

As expected the above equation overestimates the cross sections at low $t$. To remove this deficiency Kim and Rudd included the effect of the velocity distribution of the bound electrons of the target on the ionization process through the binary encounter theory (Grizinski, 1965a,b,c). According to this theory, $t$ occurring outside the square bracket of (9.7.11) changes to $t + u + 1$, with $u$ being the ratio of the average kinetic energy $U$ of the bound electron of the $i$th molecular orbital and $I_i$. Thus in the binary encounter dipole (BED) model of Kim and Rudd, the total ionization cross section for the $i$th molecular orbital is given by

$$\sigma_i = \sigma_{\text{KMD}} + \sigma_{\text{KBD}} \qquad (9.7.12)$$

where $\sigma_{\text{KMD}}$ and $\sigma_{\text{KBD}}$ are the Mott (hard collision) and the Bethe (soft collision) cross sections, respectively, in the BED model of Kim and Rudd. These cross sections are given by

$$\sigma_{\text{KMD}} = \frac{S}{t+u+1}(2 - N_i/N)[1 - 1/t - \ln t/(t+1)] \qquad (9.7.13)$$

and

$$\sigma_{\text{KBD}} = \frac{S}{t+u+1} D(t) \ln t \qquad (9.7.14)$$

A summation over all the molecular orbitals gives the total ionization cross sections $\Sigma_i \sigma_i$.

For H, He, Ne, and $H_2$ Kim and Rudd used the experimental values of $df(W,0)/dW$, compiled and recommended by Berkowitz(1979), and represented them by the following power series:

$$\frac{df(W,0)}{dW} = \frac{1}{I}\sum_{n=1}^{7} a_n \left(\frac{I}{W}\right)^n \qquad (9.7.15)$$

The values of $a_n$, $U$, and $I$ as given by these authors for H, He, and $H_2$ are shown in Table 9.3. Recently Kim et al. (2000) utilized $df(W,0)/dW$ for helium atoms obtained by the *relativistic random-phase approximation* (Johnson and Lin, 1979; Johnson and Cheng, 1979). These values are also represented by (9.7.15). The new coefficients $a_n$ are also given in the same table. With these new coefficients they have obtained excellent agreement between their theoretical $\sigma_i$ and the experimental data for the above targets.

For complex molecular targets, Kim and Rudd proposed a simpler binary encounter–Bethe (BEB) model. In this model COOS is given by

$$\frac{df(W,0)}{dW} = \frac{N_i I}{W^2}$$

or

$$\frac{df(\omega,0)}{d\omega} = \frac{N_i}{(\omega+1)^2} \qquad (9.7.16)$$

**Table 9.3** The Values of $I$, $U$, and $a_n$ for H, He and $H_2{}^a$

|        | H*          | He*         | He**       | $H_2$*   |
|--------|-------------|-------------|------------|----------|
| $I$ (eV) | 13.6057   | 24.59       | 24.587     | 15.43    |
| $U$ (eV) | 13.6057   | 39.51       | 39.51      | 25.68    |
| $a_1$  | 0           | 0           | 0          | 0        |
| $a_2$  | −2.2473 (−2) | 0          | 0          | 0        |
| $a_3$  | 1.1775      | 1.2178 (1)  | 8.24012    | 1.1262   |
| $a_4$  | −4.6264 (−1) | −2.9585 (1) | −10.4769  | 6.3982   |
| $a_5$  | 8.9064 (−2) | 3.1251 (1)  | 3.96496    | −7.8055  |
| $a_6$  | 0           | −1.2175 (1) | −0.0445976 | 2.1440   |
| $a_7$  | 0           | 0           | 0          | 0        |

$^a a(b) = a \times 10^b$, * Kim and Rudd (1994), ** Kim et al. (2000).

The symmetrized forms of $I_B(t,\omega)$ and $D(t)$ in the BEB model are given by

$$I_B(t,\omega) = \frac{S}{I} F_3(t) \left[ \frac{1}{(\omega+1)^3} + \frac{1}{(t-\omega)^3} \right]$$

and

$$D(t) = \frac{N_i}{N} \int_0^{(t-1)/2} \left[ \frac{1}{(\omega+1)^3} + \frac{1}{(t-\omega)^3} \right] d\omega$$

where $F_3(t) = N_i/(t + u + 1)$. On integration of the equation for $D(t)$ we get

$$D(t) = \frac{1}{2} \frac{N_i}{N} (1 - 1/t^2) \tag{9.7.17}$$

Now the total ionization cross section $\sigma_i$ is given by

$$\sigma_i = \sigma_{KMB} + \sigma_{KBB} \tag{9.7.18}$$

In the BEB model of Kim and Rudd, with $N_i = N$, the Mott cross section is given by

$$\sigma_{KMB} = \frac{S}{t+u+1} \left( 1 - \frac{1}{t} - \frac{\ln t}{t+1} \right) \tag{9.7.19}$$

and the Bethe cross section is

$$\sigma_{KBB} = \frac{S}{2(t+u+1)} \left( 1 - \frac{1}{t^2} \right) \ln t \tag{9.7.20}$$

Using their BEB model Kim and his associates (Kim and Rudd, 1994; Hwang et al., 1996; Kim et al., 1997; Ali et al., 1997; Nishimura et al., 1999) have calculated $\Sigma_i \sigma_i$ for a good number of molecules. They varied $E$ from $I_i$ to 1 keV (20 keV in some cases). The values of $U$ obtained by Hwang et al. (1996) are shown in Table 9.4. For most of the molecules their cross sections are in good agreement with the experimental data. They have noted that much of the success of their model is due to the replacement of $t$ by $t_{\text{eff}} = t + u + 1$, as required by the BE theory. This change in $t$ may also be justified by considering the increase in the initial kinetic energy $E$ of the incident electron by $U + I$ due to the attractive field of the target before it collides with a bound electron.

## 9.7.2   The Saksena Model

Saksena and Kushwaha (1996) started from the FBA, which includes longitudinal as well as transverse interactions. At relativistic energies the total ionization cross section is given by the sum of (7.6.15) and (7.6.30). Hence,

$$\sigma_i = \frac{4\pi a_0^2 R}{E_r} \left\{ \int_I^E \int_{\ln Q_-}^{\ln Q_+} \frac{R}{W} \frac{df(W,Q)}{dW} d(\ln Q) dW - M^2[\ln(1-\beta^2)+\beta^2] \right\} \quad (9.7.21)$$

where $E_r = 1/2\, mv^2$, $v$ being the velocity of the incident electron. The presence of the CGOS in the above equation makes the evaluation of $\sigma_i$ for the molecules difficult. On the other hand, the COOS are available for a number of molecules (Zeiss et al., 1975, 1977; Berkowitz, 1979; Gallagher et al., 1988). Using the semiphenomenological relation of Mayol and Salvat (1990), Saksena and Kushwaha expressed the CGOS in terms of the COOS, which

$$\frac{df(W,Q)}{dW} = \frac{df(W,0)}{dW}\theta(W-Q) + h(Q)\delta(W-Q) \quad (9.7.22)$$

where $\theta$ and $\delta$ are the step function and Dirac delta function, respectively, and

$$h(Q) = \int_I^Q \frac{df(W',0)}{dW'} dW' \quad (9.7.23)$$

Putting (9.7.22) into (9.7.21) and integrating over $Q$ we get

$$\sigma_i = \frac{4\pi a_0^2 R^2}{E_r} \left\{ \int_I^E \left\{ \frac{1}{W} \frac{df(W,0)}{dW} \ln\left(\frac{W}{Q_-}\right) + \frac{h(W)}{W^2} \right\} dW \right.$$
$$\left. - \frac{M^2}{R}[\ln(1-\beta^2)+\beta^2] \right\} \quad (9.7.24)$$

**Table 9.4** Values of $I$, $U$, and $N$ for Different Molecular Orbitals of Some Diatomic and Polyatomic Molecules (Hwang et al., 1996)

| Molecule | Molecular orbital | $I$ (eV) | $U$ (eV) | $N$ |
|---|---|---|---|---|
| $H_2$ | $1\sigma_g$ | 15.43 | 25.68 | 2 |
| $N_2$ | $3\sigma_g$ | 15.58 | 54.91 | 2 |
| | $1\Pi_u$ | 17.07 | 44.30 | 4 |
| | $2\sigma_u$ | 21.00 | 63.18 | 2 |
| | $2\sigma_g$ | 41.72 | 71.13 | 2 |
| CO | $5\sigma$ | 14.01 | 42.26 | 2 |
| | $1\Pi$ | 17.66 | 54.30 | 4 |
| | $4\sigma$ | 21.92 | 73.18 | 2 |
| | $3\sigma$ | 41.92 | 79.63 | 2 |
| $O_2$ | $1\Pi_g$ | 12.07 | 84.88 | 2 |
| | $3\sigma_g$ | 19.79 | 71.84 | 2 |
| | $1\Pi_u$ | 19.64 | 59.89 | 4 |
| | $2\sigma_u$ | 29.82 | 90.92 | 2 |
| | $2\sigma_g$ | 46.19 | 79.73 | 2 |
| $CH_4$ | $1t_2$ | 12.51 | 25.96 | 6 |
| | $2a_1$ | 25.73 | 33.05 | 2 |
| $NH_3$ | $3a_1$ | 10.16 | 43.25 | 2 |
| | $1e$ | 17.19 | 35.62 | 4 |
| $H_2O$ | $1b_1$ | 12.61 | 61.91 | 2 |
| | $3a_1$ | 15.57 | 59.52 | 2 |
| | $1b_2$ | 19.83 | 48.36 | 2 |
| | $2a_1$ | 36.88 | 70.71 | 2 |
| $C_2H_6$ | $1e_g$ | 11.52 | 28.17 | 4 |
| | $3a_{1g}$ | 13.90 | 32.78 | 2 |
| | $1e_u$ | 16.31 | 24.42 | 4 |
| | $2a_{2u}$ | 22.99 | 33.60 | 2 |
| | $2a_{1g}$ | 27.75 | 34.37 | 2 |
| $CO_2$ | $1\Pi_g$ | 13.77 | 64.43 | 4 |
| | $1\Pi_u$ | 19.70 | 49.97 | 4 |
| | $3\sigma_{2u}$ | 20.27 | 71.56 | 2 |
| | $4\sigma_{1g}$ | 21.62 | 74.66 | 2 |
| | $2\sigma_{2u}$ | 40.60 | 78.38 | 2 |
| | $3\sigma_{1g}$ | 42.04 | 75.72 | 2 |

In the energy range 0.1 to 2.7 MeV, Reike and Prepejchal (1972) measured $\sigma_i$ for a number of molecules and expressed their cross sections by

$$\sigma_i = \frac{4\pi a_0^2}{E_r} R\{C_{RP} + M^2[\ln\beta^2 - \ln(1-\beta^2) - \beta^2]\}  \qquad (9.7.25)$$

A comparison of (9.7.24) and (9.7.25) yields

$$C_{RP} = C'_{RP} + \int_I^E \frac{R}{W^2} h(W) dW  \qquad (9.7.26)$$

where

$$C'_{RP} = \int_I^E \frac{R}{W} \frac{df(W,0)}{dW} \ln\left(\frac{W}{\beta^2 Q_-}\right) dW  \qquad (9.7.27)$$

Using the available theoretical and experimental values of COOS, Saksena and Kushwaha have calculated $C'_{RP}$ and $C_{RP}$ for a number of molecules in the MeV region, In this energy region $C_{RP}$ is independent of $E$ and there is hardly any difference between the values of $C'_{RP}$ and $C_{RP}$. The theoretical values of $C_{RP}$, obtained by Saksena and Kushwaha at 1 MeV, are shown in Table 9.5 along with the experimental data of Reike and Prepejchal. For most of the molecules the two

**Table 9.5** The Parameter $C_{RP}$, Defined by (9.7.26) and (9.7.27), for Molecules

| Molecule | Theory Saksena and Kushwaha, (1996) | Experiment Reike and Prepejchal (1972) |
|----------|------------------------------------|---------------------------------------|
| $H_2$    | 7.60   | 8.12  |
| $N_2$    | 34.56  | 34.84 |
| $O_2$    | 41.41  | 38.84 |
| CO       | 40.77  | 35.14 |
| $H_2O$   | 29.49  | 32.26 |
| $H_2S$   | 60.68  | 42.19 |
| $CO_2$   | 54.11  | 55.92 |
| $NH_3$   | 37.62  | 34.86 |
| $CS_2$   | 82.46  | —     |
| $SF_6$   | 154.88 | —     |

sets of values of $C_{RP}$ do not differ by more than 10%. Such an agreement between theory and experiment shows the appropriateness of (9.7.22).

Equation (9.7.24) may also be written as

$$\sigma_i = \sigma_B + \sigma_M + \sigma_t \tag{9.7.28}$$

where the Bethe and the Mott cross sections are given by

$$\sigma_B = \frac{4\pi a_0^2 R}{E_r} \int_I^E \frac{R}{W} \frac{df(W,0)}{dW} \ln\left(\frac{W}{Q_-}\right) dW \tag{9.7.29}$$

and

$$\sigma_M = \frac{4\pi a_0^2 R^2}{E_r} \int_I^E \frac{h(W)}{W^2} dW \tag{9.7.30}$$

Asymptotically,

$$\sigma_B = \frac{4\pi a_0^2 R}{E_r} M^2 \ln(C_B E_r) \tag{9.7.31a}$$

where

$$M^2 \ln(C_B E_r) = \int_I^E \frac{R}{W} \frac{df(W,0)}{dW} \ln\left(\frac{W}{Q_-}\right) dW \tag{9.7.32}$$

Using (9.7.25) to (9.7.29), we also get

$$\sigma_B = \frac{4\pi a_0^2 R}{E_r} (C'_{RP} + M^2 \ln \beta^2) \tag{9.7.31b}$$

Equating (9.7.31a) to (9.731b), we obtain

$$C'_{RP} + M^2 \ln \beta^2 = M^2 \ln(C_B E_r)$$

Hence,

$$C'_{RP} = M^2 \ln(C_B E_0)$$
$$= M^2 [\ln(E_0/R) + \ln(C_B R)] \tag{9.7.33}$$

where $E_0$ is half of the rest mass energy of the electron. Since $E_0/R$ is much greater than $C_B R$, we have

$$C'_{RP}/M^2 \approx \ln(E_0/R) = 9.84 \qquad (9.7.34)$$

This value of $C'_{RP}/M^2$ is close to those obtained from the experimental data of Reike and Prepejchal.

Saksena et al. (1997a) have used (9.7.24) to calculate $\sigma_i$ for a number of molecules in the energy range 1 keV to 3 MeV. They employed the theoretical and experimental values of $df(W,0)/dW$ available for the whole molecule (summed over all the molecular orbitals) provided by Zeiss et al. (1975, 1977) and Gallagher et al. (1988). Thus they obtained $\Sigma_i \sigma_i$. Their results for $H_2$, $N_2$, $O_2$, and $H_2O$ molecules in the energy range 0.1 to 3 MeV are shown in Fig. 9.8. A comparison between the theoretical results and the experimental cross sections of Reike and Prepejchal (1972) shows excellent agreement. Figure 9.9 compares the theoretical cross sections of Saksena et al. at lower energies (1 to 20 keV) with the experimental data of Schram et al. (1965) and Shutten et al. (1966). The agreement between theory and experiment is again quite good. With such nice agreements it may be concluded that the representation of the CGOS in terms of the COOS through (9.7.22) is satisfactory for the evaluation of $\sigma_i$.

At $E < 1$ keV, the contributions of the transverse interaction and the relativistic effect become negligibly small. But the effect of exchange should be included in the evaluation of $\sigma_i$. Saksena et al. (1997b) included the effect through the Ochkur approximation. They neglected $\sigma_t$, replaced $E_r$ by $E$ in (9.7.21), and multiplied its integrand by $F_{ex}(E,Q)$ (given by 7.6.18). Further, it is well known that the FBA overestimates $\sigma_i$ at low $E$, so they also multiplied the integrand of (9.7.21) by an energy-dependent empirical factor $(1 - W/E)$.

The Bethe and Mott differential cross sections, as modified by Saksena et al. (1997b) for $E < 1$ keV, are integrated over $Q$, which gives

$$\sigma_B = \frac{4\pi a_0^2 R^2}{E} \int_I^{(E+I)/2} \frac{1}{W} \frac{df(W,0)}{dW} \left(1 - \frac{W}{E}\right) \left[ \ln\left(\frac{W}{Q_-}\right) \right.$$
$$\left. - \left(\frac{W - Q_-}{E}\right) + \frac{1}{2E^2}(W^2 - Q_-^2) \right] dW \qquad (9.7.35)$$

and

$$\sigma_M = \frac{4\pi a_0^2 R^2}{E} \int_I^{(E+I)/2} h(W)(1 - W/E) \left(\frac{1}{W^2} - \frac{1}{WE} + \frac{1}{E^2}\right) dW \qquad (9.7.36)$$

where $Q_-$ is given by the nonrelativistic equation (7.6.16), and $h(W)$ by (9.7.23) with the upper limit equal to $W$.

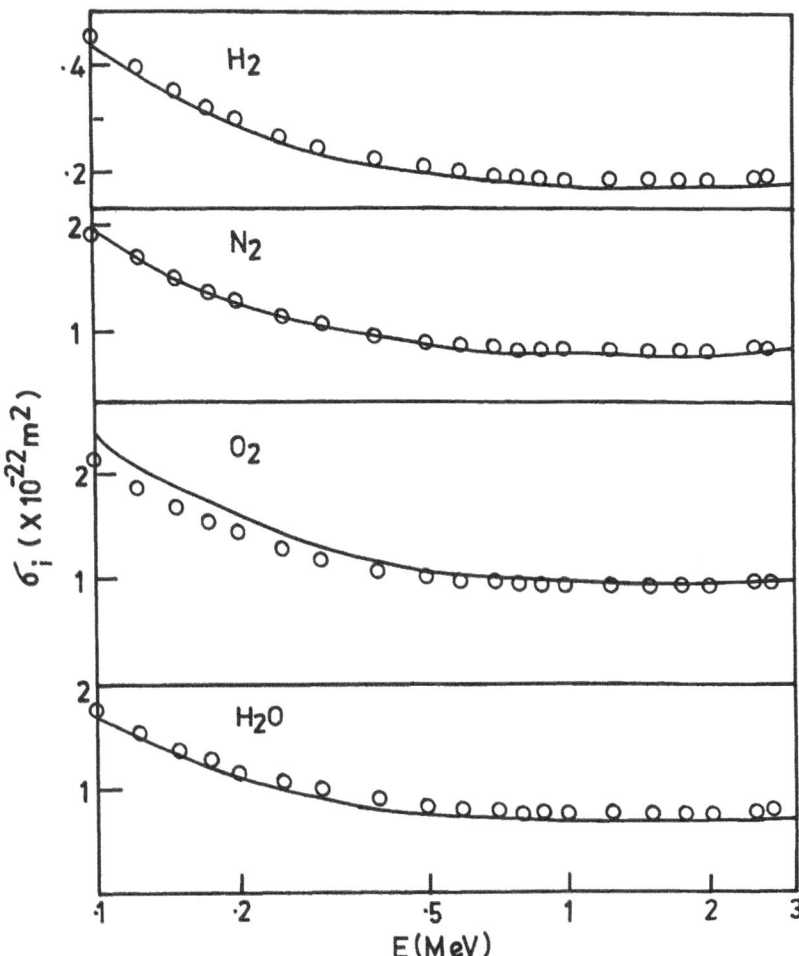

**FIGURE 9.8** Total ionization cross sections of $H_2$, $N_2$, $O_2$, and $H_2O$ due to electron impact in the energy range 0.1 to 3 MeV. Solid curves show the theoretical cross sections of Saksena et al. (1997a) and the open circles are the experimental data of Reike and Prepejchal (1972). Reproduced from "Electron impact ionization of molecules at high energies," V. Saksena, M. S. Kushwaha, and S. P. Khare, *Int. J. Mass Spectrom. Ion Proc.* **171**, L1, 1997, with permission from Elsevier Science.

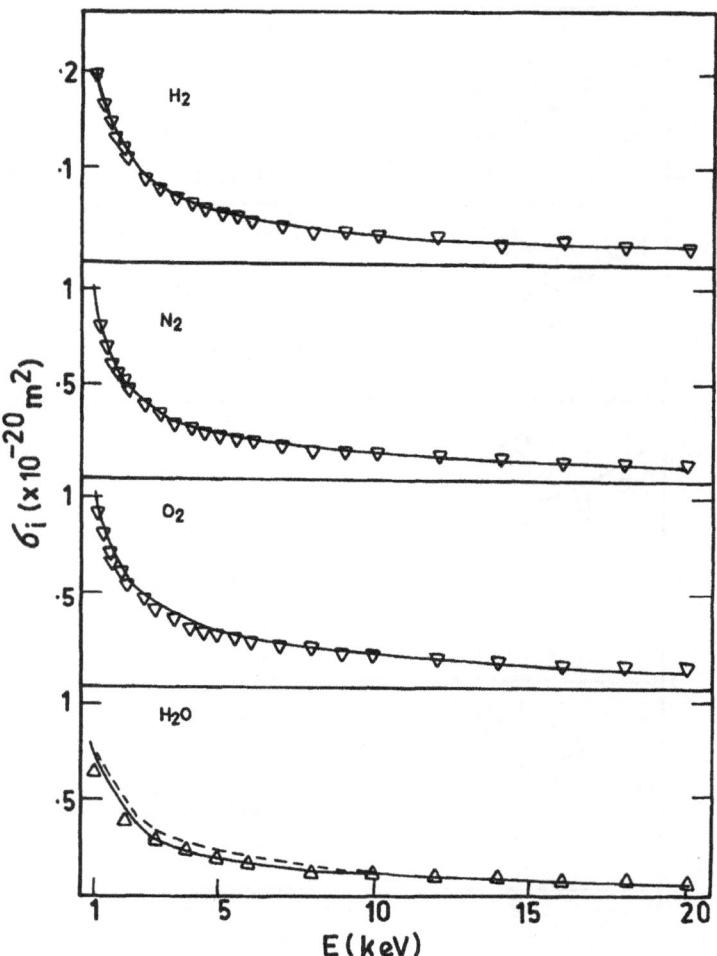

**FIGURE 9.9** Total ionization cross sections of $H_2$, $N_2$, $O_2$, and $H_2O$ due to electron impact in the energy range 1 to 20 keV. The solid and dashed curves show the theoretical cross sections of Saksena et al. (1997a) and Kim and Rudd (1994), respectively. $\nabla$ and $\Delta$ are the experimental cross sections of Schram et al. (1965) and Shutter et al. (1966), respectively. Reproduced from "Electron impact ionization of atoms and molecules," S. P. Khare and S. Tomar, in: *Trends in Atomic and Molecular Physics*, eds. K. K. Sud and U. N. Upadhayaya, p. 110, 1999.

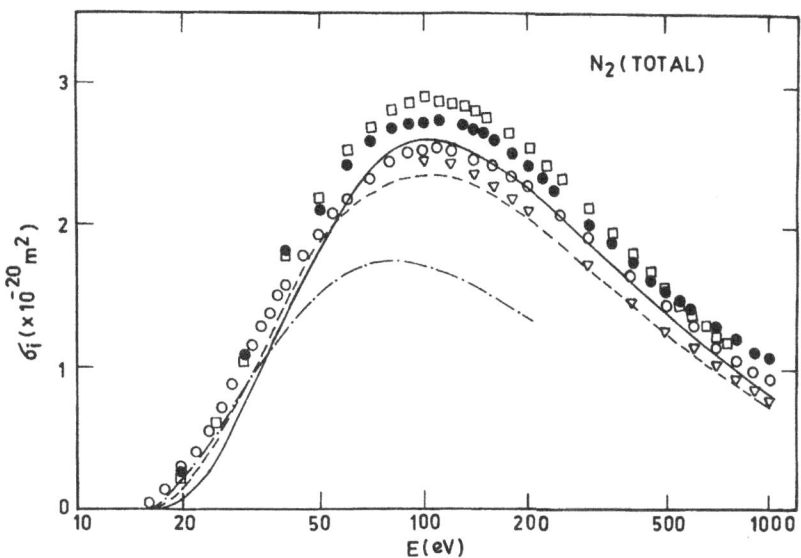

**FIGURE 9.10** Total ionization cross sections of $N_2$ due to electron impact. Theoretical: _____, Saksena et al. (1977b); – – –, Khare and Meath formula (1987), and –·–·–, Margreiter et al. (1990). Experimental: O, Rapp and Englander-Golden (1965); ▽, Schram et al. (1965, 1966); □, Tate and Smith (1932); ●, Krishnakumar and Srivastava (1990). Reproduced from "Ionization cross sections of molecules due to electron impact," V. Saksena, M. S. Kushwaha, and S. P. Khare. *Physica* B**233**, 201, 1997, with permission from Elsevier Science.

Saksena et al. (1977b) used (9.7.35) and (9.7.36) to calculate $\sigma_i$ (= $\sigma_M$ + $\sigma_B$) for $H_2$, $N_2$, $O_2$, $NH_3$, $H_2O$, and $CO_2$ in the energy range $I$ to 1 keV. Their results for $N_2$ and $O_2$ are compared with the experimental data in Figs. 9.10 and 9.11. In general, the agreement between theory and experiment is good for $E$ greater than about 50 eV. At lower $E$ the theory underestimates the cross sections. Similar agreement has been obtained for the other molecules except for $H_2$. For this molecule the Saksena model is found to underestimate the cross sections over practically the whole energy region. The Saksena model has also been applied to obtain the dissociative cross sections (Saksena et al., 1997b). Their results, shown as the ratio of the dissociative cross section $\sigma_d$ to $\sigma_i$ along with the experimental data, are shown in Fig. 9.12. The agreement between theory and experiment is again satisfactory. They have concluded that their model is quite satisfactory for $E \geq 300$ eV. For energies between 50 and 300 eV it overestimates the cross sections slightly, but for $E < 50$ eV it underestimates them for most of the molecules. This shows that at low $E$ the correction introduced by the empirical factor $(1 - W/E)$ is not adequate.

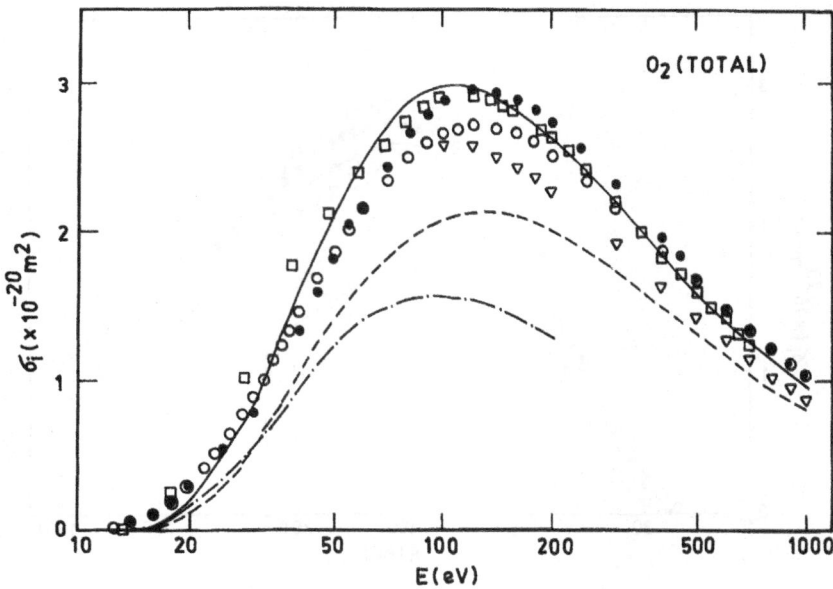

**FIGURE 9.11** Same as Fig. 9.10 but for $O_2$. ● are the experimental cross sections of Krishnakumar and Srivastava (1992). Reproduced from the same source given in Fig. 9.10, with permission from Elsevier Science.

It should be noted that the Saksena model does not require a prior knowledge of the collision parameter $C_B$ and the mixing parameter $\varepsilon_0$, which were needed in the methods proposed by Jain and Khare (1976) and Khare and Meath (1987).

### 9.7.3  The Khare model

Khare et al. (1999) have incorporated useful features of the Kim and Rudd in the Saksena model to remove the deficiency of the latter at low $E$. They removed the empirical factor $(1 - W/E)$ and replaced $E$ (which exist outside the of integral) by $E_r + U + I$ in (9.7.35) and (9.7.36). They also neglected exchange in the soft collisions. Thus in the Khare's BED model the Bethe and the Mott cross sections are given by

$$\sigma_{\text{KHBD}} = \frac{4\pi a_0^2 R^2}{E_r + U + I} \int_I^E \frac{1}{W} \frac{df(W,0)}{dW} \ln\left(\frac{W}{Q_-}\right) dW \qquad (9.7.37)$$

and

$$\sigma_{KHMD} = \frac{4\pi a_0^2 R^2}{E_r + U + I} \int\limits_I^{(E+I)/2} h(W)\left(\frac{1}{W^2} - \frac{1}{WE} + \frac{1}{E^2}\right)dW \qquad (9.7.38)$$

respectively. In this model $\sigma_i$ for the $i$th molecular orbital is given by

$$\sigma_i = \sigma_{KHBD} + \sigma_{KHMD} + \sigma_t \qquad (9.7.39)$$

where $\sigma_t$ is obtained from (7.6.30). Following Kim and Rudd, Khare and his associates also represented $df(W,0)/dW$ by (9.7.16) in their BEB model. Thus

$$h(W) = N(1 - I/W) \qquad (9.7.40)$$

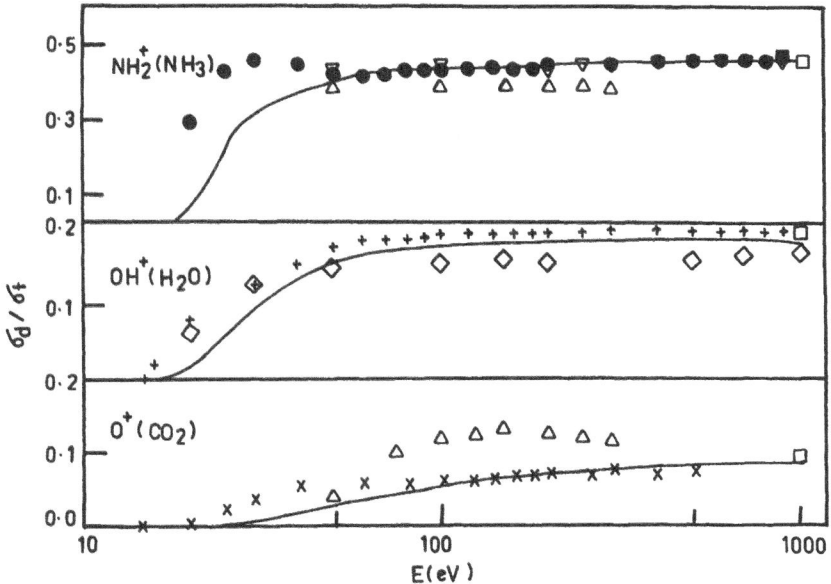

**FIGURE 9.12** The variation of the ratio $\sigma_d/\sigma_i$ with electron impact energy for a polyatomic molecule. Solid curves are the theoretical ratios of Saksena et al. (1997b). Experimental values: ●, Δ, and ∇ for $NH_2^+$ ions are from Rao and Srivastava (1992), Crowe and McConkey (1977), and Bederski et al. (1980), respectively. For $OH^+$ ions + and ◇ are from Rao et al. (1995) and Schutten et al. (1966), respectively. For $O^+$ ion X are from Orient and Srivastava (1987) and Δ are obtained by taking $\sigma_d$ from Crowe and McConkey and $\sigma_i$ from Rapp and Englander-Golden (1965). Reproduced from the same source given in Fig. 9.10, with permission from Elsevier Science.

$$\sigma_{\text{KHBB}} = \frac{SI^3}{E_r + U + I} \int_I^E \frac{1}{W^3} \ln\left(\frac{W}{Q_-}\right) dW \qquad (9.7.41)$$

and

$$\sigma_{\text{KHMD}} = \frac{SI^2}{E_r + U + I} \int_I^{(E+I)/2} (1 - I/W)\left[\frac{1}{W^2} - \frac{1}{WE} + \frac{1}{E^2}\right] dW \qquad (9.7.42)$$

In the nonrelativistic region integration over $W$ in (9.7.42) gives

$$\sigma_{\text{KHMB}} = \frac{S}{t + u + 1}\left[1 - \frac{2}{t+1} + \frac{t-1}{2t^2} + \frac{5-t^2}{2(t+1)^2} - \frac{1}{t(t+1)} - \frac{t+1}{t^2}\ln\left(\frac{t+1}{2}\right)\right] \qquad (9.7.43)$$

We may write (Khare et al. 2000)

$$\sigma_{\text{KHMB}} = \sigma_{\text{KMB}} + \sigma' \qquad (9.7.44)$$

where

$$\sigma' = \frac{S}{t + u + 1}\left[-\frac{1}{2} + \frac{2}{(t+1)^2} + \frac{1}{2}\frac{t-1}{t^2} - \frac{t+1}{t^2}\ln\left(\frac{t+1}{2}\right) + \frac{\ln t}{t+1}\right] \qquad (9.7.45)$$

It is found that $\sigma'$ is negative at all the values of $t$ and asymptotically it is half of $\sigma_{\text{KMB}}$. Hence, $\sigma_{\text{KHMD}}$ is always less than $\sigma_{\text{KMB}}$ and at large values of $t$ we have $\sigma_{\text{KHMD}} = \frac{1}{2}\sigma_{\text{KMB}}$.

We may also write (9.7.41) as

$$\sigma_{\text{KHBB}} = \sigma_{\text{KBB}} + \sigma'' \qquad (9.7.46)$$

where in the nonrelativistic region $\sigma_{\text{KBB}}$ is given by (9.7.20) and

$$\sigma'' = \frac{S}{t + u + 1} \int_I^E \frac{1}{W^3} \ln\left(\frac{W}{Q_-}\right) dW \qquad (9.7.47)$$

To compare $\sigma'$ with $\sigma''$ it is desirable to have an analytical expression for $\sigma''$ as well. On the assumption that a large contribution to the above integral comes from the small values of $W$, Khare et al. (2000) expanded $(1 - W/E)^{1/2}$ for $W \ll E$ and from (7.6.16) obtained an approximate value of $Q_-$:

$$Q_- = W^2/4E \qquad (9.7.48)$$

Putting (9.7.47) into (9.7.48) and evaluating the integral we get

$$\sigma_A'' = \frac{S}{t+u+1}[0.4431(1-1/t^2)+1/2\ \ln t/t^2]$$ (9.7.49)

The subscript $A$ on $\sigma''$ indicates that we have taken an approximate expression for $Q_-$, given by (9.7.48). The above equation shows that $\sigma_A''$ is always positive, so $\sigma_{KHBB} > \sigma_{KBB}$. Defining

$$\sigma_{KHTB} = \sigma_{KHBB} + \sigma_{KHMB}$$

and

$$\sigma_{KTB} = \sigma_{KBB} + \sigma_{KMB}$$ (9.7.50)

we get for the ratio

$$\lambda = \frac{\sigma_{KHTB}}{\sigma_{KTB}} = 1 + \frac{\sigma' + \sigma''}{\sigma_{KTB}}$$ (9.7.51)

In Table 9.6 $\lambda_A$ (with approximate $\sigma_A''$) is shown as a function of $t$. It can be seen that for $t \geq 5$ the ratio $\lambda_A$, which depends only upon $t$, does not differ from unity by more than 10%. High values of $\lambda_A$ for $t < 5$ are due to the use of the approximate expression for $Q_-$. Khare et al. (2000) employed (7.6.16) for $Q_-$ and recalculated $\lambda$ as obtained from (9.7.51). These new values of $\lambda$ lie between $1 \pm 0.03$. Hence, although the derivations of the expression for $\sigma_{KHTB}$ and $\sigma_{KBB}$

Table 9.6 Ratios $\lambda_A$ and $\lambda$ as Functions of $t$ for Any Molecular Orbital (Khare et al., 2000)

| $t$ | $\lambda_A$ | $\lambda$ |
|------|------|------|
| 1.5 | 1.64 | 0.97 |
| 2.0 | 1.37 | 1.01 |
| 3.0 | 1.18 | 1.01 |
| 4.0 | 1.11 | 1.00 |
| 5.0 | 1.07 | 1.00 |
| 6.0 | 1.05 | 1.00 |
| 8.0 | 1.03 | 0.99 |
| 10.0 | 1.02 | 0.99 |
| 20.0 | 1.00 | 0.99 |
| 40.0 | 0.99 | 0.98 |
| 60.0 | 0.99 | 0.98 |
| 80.0 | 0.98 | 0.98 |

look quite different, in effect in the BEB model the $\sigma_{KHBB}$ differ from $\sigma_{KBB}$ by $\sigma''$. Further, nearly the whole of $\sigma''$ is added to $\sigma_{KHMB}$ to obtain $\sigma_{KMB}$, so $\sigma_{KHTB}$ and $\sigma_{KTB}$ are nearly equal at every $E$ and for each molecular orbital. The BED model of Khare et al. does not involve the Bethe stopping power cross sections, which also include discrete excitations.

Khare et al. (1999) calculated the total ionization cross section $\sigma_i$ for the CH$_4$ molecule for impact energy $E$ varying from the ionization threshold to 3 MeV in their BED and BEB models as well as the cross sections in the Saksena model for comparison. The ionization of the two outermost orbitals of CH$_4$, namely, $1t_2$ and $2a_1$ were considered. The contribution of the third orbital to $\sigma_i$ is negligibly small owing to its high ionization potential (290.7 eV). Variations in $\sigma_{KHBEB}$, $\sigma_{KHMEB}$, $\sigma_{KBEB}$, and $\sigma_{KMEB}$ with $E$ for the orbital ($1t_2$) of methane are shown in Fig. 9.13. It is evident that $\sigma_{KHBEB}$ is always greater than $\sigma_{KBEB}$, but the opposite is true for $\sigma_{KHMEB}$ and $\sigma_{KMEB}$. However, $\sigma_{KHTB}$ is always very close to $\sigma_{KTB}$. The reason for this closeness has already been discussed.

In Fig. 9.14 the variations of $\sigma_{KHTD}$ and $\sigma_{KHTB}$ with $E$ in the energy range 10 eV to 20 keV are shown along with the theoretical cross sections $\sigma_{TS}$ obtained

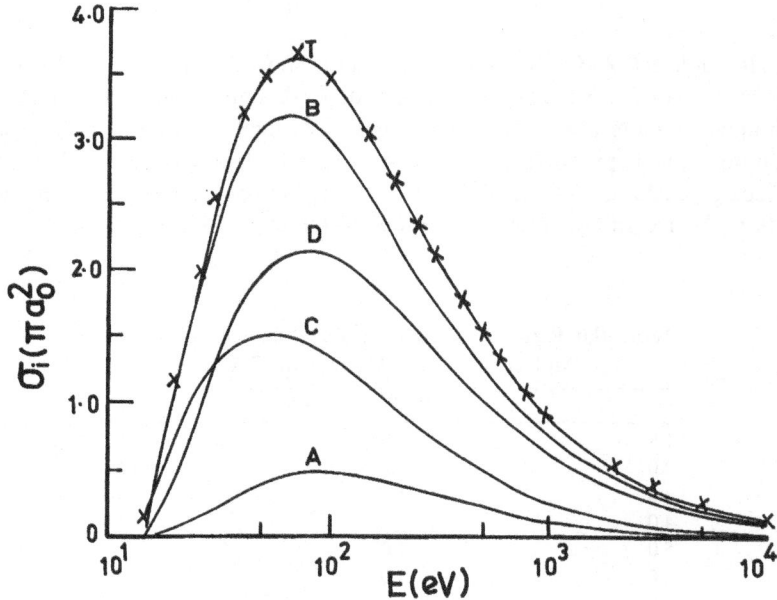

**FIGURE 9.13** Variation of ionization cross sections of the orbital ($1t_2$) of methane with electron energy $E$ in the BEB model. Curves A, B, and T are the cross sections $\sigma_{KHBEB}$, $\sigma_{KHMEB}$, and $\sigma_{KHTB}$, respectively, calculated by Khare et al. (1999). Curves C, D, and X are the cross sections of Kim et al. (1997). Reproduced from "Electron impact ionization of methane," S. P. Khare, M. K. Sharma, and S. Tomar, *J. Phys. B* **32**, 3147, 1999, with permission from the Institute of Physics Publishing, Ltd., Ltd., UK.

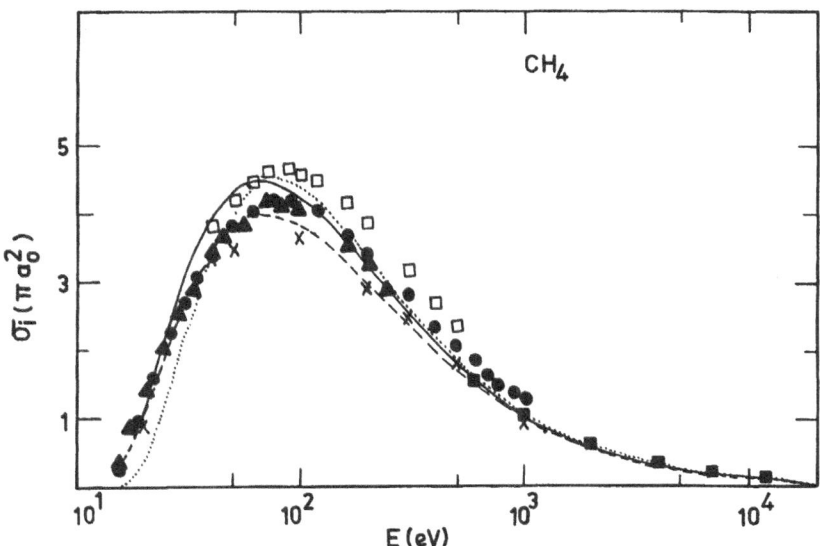

**FIGURE 9.14** Total ionization cross sections of methane due to electron impact in the energy range 10 eV to 20 keV. The solid and dashed curves show $\sigma_{KHTD}$ and $\sigma_{KHTB}$, respectively, of Khare et al. (1999), ... are the cross sections obtained by Khare et al. (1999), using the Saksena model. ●, □, ▲, and X are the experimental cross sections of Rapp and Englander-Golden (1965), Orient and Srivastava (1987), Djuric et al. (1991), and Adamczyk et al. (1966), respectively. Reproduced from the source given in Fig. 9.13, with permission from the Institute of Physics Publishing, Ltd., UK.

in the Saksena model and the experimental data. As expected $\sigma_{KHTB}$ is always greater than $\sigma_{KHBB}$. However, the difference between the two is not more than 10%, which shows that the soft collisions dominate over the hard collisions. The cross sections $\sigma_{TS}$ given by the Saksena model are in good accord with $\sigma_{KHTD}$ for $E$ greater than about 40 eV. However, at lower impact energies, the $\sigma_{KHTD}$ are greater than the $\sigma_{KHTS}$ and give better agreement with the experimental data. Thus the main shortcoming of the Saksena model, i.e., its underestimation at low $E$, has been overcome in the Khare model. Figure 9.15 shows the values of $\sigma_{KHTD}$ for $E$ ranging from 0.1 to 3 MeV. Their comparison with the experimental data of Reike and Prepejchal (1972) shows good agreement between theory and experiment. The values of $M^2$ and $C_{RP}$ obtained by Khare et al. are about 7 and 9% lower than the corresponding experimental values. Thus it may be concluded that the Khare model, obtained by making slight modifications in the Saksena model, is satisfactory for the calculation of $\sigma_i$ of the molecules for impact energy varying from the threshold potential to a few MeV.

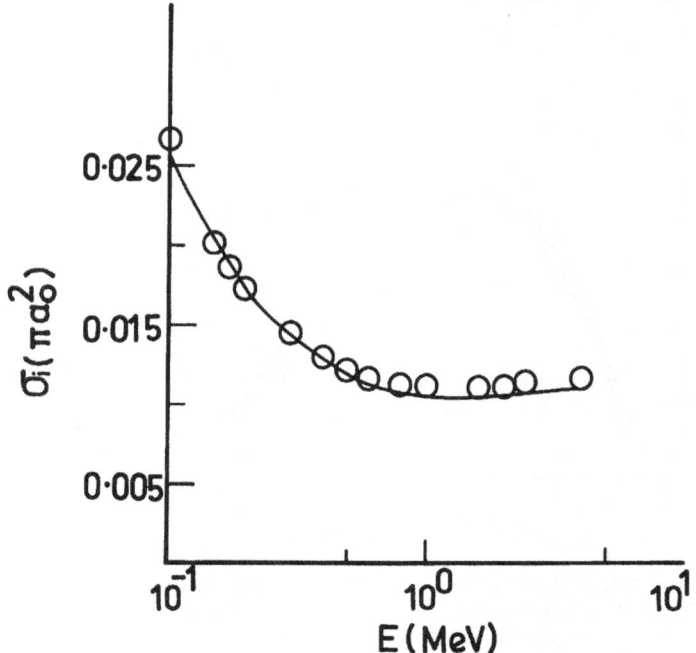

**FIGURE 9.15** Total ionization cross section of methane due to electron impact in the energy range 0.1 to 2 MeV. The solid curve represents theoretical cross sections obtained by Khare et al. (1999) their BED model and O are the experimental data of Reike and Prepejchal (1972). Reproduced from the same source given in Fig. 9.13, with permission from the Institute of Physics Publishing, Ltd., UK.

### 9.7.4   The Deutsch and Mark Model

Deutsch and Mark (1987) modified the classical formula of Grizinski (1965a, b, c) for the atomic ionization cross section and represented it by

$$\sigma = \sum_{nl} g_{nl} \pi r_{nl}^2 N_{nl} f(t) \tag{9.7.52}$$

where $r_{nl}^2$ is the mean square radius of the $(n,l)$ subshell of the atom, which has $N_{nl}$ number of electrons. The function $f(t)$ is given by

$$f(t) = \frac{d}{t}\left(\frac{t-1}{t+1}\right)^a \left\{ b + c\left(1 - \frac{1}{2t}\right)\ln\left[2.7 + (t-1)^{0.5}\right]\right\} \tag{9.7.53}$$

The parameters $a$, $b$, $c$, and $d$ have different values for $s$-, $p$-, $d$-, and $f$- electrons and are given by Deutsch et al. (2000a). The weight factors $g_{nl}$ depend on the quantum numbers $n$ and $l$ and on the ionization energy $I_{nl}$. Margreiter et al. (1990) found that the product $g_{nl}I_{nl}$ is independent of the nuclear charge $Z$ for completely filled subshells. They also extended (9.7.52) to molecules (Margreiter et al., 1990; Deutsch et al., 1994). In their method the ionization cross section of each molecular orbital is expressed in terms of the appropriate atomic weight factor $g_{A(nl)}^j$, effective occupation number $N_{A(nl)}^j$, mean square atomic ratio $r_{A(nl)}^j$, and function $f_{A(nl)}^j(t)$, where $A$ denotes the various constituent atoms of the molecule under investigation (Deutsch et al., 2000a). A Mulliken population analysis is carried out to obtain the values of $g_{A(nl)}^j$, but that does not always result in a unique representation of the molecular orbitals in terms of the atomic orbitals of the constituent atoms. For example, the three outermost molecular shells of $H_2O$ have been represented by four different atomic basis sets, which give rise to significantly different ionization cross sections for its three molecular orbitals (Deutsch et al., 2000a). The application of the Deutsch and Mark model (DM) model to 31 molecules and radicals has been discussed by Deutsch et al. (2000a). The DM model has been also applied to small clusters (Margreiter et al., 1994; Deutsch et al., 2000a). Recently the DM model was used by Deutsch et al. (2000b) to examine isomer effects in the total ionization cross section of cyclopropane and propane ($C_3H_6$). Although the theoretical cross sections are in satisfactory agreement with the experimental data (Nishimura and Tawara, 1994), the model could not reproduce the slight isomer effect obtained experimentally. This model has also been applied to the dimers $S_2$, $F_2$, $Br_2$, $I_2$, and $C_2$; trimers $O_3$ and $C_3$; and fullerenes $C_{60}$ and $C_{70}$ (Deutsch et al., 2000c). Absolute electron impact ionization cross sections of several other molecules such as AlO, $Al_2O$, $WO_x$ ($x = 1-3$), $NO_2$, BF, HX (X = F, Cl, Br, I), TMS (tetramethylsilane), HMDSO (hexamethyldisiloxane), TEOS (tetraethoxysilane) have been obtained by the DM model (Deutsch et al., 2001; Probst et al., 2001).

## 9.8 The Differential Approach for Electron–Molecule Collisions

Molecules are multicenter objects and the optical potential for the scattering of electrons by molecules, even in the SFA, is noncentral. Hence, the partial wave method discussed in the last chapter for the scattering of electrons by atoms is not appropriate for electron–molecule scattering. At intermediate and high energies, with a large number of open channels, the application of the close-coupling method becomes almost impossible, even with present day supercomputers, so such investigations have only been carried out at low impact energies (Lane, 1980; Shimamura and Takayanagi, 1984; Gianturco and Jain, 1986; Burke, 1987; Morrison, 1988; Gianturco,1995). In a number of investigations at intermediate and high energies, spherically symmetric optical potentials have been employed and the method of partial waves was used to obtain collision cross sections. A few such investigations are discussed in the present section.

Let us first consider the scattering of electrons by a potential that is separable in the spheroidal coordinate system. To obtain the phase shifts we have to expand the plane wave $F_0(r)$ and the wave function of the scattered electrons $F(r)$ in $(\lambda, \mu, \varphi)$ coordinates. The expansion of the plane wave is given by

$$F_0(r) = (2\pi)^{-3} \sum_{l,m} a_{ml} S_{ml}(C, \cos\theta_0) S_{ml}(C, \mu) f_{ml}(C, \lambda) \cos m(\varphi - \varphi_0) \quad (9.8.1)$$

where $a_{ml}$ are the expansion coefficients. The $S_{ml}$ are spheroidal harmonics and are the solutions of the following differential equation:

$$\frac{d}{d\mu}\left[(\mu^2 - 1)\frac{d}{d\mu} S_{ml}\right] + \left[A_{ml} + C^2\mu^2 - \frac{m^2}{\mu^2 - 1}\right] S_{ml} = 0 \quad (9.8.2)$$

The parameter $C$ is equal to $k_i R/2$ and $(\theta_0, \varphi_0)$ are the polar angles of the internuclear axis $R$ with respect to the vector $k_i$. As $R$ tends to zero, the spheroidal functions $S_{ml}(C, \mu)$ reduce to the associated Legendre polynomials $P_l^{|m|}(\cos\theta)$. For large values of $\lambda$ the function $f_{ml}(C, \lambda)$ is given by

$$f_{ml}(C, \lambda) \underset{\lambda\to\infty}{\sim} \frac{1}{C\lambda} \sin[\lambda C - \tfrac{1}{2}(m + l)\pi] \quad (9.8.3)$$

It is interesting to compare (9.8.1), (9.8.2), and (9.8.3) with (2.6.24), (2.6.1), and (2.6.15), respectively.

It should be noted that the interaction of the incident electron with the molecule changes $F_0(r)$ to $F(r)$. If the reduced interaction potential energy is approximated by

$$U(r) = \frac{F(\lambda)}{\lambda^2 - \mu^2} \quad (9.8.4)$$

then the expansion of $F(r)$ is given by

$$F(r) = (2\pi)^{-3/2} \sum_l \sum_m a_{ml} \exp(i\eta_l^m) S_{ml}(C, \cos\theta_0) S_{ml}(C, \mu)$$
$$\times T_{ml}(C, \lambda) \cos m(\varphi - \varphi_0) \quad (9.8.5)$$

Now the phase shift $\eta_l^m$ depends upon $l$ as well as on $m$. The function $T_{ml}$ satisfies the following one-dimensional differential equation:

$$\frac{d}{d\lambda}\left[(\lambda^2-1)\frac{d}{d\lambda}T_{ml}\right]+\left[A_{ml}+C^2\lambda^2-\frac{m^2}{\lambda^2-1}-\frac{R^2}{4}f(\lambda)\right]T_{ml}=0 \quad (9.8.6)$$

At large $\lambda$ we have

$$T_{ml}(C,\lambda)\sim\frac{1}{C\lambda}\sin[\lambda C-\tfrac{1}{2}(m+l)\pi+\eta_l^m] \quad (9.8.7)$$

and (3.1.2) modifies to

$$F_0(r)\underset{\lambda\to\infty}{\sim}A\left[e^{ik_i r}+\frac{2f(\mu,\phi;\theta_0,\varphi_0)e^{iC\lambda}}{R\lambda}\right] \quad (9.8.8)$$

The scattering amplitude is given by

$$f(\mu,\varphi;\theta_0,\varphi_0)=\frac{1}{2ik_i}\sum_m\sum_l a_{ml}(-1)^{m+l}[\exp(2i\eta_l^m)-1]$$
$$\times S_{ml}(C,\cos\theta_0)S_{ml}(C,\mu)\cos m(\varphi-\phi_0) \quad (9.8.9)$$

Thus the simple form of the scattering amplitude given by (3.9.13) changes to a more complicated expression because the interaction potential is not spherically symmetric. The scattering amplitude given by (9.8.9) is in the fixed-nuclei approximation. For a nonvibrating (i.e., with a fixed $|R|$) but rotating molecule, the averaged differential cross section is given by

$$\bar{I}(\mu,\varphi)=\frac{1}{4\pi}\int|f(\mu,\varphi;\theta_0,\varphi_0)|^2\sin\theta_0\,d\theta_0\,d\phi_0 \quad (9.8.10)$$

and the total elastic cross section $\sigma_{el}$ is obtained by integrating $\bar{I}$ over $\mu$ and $\varphi$. The partial cross sections also depend upon $l$ as well as $m$, and the $\sigma_{el}$ is given by (Stier, 1932; Fisk, 1936):

$$\sigma_{el}=\sum_m\sum_l\sigma_{lm} \quad (9.8.11)$$

where the partial cross sections are given by

$$\sigma_{lm}=\frac{4\pi}{k_i^2}(2-\delta_{om})\sin^2\eta_l^m \quad (9.8.12)$$

A comparison of (9,8.11) and (9.8.12) with (3.9.25) and (3.9.26) again shows the difference due to the replacement of a spherically symmetric potential by a non-central potential. However, as with a central potential, here too all the phase shifts except $\eta_o^o$ go to zero as $k_i \rightarrow 0$ and

$$\sigma_{\mathrm{el}} \underset{k_i \rightarrow 0}{\rightarrow} \sigma_{oo} = 4\pi a_s^2 \qquad (9.8.13)$$

where $a_s$ is the scattering length.

## 9.8.1    The Independent Atom Model

As molecules are diatomic or polyatomic objects, it is tempting to investigate electron–molecule scattering by replacing the molecule by its constituent atoms. In a number of simple investigations to obtain the DCS for a molecule, the following two approximations were employed:

1. Linear combination of atomic differential cross sections (LCADCS).

2. Linear combination of atomic scattering amplitudes (LCASA).

In the LCADCS, the molecular differential cross section $I_M(\theta)$ for a molecule having $N$ atoms is given by

$$I_M(\theta, \varphi) = \sum_{a=1}^{M} I_a(\theta, \varphi) \qquad (9.8.14)$$

where $I_a(\theta, \varphi)$ is the differential cross section for the scattering of the incident electron by the $a$th atom of the molecule. The above equation follows the additivity rule (AR) at the macroscopic level. In this approximation we represent $I_M(\theta, \varphi)$ by $I_{AR}(\theta, \varphi)$. On the other hand, in the LCASA, known as the independent atom model (IAM), the molecular scattering amplitude $f_M(\theta)$ is taken as a linear combination of the atomic scattering amplitude $f_a(\theta)$:

$$f_M(\theta) = f_{\mathrm{IAM}}(\theta) = \sum_{a=1}^{N} C_a f_a(\theta) \qquad (9.8.15)$$

where $C_a$ are the expansion coefficients and are determined in the following manner.

Let the incident electrons traveling in the $z$ direction with momentum $\hbar k_i$ be scattered in the direction $(\theta, \varphi)$ and the final momentum be $\hbar k_f$. Now, asymptotically, the scattered wave due to the $a$th atom with the $i$th nucleus at the origin, is given by

$$\psi \sim A\left[e^{ik_i \cdot (|r - r_a|)} + f_a(\theta, \varphi)\frac{e^{ik_i|r - r_a|}}{|r - r_a|}\right] \tag{9.8.16}$$

where $r$ and $r_a$ are the coordinates of the point of observation $P$ and the $a$th nucleus, respectively (see Fig. 9.16). If we had taken $O$ as the center, the plane wave would have been $Ae^{ik_i \cdot r}$, so to write the above equation in the same form we multiply it by $e^{ik_i \cdot r_a}$ and get

$$\psi \sim A\left[e^{ik_i r} + f_a(\theta, \varphi)\frac{e^{ik_i|r - r_a|}}{|r - r_a|}e^{ik_i r_a}\right] \tag{9.8.17}$$

Since $r \gg r_i$ the above equation reduces to

$$\psi \sim A\left[e^{ik_i r} + f_a(\theta, \varphi)e^{ik_i r_a}\frac{e^{ik_i r}}{r}\right] \tag{9.8.18}$$

Comparing (9.8.15) for the $a$th with (9.8.18) we obtain

$$C_a = e^{iK \cdot r_a} \tag{9.8.19}$$

Thus the expansion coefficients are only phase terms. Finally, the molecular differential cross section in the IAM for elastic scattering is given by

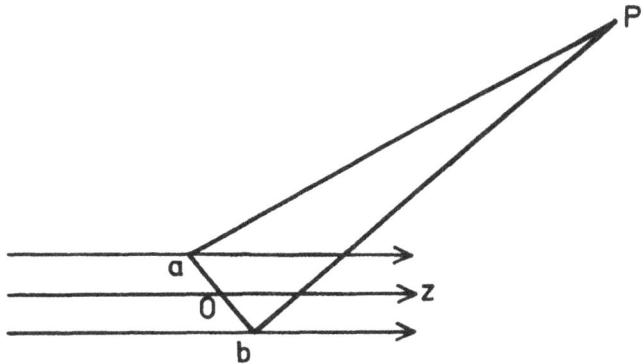

**FIGURE 9.16** Scattering of electrons by two atomic centers a and b.

$$I_{IAM} = \sum_{a,b} f_a^*(\theta, \varphi) f_b(\theta, \varphi) e^{iK \cdot (r_b - r_a)}$$

$$= I_{AR} + \sum_{a \neq b} f_a^*(\theta, \varphi) f_b(\theta, \varphi) e^{iK \cdot (r_b - r_a)} \qquad (9.8.20)$$

We note from (9.8.20) that to obtain $I_{AR}$, the differential cross sections $I_a$ are added (incoherent addition), but that to obtain $I_{IAM}$, the scattering amplitudes $f_a$ multiplied by a phase vector $e^{ik \cdot r_a}$ are added (coherent addition). Thus in the IAM interference of scattering waves originating from the different atoms is taken into account and the geometry of the molecule enters into the calculation through the internuclear axis $R$. These features are neglected in the additivity rule. Both these approximations assume that:

1. Each atom scatters independently.

2. Any redistribution of atomic electrons due to molecular binding is unimportant, so that each atom scatters as if it were free.

3. Multiple scattering within the molecule is negligible.

For the above assumptions to be valid the de Broglie wavelength of the incident electron should be small in comparison to the inner atomic distances. Hence, both approximations are high-energy approximations. Equation (9.8.20) is for a fixed nuclear axis. Since the molecule rotates, $I_{IAM}$ $(\theta, \varphi)$ is *averaged* over all the orientations of the molecular axis. We choose $K$ as the axis of reference, integrate (9.8.20) over all possible orientations of $R$, and divide the result by $4\pi$ to get the average value of the molecular differential cross section. Thus

$$\bar{I}_{IAM}(\theta, \varphi) = \left[ \sum_a |f_a|^2 + \sum_{a \neq b} f_a^*(\theta, \varphi) f_b(\theta, \varphi) \frac{\sin K|r_a - r_b|}{K|r_a - r_b|} \right] \qquad (9.8.21)$$

The $f_a$ are usually obtained by using the partial wave method with some suitable optical potential, as discussed in the Chapter 8.

Jain and Khare (1977) investigated the elastic scattering of electrons by the ground state hydrogen molecule in IAM. However, to take account of valence–bond distortion they took a hydrogenic atom with $Z = 1.193$ instead of a hydrogen atom ($Z = 1$). Hence, from (9.8.21),

$$\bar{I}_{H_2} = 2 I_H(Z = 1.19)\left(1 + \frac{\sin KR}{KR}\right) \qquad (9.8.22)$$

They solved the radical Schrödinger equation due to a central optical potential. The optical potential was taken as the sum of a static field, a local exchange

**Table 9.7** Comparison of the Ratios $\lambda = I_{H_2}^{Exp}/I_H^{Th}$ ($Z = 1.193$) with $f(KR) = 2(1 + \sin KR/KR)$ for the Elastic Scattering of 100-eV Electrons by the Hydrogen Molecule

| $\theta$ (deg) | $\lambda$ | $f(KR)$ | $\theta$ (deg) | $\lambda$ | $f(KR)$ |
|---|---|---|---|---|---|
| 5 | 4.31 | 3.96 | 70 | 1.36 | 1.57 |
| 10 | 4.39 | 3.86 | 80 | 1.38 | 1.60 |
| 20 | 3.53 | 3.47 | 90 | 1.48 | 1.71 |
| 30 | 2.82 | 2.94 | 100 | 1.62 | 1.85 |
| 40 | 2.23 | 2.40 | 110 | 1.70 | 1.98 |
| 50 | 1.58 | 1.96 | 120 | 1.80 | 2.09 |
| 60 | 1.47 | 1.68 | 130 | 1.87 | 2.16 |

potential with $Z = 1.193$, and an energy-ependent polarization potential. To compare their results with the experimental data they obtained the ratios $\lambda = I_{H_2}^{Ex}/I_H^{Th}$. ($Z = 1.193$), where $I_{H_2}^{Ex}$ is the experimental DCS for hydrogen molecules. These ratios are compared with $f(KR) = 2(1 + \sin KR / KR)$. Jain and Khare used the experimental values of van Wingerden et al. (1977) for $\theta \le 50°$ and those of Fink et al. (1975) (renormalized by van Wingerden et al.) for $I_{H_2}^{Ex}$. Their values of $\lambda$ at 100 eV are shown in Table 9.7 along with $f(KR)$. The table shows that the differences between the ratio $\lambda$ and $f(KR)$ are less than 15% at all angles except 50°, where the difference is about 20%. These differences are close to the accuracy of the experimental data for $I_{H_2}$. We also note oscillations both in $\lambda$ and $f(KR)$ as a function of $\theta$; these are due to the diffraction of the electron waves by the two-center molecular object.

To include the effect of the vibration of the molecule on the DCS we take

$$\bar{I}_M(\theta) = \sum_{a,b}^{N} f_a^*(\theta) f_b(\theta) \int P_{ab}(R) \frac{\sin KR}{KR} dR \qquad (9.8.23)$$

where the function $P_{ab}(R)$ is the probability that the two nuclei are separated by a distance $R$. A suitable expression for $P_{ab}(R)$ is given by Kuchitsu and Bartell (1961). Equation (9.8.23) has been utilized by Khare and Raj (1979) to investigate the elastic scattering of electrons by the heteronuclear CO molecule. Assuming that the vibrational distribution function of the molecule is harmonic, we get from (9.8.23)

$$\bar{I}_{CO}(\theta) = I_C(\theta) + I_O(\theta) + \left[ f_C^*(\theta) f_O(\theta) + f_C(\theta) f_O^*(\theta) \right]$$
$$\times \exp(-\tfrac{1}{2} l_e^2 K^2) \{ \sin[K(R - l_e^2/R)]/KR \} \qquad (9.8.24)$$

where $l_e$ is the mean vibrational amplitude. Khare and Raj (1979) employed the partial wave method to calculate $f_C(\theta)$ and $f_O(\theta)$. The optical potentials were taken to be the sum of the static potential and an energy-dependent polarization potential $V_{dp}(r)$. For the static potential, Strand and Bonham's (1964) equation [Eq. (8.3.12)] was utilized and $V_{dp}(r)$ was represented by Eq. (7.7.37). Their results for $E = 300\,eV$ are shown in Fig. 9.17 along with the experimental data of Bromberg (1970). The agreement between theory and experiment is satisfactory. For a better comparison the ratio $\lambda = I_{CO} / (I_C + I_O)$ for $E = 400\,eV$ is shown as a function of $\theta$ in Fig. 9.18.

The experimental points in Fig. 9.18 have been obtained by dividing experimental $I_{CO}$ by the theoretical values of $I_C$ plus $I_O$. In this figure the shape of the theoretical curve is in very good agreement with the experimental data. Almost all the maxima and minima are faithfully reproduced by the theory. Quantitatively, the theory overestimates the value of $\lambda$ for $\theta$ greater than about $10°$. Similar investigations have been carried out by Raj (1991a,b) and Khare and Raj (1982) for $O_2$, $CO_2$, and $CF_4$ molecules.

The above investigations do not include exchange and absorption effects. These two effects were included by Khare et al. (1994c) in the elastic scattering of electrons by the $CF_4$ molecule at intermediate energies. They employed $V_{ex}(r)$, given by Riley and Truhlar (1976). For the imaginary part of the optical potential $V_{abs}(r)$, they employed the nonempirical expression derived from a quasi-free scattering model by Staszewska et al. (1983).

The theoretical cross sections of Raj (1991a) and Khare et al. (1994c) for $E = 200$ and $300\,eV$ are shown in Fig. 9.19 along with the experimental data of Sakae et al. (1989). The figure shows that the two theoretical curves have almost the same nature but the agreement between theory and experiment improves significantly when the exchange and absorption effects are included. However, at higher energies the values of the theoretical cross sections are lower than the experimental data [see Fig. 3 of Khare et al. (1994c)].

Recently Raj and Kumar (2001) pointed out that in general the absorption potential of Staszewska et al. (1983) underestimates the elastic DCS at intermediate and large scattering angles. Further, the underestimation becomes worse with the increase in $E$. Such behavior is contrary to expectation. Underestimation has also been noted for the integrated elastic cross section $\sigma_{el}$ and the momentum cross section $\sigma_m$. To improve the agreement between theory and experiment Raj and Kumar (2001) divided the absorption potential of Staszewska et al. (1983) [given by (8.7.8)] by $k$. They used this modified absorption potential in their IAM calculation to obtain the DCS, $\sigma_{el}$, and $\sigma_m$ for the elastic scattering of electrons by $O_2$ molecules in the energy range from 300 to 1000 eV. Their cross sections with the modified absorption potential are in much better agreement with experimental data (see Figs. 1(a) to (d) and Table 1 of their paper). It may be noted that these methods neglect multiple scattering within the molecule. A method to include these scattering terms was provided by Hayashi and Kuchitshu (1976).

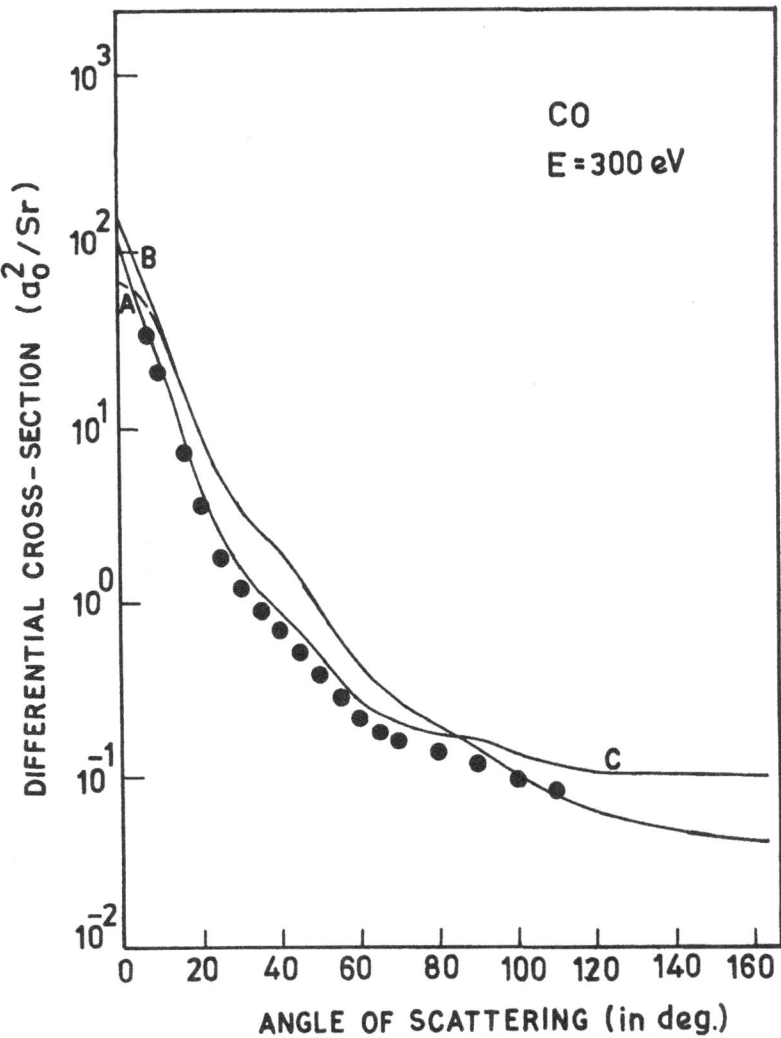

**FIGURE 9.17** Differential cross sections for the elastic scattering of 300-eV electrons from a CO molecule obtained by Raj (1981) in the independent atom model. For atoms, the FBA (curve A), the PWA (curve B), and the SFPA (curve C) were employed. ● represents the experimental data of Bromberg (1970).

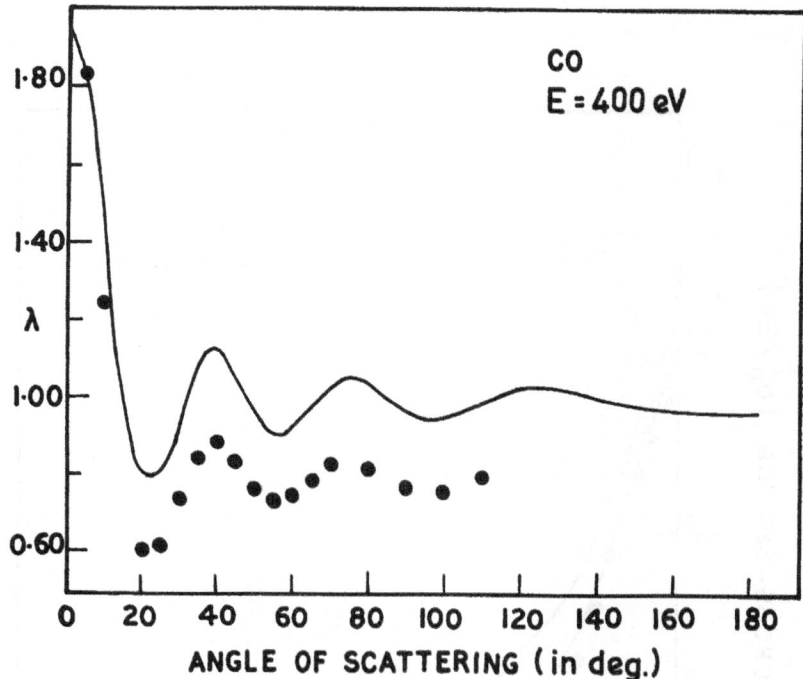

**FIGURE 9.18** Variation of the ratio $\lambda = \bar{I}_{CO}(\theta)/(I_C(\theta) + I_O(\theta))$ at 400 eV. For the solid curve $I_C(\theta)$, $I_O(\theta)$, and $\bar{I}_{CO}(\theta)$ are theoretical. For ●, $I_C(\theta)$ and $I_O(\theta)$ are theoretical but $\bar{I}_{CO}(\theta)$ are the experimental data of Dubois and Rudd (1976). All the theoretical cross sections were obtained by Raj (1981).

To obtain the total collision cross section $\sigma_T$, the optical theorem is utilized. For this we require the imaginary part of the elastic scattering amplitude in the forward direction. In this direction $\exp(iK \cdot r_i) f_i(\theta, \varphi)$ is equal to $f_i(\theta)$. Thus the total cross section $\sigma_T$, obtained by the additivity rule (AR) and from the independent atom model (IAM) are the same.

The AR approach and its modifications were used by Joshipura and his associates (Joshipura and Patel, 1994, 1996; Joshipura and Vinodkumar (1997a,b); Joshipura (1998); Joshipura and Vinodkumar (1999); Joshipura et al. (1999, 2001) for a good number of molecules and radicals for electron energies varying from about 20 eV to a few keV. They started with the atomic charge density to construct all three short-range potentials, namely the static potential $V_{SF}$, the local exchange potential $V_{ex}$, and the absorption potential $V_{ab}$. For $V_{ex}$ the Hara free-electron gas model was employed and for $V_{ab}$ Eq. (8.7.8), derived by Staszewska et al. (1983), was used. For $E > 100$ eV, (7.7.37) or only its dipole part represented the long-range polarization potential $V_{pol}$. At lower impact energies, (7.7.37) overestimates $V_{pol}$, so it was replaced by the correlation polarization potential (Padial and Norcross, 1983). The sum of all four potentials was the complex optical potential $V_{op}(r)$. Complex phase shifts $\eta_l$ for various values of $l$ were obtained for this potential, and from (3.2.18) and (3.9.13) for the $j$th atom of the molecule we have

$$\sigma_T^j = \frac{4\pi}{k^2} \operatorname{Im} f_j(\theta = 0)$$

$$= \frac{4\pi}{k^2} \operatorname{Im} \sum_l (2l+1)\exp(i\,\eta_{jl})\sin\eta_{jl}$$

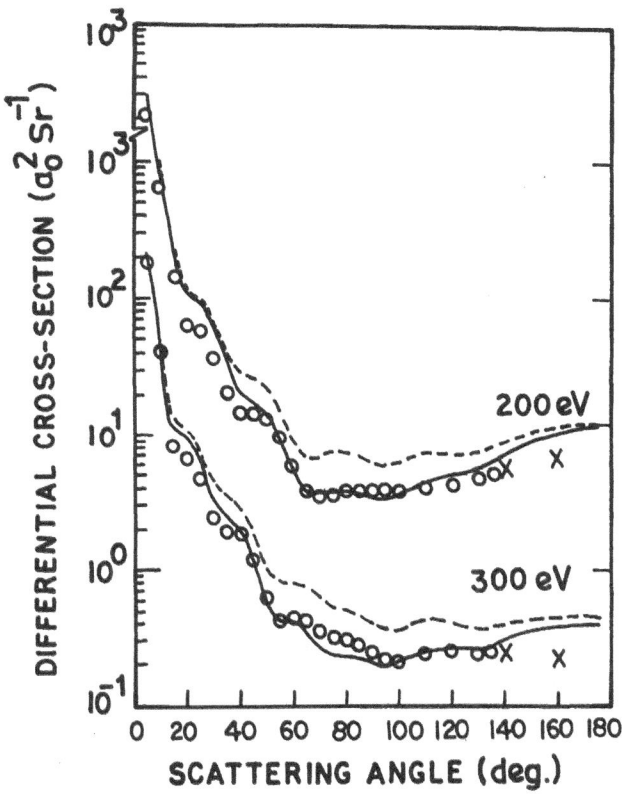

**FIGURE 9.19** Angular dependence of the differential cross section for e–CF$_4$ elastic scattering at 200- and 300-eV impact energies. Theory: solid curve, Khare et al. (1994c); broken curve, Raj (1991a). Experiment: open circles, Sakae et al. (1989); crosses, extrapolated data. Reproduced from "Absorption effects in the elastic scattering of electrons by the CF$_4$ molecule at intermediate energies," S.P. Khare, D. Raj and P. Sinha, *J. Phys. B* **27**, 2569, 1994, with permission from the Institute of Physics Publishing, Ltd., UK.

Finally, for the molecule

$$\sigma_T^M = \sum_j \sigma_T^j$$

The values of $\sigma_T^M$ for a few molecules studied by Joshipura and his associates are given in Table 9.8. Sun et al. (1994), Jiang et al. (1995), and Liu and Sun (1996) also employed the AR approach.

### 9.8.2  Modified Additivity Rule

It should be noted that both in the AR and the IAM, atomic polarizabilities are employed in the construction of the long-range polarization potential. Quite often the polarizabilities of the constituent atoms are quite different from the molecular polarizability. For example, the average spherical dipole polarizability of the CO molecule is $13.17a_0^3$, whereas the polarizabilities of the free atoms C and O are $14.17a_0^3$ and $5.2a_0^3$, respectively. Thus the polarization effect is not adequately treated in the AR and in the IAM. This effect contributes significantly in the forward direction, which is used to evaluate $\sigma_T$. Hence, Joshipura and Patel (1996) proposed the modified additivity rule (MAR), in which the spherical part of the electron–molecule interaction is taken as

$$V^{SP} = V_{SR} + V_{LR} \qquad (9.8.25)$$

where the short-range interaction potential $V_{SR}$ is

$$V_{SR} = \sum_j V_{SR}^j \qquad (9.8.26)$$

and $V_{SR}^j$ is the sum of the static field, exchange, and absorption potentials for the $j$th atom of the molecule. The long-range potential $V_{LR}$ is due to the polarization of the molecule. Joshipura and Patel took only the dipole part of the polarization potential given by (7.7.37) for $V_{LR}$, with $\alpha_d$ as the average spherically symmetric dipole polarizability of the molecule. Using the partial wave method the imaginary parts of the elastic scattering amplitudes one obtains $f_{SR}(\theta = 0)$ due to $V_{SR}^j$ and $f_{LR}^{pol}(\theta = 0)$ due to $V_{LR}$. These amplitudes are converted into $\sigma_T^j$ and $\sigma_T^{pol}$ with the help of the optical theorem. Finally in the MAR, the total collision cross section for the scattering of electrons by the spherically symmetric potential of the molecule is given by

$$\sigma_T^{SP} = \sum_j \sigma_T^j + \sigma_T^{pol} \qquad (9.8.27)$$

It should be noted that above equation considers only the spherical part of the interaction. At the low impact energies the rotational excitation of heteronuclear molecules, involving the permanent dipole moment, becomes important. For example, Jain (1988) found that for $NH_3$ the rotational excitation cross section $\sigma_{0\rightarrow1}$ from $J = 0$ to $J = 1$ is about 17% of the measured cross section $\sigma_T$ at 100 eV. Joshipura and his associates included this dipole contribution in (9.8.27) through the FBA. Due to the dipole moment $D$, the asymptotic dipole potential is equal to $-D \cdot r/r^2$. For this potential, the excitation cross section $\sigma_{0\rightarrow1}$ in the FBA is

$$\sigma_{0\rightarrow1} = \frac{8\pi D^2}{3k_i^2} \ln\left(\frac{k_i - k_f}{k_i + k_f}\right) \qquad (9.8.28)$$

This contribution is added to (9.8.27). Hence, for polar molecules the total cross section, in MAR, is given by

$$\sigma_T = \sum_j \sigma_T^j + \sigma_T^{\text{pol}} + \sigma_{0\rightarrow1} \qquad (9.8.29)$$

We note that $\sigma_T^j$ is due to the constituent atoms but $\sigma_T^{\text{pol}}$ and $\sigma_{0\rightarrow1}$ depend upon the molecular properties such as dipole polarizability, dipole moment, and ionization potential. Thus the MAR is a better model than the AR, and it has been applied by Joshipura and his associates to a number of molecules. Some of these results are shown in Table 9.8 and will be discussed further on.

### 9.8.3 The Single-Center Charge Density Method

Jain (1986, 1987, 1988) employed the spherical complex-optical potential (SCOP) to calculate $\sigma_{\text{el}}$ and $\sigma_{\text{ab}}$ in the intermediate- and high-energy range for electron–molecule collisions. Encouraged by its success Jain and Baluja (1992) extended it to many polar as well as nonpolar molecules. They considered those molecules for which the molecular wave functions are available at the Hartre–Fock level. Using these wave functions the charge density $\rho(r)$ was obtained and utilized to construct the static field, local exchange, polarization, and absorption potentials. The sum of these four potentials constituted the optical potential. Due to the multicenter nature of the molecules their $V_{\text{op}}(r)$ was non-spherical. Jain and Baluja expanded $V_{\text{op}}(r)$ around the center of mass of the molecule and considered only the spherical term of the expansion. With this complex spherical interaction potential, the scattering matrix $S_l$ for each partial wave was obtained. Finally (3.9.19), (3.9.23), and (3.9.24) were used to find $\sigma_{\text{el}}^s$, $\sigma_{\text{ab}}^s$, and $\sigma_T^s$, respectively. They employed the FBA to include the contribution of the dipole and quadrupole terms obtained in the multipole expansion of $V_{\text{op}}(r)$. The cross sections $\sigma_{0\to1}$ and $\sigma_{0\to2}$ due to anisotropic dipole and quadrupole terms were added incoherently to $\sigma_T^s$ to get $\sigma_T$. Thus

$$\sigma_T = \sigma_T^s + \sigma_{0\to1} + \sigma_{0\to2} \qquad (9.8.30)$$

For polar molecules the contribution of the anisotropic terms becomes quite important for $E < 100\,\text{eV}$. Hence, the incoherent addition is likely to introduce error in $\sigma_T$ at low energies. The *SCOP* method has been applied to $H_2$, $Li_2$, HF, $CH_4$, $N_2$, CO, $C_2H_2$, HCN, $O_2$, HCl, $H_2S$, $PH_3$, $SiH_4$, and $CO_2$ molecules in the energy range from $10\,\text{eV}$ to $5\,\text{keV}$. The results will be discussed a little later.

Kumar et al. (1995) and Jain and Tripathi (1997) followed the procedure of Jain et al. (1991) to investigate the scattering of electrons by $GeH_4$ (germane) and $SiH_4$ (silane) molecules, respectively. They also employed the molecular charge density, which was obtained by using nonrelativistic multicenter molecular wave functions at the Hartree-Fock level, to construct their spherically symmetric $V_{SF}, V_{ex}, V_{pol}$, and $V_{ab}$ potentials. However, as they solved the Dirac equation instead of the nonrelativistic Schrödinger equation, the spin-orbit interaction was automatically included. They calculated the elastic DCS, $\sigma_{el}$, $\sigma_m$ (the momentum cross section), and $\sigma_T$ for the energy of an incident electron ranging from a few to several hundred eV. For the heavier $GeH_4$ molecule they took $V_{ab} = 0$. Thus for this molecule their $\sigma_{el}$ and $\sigma_T$ are equal. They also obtained the polarization parameters $P$, $T$, and $U$.

Figure 9.20 shows the theoretical elastic DCS of electrons scattered by $GeH_4$ obtained by Kumar et al. (1995), by Jain et al. (1991) in their SFPE model, and by Dillon et al. (1993) with their continuum multiple scattering (CMS) model, along with the experimental data. It is evident from the figure that the theoretical cross sections of Dillon et al. are in best agreement with the experimental data. The agreement between the theoretical cross sections of Kumar et al. and the experimental data is also quite satisfactory. However, these cross sections differ strongly from the CMS cross sections in the backward direction. The SFPE model of Jain et al. (1991) is found to underestimate the cross sections over most of the angular range. Their model does not include the spin–orbit interaction. Its inclusion by Kumar et al. has improved the agreement between theory and experiment. The theoretical curves of $P$, $T$, and $U$ obtained by Kumar et al. also exhibit structure, which is in accord with expectations.

Kumar et al. (1995) have compared their DCS, $P$, $T$, and $U$ of $GeH_4$ with those of the Ge atom and isoelectronic Kr atom. They noted that their $GeH_4$ results are close to those of the Ge atom but differ significantly from those of the isoelectronic Kr atom. From their observations they concluded that for the heavy molecule $GeH_4$ the four hydrogen atoms hardly contribute to the scattering process.

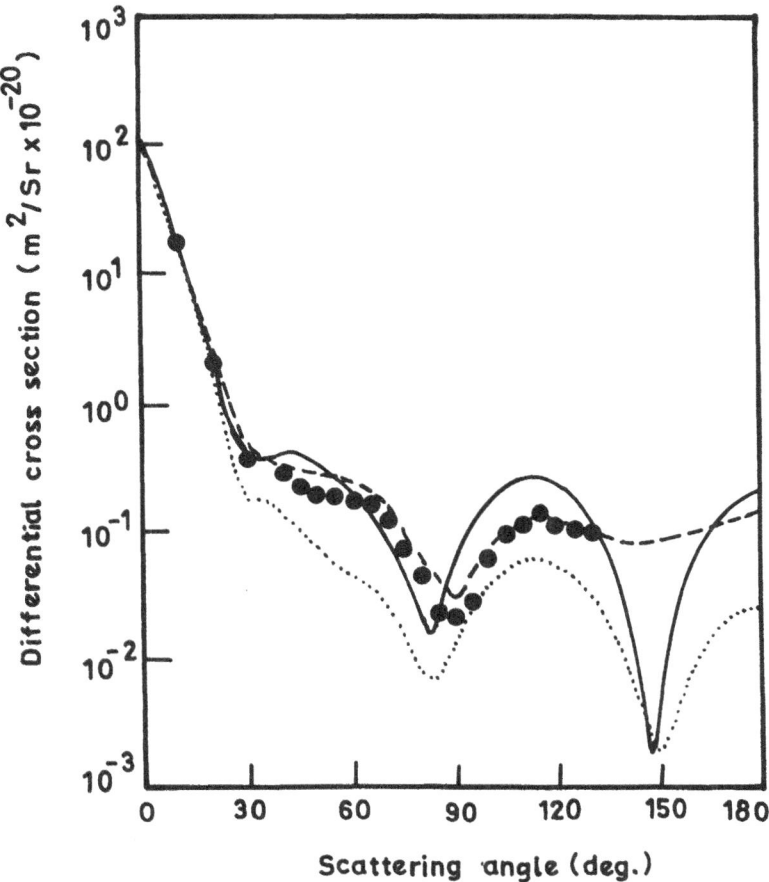

**FIGURE 9.20** Differential cross section for elastic scattering of electrons from the $GeH_4$ molecule at 100 eV. Calculations: ——— Kumar et al. (1995); . . . . , SFPE results from Jain et al. (1991), - - - - CMS results from Dillon et al. (1993). Experiment: ●, Dillon et al. (1993). Reproduced from "Spin polarization and cross sections of electrons elastically scattered from germane molecules," P. Kumar, A. K. Jain, and A. N. Tripathi, *J. Phys. B* **28**, L387, 1995, with permission from the Institute of Physics Publishing, Ltd., UK.

The results obtained by Jain and Tripathi (1997), Tanaka et al. (1990) and Jain (1987) for $SiH_4$ molecule are similar to those for the $GeH_4$ molecule. However, for the relatively lighter molecule $SiH_4$, the four covalent hydrogen bonds introduce additional features into the DCS and polarization parameters (Jain and Tripathi, 1997).

The single-center method was also utilized by Joshipura and Vinodkumar (1997a), but they started with the atomic charge density rather than the molecular charge density to calculate $\sigma_T$ for HF, OH, NH, and CH hydrides, which are highly reactive molecules. Following Watson (1958) they expanded the charge density of the hydrogen atom about the nucleus of the heavier atom. To take an approximate account of the valence bond effect they represented the hydrogen atom by (6.11.1) with $Z_s = \sqrt{I_M}$, where $I_M$ is the ionization potential of the molecule in Rydberg units. The spherical part of the expanded charge density $\rho_H(r; Z_s, R)$ was added to the charge density of the heavier atom A to obtain the molecular charge density $\rho_M(r; Z_s, R)$. Thus

$$\rho_M(r; Z_s, R) = \rho_A(r) + \rho_H(r; Z_s, R) \tag{9.8.31}$$

where $R$ is the internuclear distance. This charge density was utilized to construct the spherically symmetric and real static field, the local exchange potential, and the imaginary (absorption) potential. For $E > 100\,eV$ the polarization potential was represented by the dipole part of (7.7.37) with the molecular dipole polarizability. At lower energies an independent correlation polarization potential (Jiang et al., 1995) was used. The sum of the above four potentials was their spherically symmetric optical potential. With this potential the method of partial waves was used to obtain $\sigma_T^s$. To this cross section $\sigma_{0\to1}$, obtained in the FBA, was added to get $\sigma_T$:

$$\sigma_T = \sigma_T^s + \sigma_{0\to1} \tag{9.8.32}$$

Joshipura and Vinodkumar (1997a) used the above SC method to calculate $\sigma_T$ for HF, OH, NH, and CH molecules in the energy range 30–2000 eV. Their results are shown in Table 9.8.

Joshipura and Vinodkumar (1999) modified their SC method for molecules having two heavy atoms A and B and a number of light atoms for example $AH_nBH_m$. In such cases the molecule is divided into two groups. For the above molecule, $AH_n$ and $BH_m$ are taken as the two groups. Now the charge densities of the $n$ hydrogen atoms are expanded about the atom A and those of $m$ hydrogen atoms are expanded about the atom B. Then from (9.8.31),

$$\rho_{AH_n}(r; Z_s, R_{A-H}) = \rho_A(r) + n\rho_H(r; Z_s, R_{A-H}) \tag{9.8.33a}$$

and

$$\rho_{BH_m}(r; Z_s, R_{B-H}) = \rho_B(r) + m\rho_H(r; Z_s, R_{B-H}) \tag{9.8.33b}$$

where $R_{A-H}$ and $R_{B-H}$ are the bond lengths for A–H and B–H combinations, respectively. The bond length $R_{A-B}$ is usually greater than $R_{A-H}$ and $R_{B-H}$, which may be different from each other. Now using the SC approach $\sigma_T(AH_n)$ and $\sigma_T(BH_m)$ are calculated. Finally, following the MAR approach we get

$$\sigma_T(AH_nBH_m) = \sigma_T(AH_n) + \sigma_T(BH_m) \qquad (9.8.34)$$

Since this method combines the MAR and the SC approaches, it is known as the MAR–SC approach. Joshipura and Vinodkumar (1999) have employed the MAR as well as the MAR–SC methods to obtain $\sigma_T$ for $C_2H_2$, and $C_2H_4$, and $CH_3X$ molecules (X = $CH_3$, $NH_2$, OH, and F) due to electron impact in the energy range from 30 to 1000 eV. Their values are shown in Table 9.8. It may be noted that for $CH_3X$ molecules $n$ is always 3 but $m$ = 3, 2, 1, and 0 for $CH_3$, $NH_3$, OH, and F groups, respectively.

In both the above methods $\rho_H$, in (9.8.31) and (9.8.33) is the charge density of a free hydrogen atom. However, in the formation of the molecular covalent bond A–H there is a readjustment of the electronic charge. To take this readjustment into account Joshipura et al. (1999) modified (9.8.33) to

$$\rho'_{AH_n}(r, R_{A-H}) = f_A\rho_A(r) + nf_H\rho_H(r, R_{A-H}) \qquad (9.8.35)$$

where

$$f_A = 1 + nN(H, A)/N(A) \qquad (9.8.36)$$

and

$$f_H = 1 - N(H, A)/N(H) \qquad (9.8.37)$$

where $N(A)$ and $N(H)$ are the number of electrons in the free atoms A and H, respectively, and $N(H, A)$ is the number of electrons transferred from each hydrogen atom to atom A. It is easy to verify that $f_A$ and $f_H$, given by (9.8.33), ensure the conservation of electrons in the molecule. $N(H, A)$ have been tabulated by Bader (1990) for a number of atoms. Using this MAR–SCCT (modified additivity rule–single center-charge transfer) approach. Joshipura et al. (1999) calculated $\sigma_T$ for $CH_4$, $SiH_4$, $F_2$, $H_2S$, and $C_2H_6$ molecules. Some of their cross sections are shown in Table 9.8.

Let us now consider Table 9.8, where the $\sigma_T$ obtained by the AR, MAR, SC, AR–SC, MAR–SCCT, and SCOP methods are shown in the energy range 100–1000 eV. Representative experimental cross sections are also shown for comparison. The table shows that for $O_2$, $N_2$, CO, and $CO_2$ molecules the $\sigma_T$ (AR) are smaller than the $\sigma_T$ (SCOP) and closer to the experimental data. However, for $SiH_4$, $CH_4$ and HF molecules the $\sigma_T$ (AR) are greater than the $\sigma_T$ (SCOP). For $SiH_4$ the SCOP method yields better cross sections. But for $CH_4$, at $E$ = 100 and 300 eV the $\sigma_T$ (Exp.) lie between the $\sigma_T$ (SCOP) and the $\sigma_T$ (AR) and at $E$ = 500 and 700 eV, the $\sigma_T$ (AR) are closer to the experimental data. No experimental data exists for the HF molecule.

A comparison between $\sigma_T$ and (AR) and $\sigma_T$ (MAR) is possible for the CO, NO, $NO_2$, $CO_2$, $NH_3$, $CH_4$, $C_2H_2$, $CH_3OH$, $C_2H_4$, $CH_3F$, HF, OH, CH, and NH molecules. For all of them the $\sigma_T$ (MAR) are less than the $\sigma_T$ (AR) at low values of the impact energy but are practically the same at high $E$. It is also to be noted that for almost all the molecules, the $\sigma_T$ (MAR) are closer to the experimental data in comparison of to the $\sigma_T$ (AR). This shows that the effect replacing the atomic polarizabilities by the molecular polarizability reduces the value of $\sigma_T$ and, in general, yields better agreement with experiment. Thus out of the AR, MAR, and SCOP methods, the MAR method seems to give best cross sections.

Let us now compare $\sigma_T$ (MAR) with $\sigma_T$ (MAR–SC). Values for these two sets of the cross sections are available for $C_2H_2$, $CH_3OH$, $C_2H_4$, $CH_3F$, and $CH_3NH_2$. In all these cases we find that the $\sigma_T$ (MAR–SC) are smaller than the $\sigma_T$ (MAR). For all the molecules, with the exception of $C_2H_4$ and $CH_3OH$, the $\sigma_T$ (MAR–SC) are in better agreement with the experimental data in comparison with the $\sigma_T$ (MAR). It is noted that in a molecule such as $C_2H_6$, where the C–C bond length is appreciably greater than the C–H bond length the MAR–SC method is quite successful. Thus for the energy range covered in Table 9.8 we may conclude that, in general, the overestimation of the cross sections by the different theoretical method decreases as we move from the AR method to the MAR method and then to the MAR–SC method.

Joshipura et al. (1999) have calculated $\sigma_T$ for $CH_4$, $SiH_4$, $F_2$, $H_2S$, and $C_2H_2$ molecules in their MAR–SCCT model. As already discussed, this model considers partial charge transfer from the hydrogen atom to the heavier atom of the molecule during the formation of the molecular bonds. A comparison of the $\sigma_T$ (MAR) and $\sigma_T$(MAR–SCCT) shows that the use of single-center charge density with charge transfer reduces the cross sections and gives better agreement with the experimental data.

## 9.8.4    The Two-Parameter Fit for the Total Collision Cross Section $\sigma_T$

Trajmar et al. (1983), Christophorou (1984), Stein and Kauppila (1986), Sueoka (1987), Szmytkowski (1989), and Jain and Baluja (1992) have summarized the experimental data on electron–molecule systems. Recently Szmytkowski et al. (1997) and Karwasz et al. (1999) measured $\sigma_T$ for a number of molecules. For quite some time there have been attempts to correlate $\sigma_T$ with some microscopic target properties. Floeder et al. (1985) measured $\sigma_T$ for a series of hydrocarbons and showed that for $E$ between 100 and 400 eV the cross section increases linearly with the number of molecular electrons. This correlation was supported by Jain and Baluja (1992), who calculated $\sigma_T$ for a large number of molecules using their SCOP method. According to this correlation $\sigma_T$ for isoelectronic molecules should be the same for any given energy in the

**Table 9.8** Total Collision Cross Sections for $\sigma_T$ (in $10^{-20} m^2$) for the Scattering of Electrons by Molecules

(a)

| E (eV) | $O_2$ | | | $N_2$ | | | CN | $C_2$ |
|---|---|---|---|---|---|---|---|---|
| | Theory | | Expt.[c] | Theory | | Expt. | Theory | Theory |
| | AR[a] | SCOP[b] | | AR[a] | SCOP[b] | | AR[a] | AR[a] |
| 100 | 9.84 | 14.59 | 8.57 | 10.28 | 13.49 | 8.64[d] | 10.87 | 11.46 |
| 300 | 5.12 | 6.87 | 5.14 | 5.08 | 6.03 | 4.86[d] | 5.3 | 5.52 |
| 500 | 3.62 | 4.82 | 3.66 | 3.42 | 4.18 | 3.46[d] | 3.62 | 3.82 |
| 700 | 2.86 | 3.72 | 2.88 | 2.66 | 3.20 | 2.76[c] | 2.8 | 2.94 |
| 1000 | 2.22 | 2.77 | 2.10 | 2.02 | 2.36 | 2.10[c] | 2.11 | 2.2 |

(b)

| E (eV) | CO | | | | NO | | |
|---|---|---|---|---|---|---|---|
| | Theory | | | Expt.[l] | Theory | | Expt.[c] |
| | AR[a] | MAR[e] | SCOP[b] | | AR[a] | MAR[e] | |
| 100 | 10.65 | 9.95 | 13.91 | 9.01 | 10.06 | 8.89 | 8.46 |
| 300 | 5.32 | 5.27 | 6.22 | 4.85 | 5.1 | 4.91 | 5.18 |
| 500 | 3.72 | 3.72 | 4.32 | 3.56 | 3.52 | 3.52 | 5.60 |
| 700 | 2.9 | — | 3.31 | 2.8 | 2.76 | — | — |
| 1000 | 2.21 | 2.21 | 2.45 | 2.2 | 2.12 | 2.12 | 2.11 |

(c)

| E (eV) | $NO_2$ | | | $CO_2$ | | | |
|---|---|---|---|---|---|---|---|
| | Theory | | Expt.[f] | Theory | | | Expt. |
| | AR[a] | MAR[e] | | AR[a] | MAR[e] | SCOP[b] | |
| 100 | 14.98 | 13.35 | 11.9 | 15.57 | 14.45 | 19.74 | 12.9[g] | 
| 300 | 7.66 | 7.42 | 7.2 | 7.88 | 7.77 | 10.58 | 7.12[g] |
| 500 | 5.33 | 5.35 | 5.28 | 5.53 | 5.56 | 7.34 | 5.16[g] 5.12[h] |
| 700 | 4.19 | 4.19 | 4.2 | 4.33 | 4.33 | 5.64 | 4.05[h] |
| 1000 | 3.23 | 3.23 | 3.2 | 3.32 | 3.32 | 4.20 | 3.16[h] |

(d)

| E (eV) | $SiH_4$ | | | | $NH_3$ | | |
|---|---|---|---|---|---|---|---|
| | Theory | | | Exp.[k] | Theory | | Exp.[k] |
| | AR[i] | MAR–SCCT[j] | SCOP[b] | | AR[j] | MAR[e] | |
| 100 | 16.51 | 16.75 | 12.80 | 14.70 | 11.26 | 9.57 | 8.54 |
| 300 | 7.62 | 8.45 | 6.32 | 7.92 | 5.09 | 4.32 | 4.25 |
| 500 | 5.11 | 5.89 | 4.36 | 5.50 | 3.39 | 2.89 | 2.94 |
| 700 | 3.76 | 4.60 | 3.40 | 4.14 | 2.59 | 2.19 | 2.20 |
| 1000 | — | — | 2.61 | — | 1.96 | 1.56 | 1.61 |

*(Continued)*

**Table 9.8** (*Continued*)

(e)

| E (eV) | CH₄ | | | | Exp.ⁿ | C₂H₆ | Exp.ᵒ |
|---|---|---|---|---|---|---|---|
| | Theory | | | | | Theory | |
| | AR$^i$ | MAR$^m$ | MAR–SCCT$^j$ | SCOP$^b$ | | MAR$^m$ | |
| 100 | 11.40 | 10.6 | 9.38 | 7.75 | 9.00 | 16.32 | 15.4 |
| 300 | 5.02 | 4.7 | 4.89 | 3.85 | 4.76 | 7.44 | 10.5⁺ |
| 500 | 3.22 | 3.2 | 3.46 | 2.61 | 3.18 | 4.90 | 9.26⁺⁺ |
| 700 | 2.33 | 2.47 | 2.72 | 1.99 | 3.13 | 3.86 | — |

(f)

| E (eV) | H₂O | | H₂S | |
|---|---|---|---|---|
| | Theory | | Theory | |
| | AR | MAR–SC$^m$ | MAR$^m$ | SCOP$^b$ |
| 100 | 9.0 | 7.63 | 10.51 | 11.03 |
| 300 | 4.26 | 3.46 | 5.16 | 5.55 |
| 500 | 2.93 | 2.42 | 3.62 | 3.94 |
| 700 | 2.27 | 1.90 | 2.87 | 3.10 |
| 1000 | 1.71 | 1.42 | 2.3 | 2.40 |

(g)

| E (eV) | C₂H₂ | | | | Exp.$^l$ | CH₃OH | | | Exp.ᵒ |
|---|---|---|---|---|---|---|---|---|---|
| | Theory | | | | | Theory | | | |
| | AR$^s$ | MAR$^s$ | MAR$^s$–SC | SCOP$^b$ | | AR$^s$ | MAR$^s$ | MAR$^s$–SC | |
| 100 | 13.18 | 12.95 | 11.82 | 15.58 | 9.3 | 14.16 | 13.87 | 13.22 | 12.4 |
| 300 | 6.33 | 6.3 | 5.56 | 7.21 | 5.3 | 6.83 | 6.79 | 6.52 | 8.81⁺ |
| 500 | 4.31 | 4.32 | 3.76 | 4.89 | — | 4.69 | 4.68 | 4.48 | 7.55⁺⁺ |
| 700 | 3.32 | 3.32 | 2.9 | 3.71 | — | 3.62 | 3.62 | 3.50 | — |
| 1000 | 2.49 | 2.50 | 2.2 | 2.72 | — | 2.73 | 2.73 | 2.11 | — |

(h)

| E (eV) | C₂H₄ | | | | Exp.$^p$ | CH₃F | | | Exp.$^q$ |
|---|---|---|---|---|---|---|---|---|---|
| | Theory | | | | | Theory | | | |
| | AR$^s$ | MAR$^s$ | MAR$^s$–SC | Jiang et al.$^j$ | | AR$^s$ | MAR$^s$ | MAR$^s$–SC | |
| 100 | 15.24 | 14.78 | 14.32 | 8.18 | 17.18 | 13.72 | 11.8 | 11.6 | 11.8 |
| 300 | 7.096 | 7.06 | 6.708 | 4.06 | 7.96 | 6.409 | 5.76 | 5.74 | |
| 500 | 4.784 | 4.78 | 4.556 | 2.76 | 5.20 | 4.45 | 3.98 | 3.98 | |
| 700 | 3.664 | 3.66 | 3.538 | 2.1 | 3.82 | 3.46 | 3.1 | 3.09 | |
| 1000 | 2.732 | 2.73 | 2.628 | 1.5 | 2.49 | 2.59 | 2.33 | 2.33 | |

(i)

| E (eV) | HF | | | | OH | | | CH | |
|---|---|---|---|---|---|---|---|---|---|
| | Theory | | | | Theory | | | Theory | |
| | AR$^s$ | MAR$^s$ | SC$^s$ | SCOP$^b$ | AR$^s$ | MAR$^s$ | SC$^s$ | AR$^s$ | SC$^s$ |
| | | | (with anistropic terms) | | | | | | |
| 100 | 6.82 | 5.76 | 5.59 | 5.96 | 6.39 | 5.46 | 6.22 | 7.47 | 6.95 |
| 300 | 3.36 | 2.84 | 2.86 | 3.07 | 3.37 | 2.89 | 3.05 | 3.64 | 3.27 |
| 500 | 2.35 | 1.88 | 1.90 | 2.16 | 2.40 | 2.06 | 2.14 | 2.49 | 2.20 |
| 700 | 1.84 | 1.55 | 1.57 | 1.70 | 1.89 | 1.63 | 1.67 | 1.93 | 1.68 |
| 1000 | 1.39 | 1.19 | 1.20 | 1.30 | 1.45 | 1.28 | 1.27 | 1.45 | 1.25 |

**Table 9.8**  (*Continued*)

(j)

| $E$ (eV) | NH | | |
|---|---|---|---|
| | Theory | | |
| | $AR^s$ | $MAR^s$ | $SC^s$ |
| 100 | 6.51 | 6.33 | 5.40 |
| 300 | 3.30 | 3.04 | 2.80 |
| 500 | 2.30 | 2.07 | 1.94 |
| 700 | 1.79 | 1.6 | 1.52 |
| 1000 | 1.35 | 1.20 | 1.11 |

[+] at 200 eV   [++] at 250 eV.

[a] Joshipura. and Patel (1994); [b] Jain and Baluja (1992); [c] Dalba et al. (1980); [d] Hoffman et al. (1982); [e] Joshipura and Patel (1996); [f] Zecca et al. (1995); [g] Szmytkowski et al. (1987); [h] Garcia and Manero (1996); [i] Jiang et al. (1995); [j] Joshipura et al. (1999); [k] Zecca et al. (1992); [l] Kwan et al. (1983); [m] Joshipura (1998); [n] Zecca et al. (1991); [o] Szmytkowski and Krystofowicz (1995); [p] Sueoka and Mori (1986); [q] Krystofowicz and Szmytkowski (1995); [s] Joshipura and Vinodkumar (1999); [t] Sueoka and Mori (1989).

intermediate energy range. Szmytkowski (1989) analyzed $\sigma_T$ due to electron and positron bombardments at 50 and 100 eV for a large number of atoms and molecules. He observed that at a given energy, $\sigma_T$ of targets with higher dipole polarizability are higher. Szmytkowski and his associates measured $\sigma_T$ for 18 electron targets including Ar, $H_2S$, $CH_3F$, $C_2H_6$, $CH_3NH_2$, $CH_3OH$ (Szmytkowski and Maciago, 1986; Krzystofowicz and Szmytkowski, 1995; Szmytkowski and Krzystofowicz, 1995). All these targets are isoelectronic. Szmytkowski and Krzystofowicz observed that even for intermediate energies, targets of higher polarizability have higher values of $\sigma_T$. Correlation of $\sigma_T$ with the diamagnetic susceptibility (Szmytkowski, 1989) has also been noted. Furthermore, it is well known that at intermediate and high energies $\sigma_T$ decreases with an increase in the impact energy $E$.

Zecca and his associates (Zecca et al., 1992; Karwasz et al., 1993) tried to parametrize $\sigma_T$. They noted that $\sigma_T$ measured in their laboratory in the energy range 100–4000 eV can be reproduced within the experimental error at incident energy $E$ by the following equation:

$$\sigma_T(E) = \frac{\sigma_0}{1 + \sigma_0 \, E/b} \tag{9.8.38}$$

where $\sigma_0$ and $b$ are two adjustable parameters for each target. It is evident from the above equation that $\sigma_0$ is equal to $\sigma_T(E)$ at $E = 0$ and that in the asymptotic region $\sigma_T(E)$ varies linearly with $E^{-1}$, with $b$ as the slope of the straight line. It was also observed that $\sigma_T$ depends upon the dipole polarizability of the molecule. According to Joshipura (1998) $\sigma_T$ varies as $\sqrt{(\alpha/E)}$ and the constant of proportionality depends upon the size of the molecule. Karwasz et al. (1999)

examined the variation of $\sigma_T$ with $\sqrt{\alpha}$ for 20 molecules. They noted that the parameter $\sigma_0$ could be well approximated by the expression

$$\sigma_0 = 20\left(\frac{\sqrt{\alpha}}{a_0^{3/2}} - 1\right)a_0^2 \qquad (9.8.39)$$

Thus using the known values of $\alpha$ given in the *CRC Handbook of Chemistry and Physics* (Ed. Lide, 1990) the value of $\sigma_0$ can be obtained. To determine the value of $b$, occurring in (9.8.38), Karawasz et al. started with the following additivity relation:

$$b(A_l B_n C_m) = lb(A) + nb(B) + mb(C) \qquad (9.8.40)$$

where the molecule $A_l B_n C_m$ contains $l$, $m$, and $n$ number of A, B, and C atoms, respectively. The above relation along with the experimental values of $\sigma_T$ of the molecules were employed to determine $b$ for atoms. For example the experimental values of $\sigma_T$ for $H_2$ were used in (9.8.38) to obtain $b(H_2)$. From (9.8.40), $b(H)$ is equal to $0.5b(H_2)$. Thus $b(H)$ is determined; $b(CH_4)$ is obtained with the help of the experimental values of $\sigma_T$ ($CH_4$). To obtain $b$ for the carbon atom following relation is employed:

$$b(C) = b(CH_4) - 4b(H)$$

A similar procedure has been used to determine $b$ for other atoms. Table 9.9 shows the values of $b$ derived by Karwasz et al. (1999) for a few atoms.

These values of $b$ were then used in (9.8.40) to determine $b$ for the molecule of interest. For example,

$$b(CH_3Cl) = b(C) + 3b(H) + b(Cl)$$

**Table 9.9** Semiemperical Values of Atomic Cross-Section Parameter b (in Units of $10^{-20}\,m^2\,keV$) Derived by Karwasz et al. (1999)

| Atom | $b$ |
|------|------|
| H | 0.22 |
| C | 1.01 |
| F | 1.38 |
| Si | 2.57 |
| S | 2.94 |
| Cl | 3.20 |

Using Table 9.9 we get $b(CH_3Cl)$ equal to $4.87 \times 10^{-20} \, m^2 \, keV$. Using the known value of $\alpha$ in (9.8.39), we can obtain the value of $\sigma_0$ for the molecule. Finally the calculated values of $\sigma_0$ and $b$ are put into (9.8.38) and $\sigma_T$ for the molecule at a particular energy $E$ can be determined. Karwasz et al. have remarked that their model produces the experimental values of $\sigma_T$ fairly well, better than those given by the theory of Jiang et al. (1995).

A way to include the effects of molecular geometry in the calculation of $\sigma_T$ for linear molecules has been proposed by Jiang et al. (1997a, and b). Zecca et al. (1999) expressed $\sigma_T$ in term of the atomic cross sections $\sigma_A$ by the relation

$$\sigma_T = \sum_A k_A \sigma_A \qquad (9.8.41)$$

where $k_A$ is the energy-dependent expansion coefficient. Its value depends upon the geometry of the molecule and it goes to unity at high values of $E$. Thus at high $E$ the method of Zecca et al. gives the same molecular cross section as obtained by use of the AR. For diatomic molecules the following three cases have been considered: the two atoms are (i) completely separate, (ii) partially overlap, (iii) completely overlap. For all the three cases the procedure to obtain $k_A$ has been given. This geometrical AR model is found to give fairly good values of $\sigma_T$ for the linear molecules NO, $N_2O$, and $CO_2$ for energies as low as 50 eV. Using this model Zecca et al. have also calculated $\sigma_T$ for $NO_2$ (a bent molecule) and $CH_4$.

We conclude this chapter with by noting that for the light molecules the AR method gives reasonable molecular cross sections for $E \geq 100$ eV. The geometrical AR, MAR, SCOP, SC, MAR–SC, and MAR–SCCT models are conceptually better than the AR, but they all overestimate $\sigma_T$ at lower energies. For the heavier molecule one is required to consider still higher values of $E$ to get satisfactory results using these methods. A lot more effort is needed to calculate reliable molecular cross sections at low impact energies.

## Questions and Problems

9.1 (a) Why is it more difficult to calculate cross sections for electron–molecule collisions in comparison with those for electron–atom collisions?

(b) What is the Born–Oppenheimer approximation? Discuss its utility and validity in the construction of the molecular wave functions?

9.2 Differentiate among: (a) $\Sigma$ and $\Pi$ states; (b) gerade and ungerade states; (c) + and − states for a diatomic molecule.

9.3 Pauling and Wilson (1935) represented the ground state of the hydrogen molecule in a separate form and took

$$\psi(r_2, r_3) = v(r_2)v(r_3)$$

where $v(r) = N[\exp(-Zr_A) + \exp(-Zr_B)]$ are one-electron normalized orbitals and extend to both nuclei A and B. The variational principle gives $Z = 1.193a_0^{-1}$ and the internuclear distance $R_{AB} = 1.33a_0$. Show that $v(r)$ represents a $\Sigma$ state and that

$$N^2 = \frac{Z^3}{2\pi}(1+S)^{-1}$$

Show also that the overlap integral is

$$S = \left(1 + Z R_{AB} + \tfrac{1}{3} Z^2 R_{AB}^2\right)e^{-Z R_{AB}}$$

Further show that in the separated atom limit of the hydrogen molecule the static potential for $e$–$H_2$ elastic scattering is

$$V_{st}(H_2) = V_{st}^A(H, Z) + V_{st}^B(H, Z)$$

where $V_{st}^A$ (H, Z) and $V_{st}^B$ (H, Z) are the static potential for the scattering of electrons by hydrogenic atoms situated at A and B, respectively.

9.4 One center wave function of Carter et al. (1958) for the ground state hydrogen molecule is given by

$$\psi(r_2, r_3) = v(r_2)v(r_3)$$

with normalized $v(r)$ as

$$v(r) = \frac{N}{\sqrt{\pi}}[\exp(-qr) + p(1 + sr)\exp(-tqr)]$$

where $q = 1.0837a_0^{-1}$, $p = -0.4585$, $s = 1.196a_0^{-1}$, and $t = 4.1524$. Derive an expression for $N^2$ and calculate its value.

9.5 List all the assumptions that are made in the derivations of (9.3.9) and (9.5.10) and discuss their validity.

9.6 Take again for the hydrogen molecule

$$\psi(r_2, r_3) = v(r_2)v(r_3)$$

and assume that one-center normalized molecular $v(r)$ are spherically symmetric. Using the above wave function for $H_2$ show that the averaged differential cross section $\bar{I}(H_2)$ for the elastic scattering of electrons by hydrogen molecule in the FBA is

$$\bar{I}(H_2) = \frac{16}{K^4 a_0^2}\left[\frac{1}{2}\left(1 + \frac{\sin KR}{KR}\right) - 4J\frac{\sin(KR/2)}{KR} + J^2\right]$$

where $J = \langle v(r)|e^{iK\cdot r}|v(r)\rangle$ and $R$ is the internuclear distance. Further show that $\bar{I}(H_2)$ is finite in the forward direction.

9.7 In the independent atom model (IAM) the elastic differential cross section $\bar{I}(H_2)$ is given by (9.8.23). To include exchange in $I_H$ we consider triplet and singlet scatterings for the $e$–H system [see Eq. (7.4.17)]. Show that to give $\bar{I}_{H_2}$ in terms of $I_H(Z)$, it is physically incorrect to include singlet scattering in $I_H(Z)$. Further consider the united atom limit of $H_2$ to show that a proper form of $f_H(Z)$ is

$$f_H(Z) = f_d - g/2$$

where $f_d$ and $g$ are the direct- and exchange-scattering amplitudes, respectively, for elastic scattering of electrons by a hydrogenic atom. Hence, in the IAM

$$\bar{I}_{H_2} = 2|f_d - g/2|^2\left(1 + \frac{\sin KR}{KR}\right)$$

9.8 Using (9.7.7), (9.7.9), and (9.7.15) derive expressions for $D(t)$ and $N_i$ in terms of the coefficients $a_n$. Further, take the values of $a_n$ from Table 9.3 and show that for the $H_2$ molecule

$$N_i = 1.173$$

and

$$D(t) = 0.3856 - \frac{4}{(t+1)^3}\left(0.3754 + \frac{3.199}{t+1} - \frac{6.244}{(t+1)^2} + \frac{2.859}{(t+1)^3}\right)$$

Calculate the ionization cross section $\sigma_i$ for $H_2$ due to electron impact in the BED model of Kim and Rudd (1944) for $t$ varying from 1 to 20. Compare your cross sections with the experimental data of Rapp and Englander-Golden (1965).

9.9 Compare the MAR–SC (modified additivity rule-single-center) approach of Joshipura and Vinodkumar (1999) with the SCOP (single-center optical potential) approach of Jain and Baluja (1992). Point out their similarities and differences. Why do these approaches fail at low impact energies?

9.10 Using dipole polarizabilities and the total collision cross sections for $H_2$, $CH_4$, and HF molecules, how will you obtain the parameter $b$, occurring in (9.5.38), for H, C, and F atoms.

Use the values of $b$ given in the Table 9.9 and the dipole polarizability (equal to $3.29 \times 10^{-30}\,m^3$) of the $CH_3F$ molecule to calculate $\sigma_T$ for this molecule due to electron impact in energy range 100–1000 eV. Compare your results with those given in Table 9.8(h).

# References

Adamczyk B, Boerboom A J H, Schram B L, and Kistemaker J (1966), *J. Chem. Phys.* **44**: 4640.

Ali M A, Kim Y K, Hwang W, Weinberger N M, and Rudd M E (1997), *J. Chem. Phys.* **106**: 9602.

Anderson N, Gallagher J W, and Hertel I V (1988), *Phys. Rep.* **165**: 1.

Arfken G (1968), *Mathematical Methods for Physicists* (Academic, New York).

Bader R F W (1990), *Atoms in Molecules* (Clarendon, Oxford).

Bartschet K (1993), in: AIP *XVIII International Conference on the Physics of Electronic and Atomic Collisions*, Proceedings No. 295, eds. T Andersen, B Fastrup, F Folkmann, H Knudsen, and N Andesen (AIP Press, New York), p. 251.

Bartschet K and Madison D H (1988), *J. Phys. B* **21**: 2621.

Baum G, Raith W, and Schroder W (1988a), *J. Phys. B* **21**: L501.

Baum G, Raith W, and Steidl H (1988b), *Z. Phys. D* **10**: 171.

Baum G, Moede M, Raith W, and Schroder W (1985), *J. Phys. B* **18**: 531.

Becker K, Crowe A, and McConkey J W (1992), *J. Phys. B* **25**: 3885.

Bederski K, Wojcik L, and Adamczyk B (1980), *Int. J. Mass Spectrom. Ion Phys.* **35**: 171.

Beijers J P M, Madison D H, van Eck J, and Heideman H G M (1987), *J. Phys. B* **20**: 167.

Bell K L and Moiseiwitch B L (1963), *Proc. Roy. Soc. A* **276**: 346.

Berkowitz J (1979), *Photoabsorption, Photoionization and Photoelectron Spectroscopy* (Academic, New York).

Bethe H A (1930), *Ann. Phys.* **5**: 325.

Bhatia A K, Schneider B I, and Temkin A (1993), *Phys. Rev. Letts.* **70**: 1936.

Blaauw H J, de Heer F J, Wagenaar R W, and Barends D H (1980), *J. Phys. B* **13**: 359.

Blum K and Kleinpoppen H (1979), *Phys. Rep.* **52**: 203.

Brenton A G, Dutton J, Harries F M, Jones R A, and Lewis D M (1977), *J. Phys. B* **10**: 2699.

Brenton A G, Dutton J, Harries F M (1978), *J. Phys. B* **11**: L15.

Brion C E (1985), *Comment At. Mol. Phys.* **16**: 249.

Bromberg J P (1969), *J. Chem. Phys.* **50**: 3906.

Bromberg J P (1970), *J. Chem. Phys.* **52**: 1243.

Bromberg J P (1974), *J. Chem. Phys.* **60**: 1717.

Brown C J and Humperston J W (1985), *J. Phys. B* **18**: L401.

Buhring W (1968), *Z. Physik* **208**: 286.

Burke P G (1968), in: *Advances in Atomic and Molecular Physics*, Vol. 4, eds. D R Bates and I Estermann (Academic, New York), p. 173.

Burke P G (1977), *Potential Scattering in Atomic Physics* (Plenum, New York).

Burke P G (1985), in: *Fundamental Processes in Atomic Collision Theory*, eds. H Kleinpoppen and J S Briggs (Plenum, New York).

Burke P G (1987), in: *Atomic Physics*, Vol. 10, eds. H. Nirumi and I Shimamura (Elsevier Science), p. 243.

Burke P G (1993), in: AIP *XVIII International Conference on the Physics of Electronic and Atomic Collisions*, Proceedings No. 295, eds. T Andersen, B Fastrup, F Folkmann, H Knudsen and N Andesen (AIP Press, New York), p. 26.

Burke P G and Eissener W (1983), in: *Atoms in Astrophysics*, eds. P G Burke, W B Wissener, D G Hummer and I C Percival (Plenum, New York), p. 1.

Burke P G and Williams J F (1977), *Phys. Rep. C* **34**: 325.

Burke P G, Gallaher D F, and Geltman S (1969), *J. Phys. B* **2**: 1142.

Burke P G, Hibbert A, and Robb D W (1971), *J. Phys. B* **4**: 153.

Byron F W and Joachain C J (1973a), *Phys. Rev. A* **8**: 1267.

Byron F W and Joachain C J (1973b), *Phys. Rev. A* **8**: 3266.

Byron F W and Joachain C J (1974a), *Phys. Rev. A* **9**: 2259.

Byron F W and Joachain C J (1974b), *J. Phys. B* **7**: L212.

Byron F W and Joachain C J (1975), *J. Phys. B* **8**: L284.

Byron F W and Joachain C J (1977a), *Phys. Rep. C* **34**: 235.

Byron F W and Joachain C J (1977b), *Phys. Rev. A* **15**: 128.

Byron F W, Joachain C J, and Potvliege R M (1982), *J. Phys. B* **15**: 3915.

Byron F W and Joachain C J, and Potvliege R M (1985), *J. Phys. B* **18**: 1637.

Callaway J and Oza D H (1984), *Phys. Rev. A* **29**: 2416.

Catalan G and Roberts M J (1979), *J. Phys. B* **12**: 3947.

Carter C, March N H, and Vincent D (1958), *Proc. Phys. Soc. (Lon.)* **71**: 2.

Cartwright D C and Kupperman A (1967), *Phys. Rev.* **163**: 86.

Champion C, Hansesen J, and Hervieux P A (2001), *Phys. Rev. A* **63**: 052720.

Charlton M, Griffith T C, Heyland C R, and Wright G L (1980), *J. Phys. B* **13**: L353.

Christophorou L G (1971), *Atomic and Molecular Radiation Physics* (Wiley, New York).

Christophorou L G (ed.) (1984), *Electron Molecule Interactions and Their Applications*, Vols. 1 and 2 (Academic, New York).

Chung S and Lin C C (1978), *Phys. Rev. A* **17**: 1875.

Coleman P G, Griffith T C, Heyland G R, and Twomey T R (1976), *Appl. Phys.* **11**: 321.

Corrigan S J B (1965), *J. Chem. Phys.* **43**: 4381.

*CRC Handbook of Chemistry and Physics* (1994), ed. Lide (CRC Press, Boca Raton, Fl).

Crooks G B and Rudd M E (1972), *Bull. Am. Phys. Soc.* **17**: 131.

Crowe A and McConkey J W (1977), *Int. J. Mass Spectrom. Ion Phys.* **24**: 181.

Crowe A, Nogueira J C, and Liew Y C (1983), *J. Phys. B* **16** 481.

Dalba G , Fornnasini P, Lazzizzera I, Ranieri G, and Zecca A (1980), *J. Phys. B* **13**: 2839.

Dalgarno A (1979), in: *Atomic and Molecular Physics from Atmospheric Physics*, Vol. 15, eds. D R Bates and B Bederson (Academic, New York).

Dalgarno A and Lewis J T (1955), *Proc. Roy. Soc. (Lon.) A* **233**: 70.

Damburg R J and Karule E (1967), *Proc. Phys. Soc.* **90**: 637.

Das J N and Biswas A K (1980), *Phys. Lett. A* **78**: 319.

Das J N and Biswas A K (1981), *J. Phys. B* **14**: 1363.

Das J N and Saha N (1981), *J. Phys. B* **14**: 2657.

Davidson E R and Jones L L (1962), *J. Chem. Phys.* **37**: 2966.

Davis D V, Mistry V D, and Quarles C A (1972), *Phys. Lett. A* **38**: 169.

de Heer F J, McDowell M R C, and Wagenaar R W (1977), *J. Phys. B* **10**: 1945.

de Heer F J, Jansen R J H, and van der Kay W (1979), *J. Phys. B* **12**: 979.

Deutsch H and Mark T D (1987), *Int. J. Mass Spectrom. Ion Proc.* **79**: R1.

Deutsch H, Margreiter D, and Mark T D (1994), *Z. Phys. D* **29**: 31.

Deutsch H, Becker K, Matt S, and Mark T D (2000a), *Int. J. Mass Spectrom.* **197**: 37.

Deutsch H, Becker K, Janev R K, Probst M, and Mark T D (2000b), *J. Phys. B* **33**: L865.

Deutsch H, Becker K, and Mark T D (2000c), *Eur. Phys. J. D* **12**: 283.

Deutsch H, Hilpert K, Becker K, Probst M, and Mark T D (2001), *J. Appl. Phys.* **89**: 1915.

Dewangan D P (1976), *Phys. Lett. A* **56**: 279.

Dewangan D P and Walters H R J (1977), *J. Phys. B* **10**: 637.

Dillon M A, Boesten L, Tanaka H, Kimura M, and Sato H (1993), *J. Phys. B* **26**: 3147.

Djuric N L, Cadez I M, and Kurepa M V (1991), *Int. J. Mass Spectrom. Ion Proc.* **108**: R1.

Dorn A, Elliot A, Lower J, Mazevet S F, McEachran R P, McCarthy I E, and Weigold E (1998), *J. Phys. B* **31**: 547.

Dubois R D and Rudd M E (1975), *J. Phys. B* **8**: 1474.

Dubois R D and Rudd M E (1976), *J. Phys. B* **9**: 2657.

Eminyan M, MacAdam K B, Slevin J, and Kleinpoppen H (1973), *Phys. Rev. Lett.* **31**: 576.

Eminyan M, MacAdam K B, Slevin J, and Kleinpoppen H (1974), *J. Phys. B* **7**: 1519.

Ermolaev A M and Walters H R J (1979), *J. Phys. B* **12**: L779.

Fabrikant I I (1980), *J. Phys. B* **13**: 603.

Fano U (1963), *Ann. Rev. Nucl. Sci.* **13**: 1.

Fano U (1969), *Phys. Rev.* **178**: 131.

Fink M, Jost K, and Herrmann D (1975), *Phys. Rev. A* **12**: 1374.

Fisk J B (1936), *Phys. Rev.* **49**: 167.

Fletcher G D, Alguard M J, Gay T J, Hughes V W , Wainwright P F, Lubell M S, and Raith W (1985), *Phys. Rev. A* **31**: 2854.

Fliescher R L, Price P B, and Walker R M (1975), *Nuclear Tracks in Solids* (University of California Press, Berkeley).

Floeder K, Fromme D, Raith W, Schwab A, and Sinapius G (1985), *J. Phys. B* **18**: 3347.

Fon W C and Berrington K A (1981), *J. Phys. B* **14**: 323.

Ford A L and Browne J C (1973), *Chem. Phys. Lett.* **20**: 284.

Furness J B and McCarthy I E (1973), *J. Phys. B* **6**: L42 and 2280.

Gallagher J W, Brion C E, Samson J A R, and Langhoff F W (1988), *J. Phys. Chem. Ref. Data.* **17**: 9.

Garcia G and Manero F (1996), *Phys. Rev. A* **53**: 250.

Geiger J (1964), *Z. Physik* **181**: 413.

Gerjoy E and Krall N A (1960), *Phys. Rev* **119**: 705.

Ghosh A S, Mukherjee M, and Basu M (1990), in: AIP *XVI International Conference on the Physics of Electronic and Atomic Collisions*, Proceedings No. 203, eds. A. Dalgarno, P M Koch, M S Lubell, and T B Lucatorto (AIP Press, New York), p. 633.

Gianturco F A (1995), in: AIP *XIX International Conference on the Physics of Electronic and Atomic Collisions*, Proceedings No. 360, eds. L J Dube, J B A Mitchell, J M McConkey, and C E Brion (AIP Press, New York), p. 211.

Gianturco F A and Jain A (1986), *Phys. Rep.* **143**: 347.

Gien T T (1976), *J. Phys. B* **9**: 3203.

Gien T T (1977a), *Phys. Rev. A* **16**: 123.

Gien T T (1977b), *Phys. Lett. A* **61**: 299.

Glauber R J (1959), *Lectures in Theoretical Physics* Vol. 1, eds. W E Brittin and I G Dunham (Interscience, New York), p. 315.

Golden D E (1978), *Adv. Atom Mole. Phys.* **14**: 1.

Gregory D and Fink M (1974), *At. Data Nucl. Data Tables* **14**: 39.

Green L C, Mulder M M, Lewis M N, and Woll J W (1954), *Phys. Rev.* **93**: 757.

Griffith T C and Heyland G R (1978), *Phys. Rep.* **39**: 169.

Grizinski M (1965a), *Phys. Rev. A* **138**: 305.

Grizinski M (1965b), *Phys. Rev. A* **138**: 322.

Grizinski M (1965c), *Phys. Rev. A* **138**: 336.

Gupta P and Khare S P (1978), *J. Chem. Phys.* **68**: 2193.

Gupta S C and Rees J A (1975a), *J. Phys. B* **8**: 417.

Gupta S C and Rees J A (1975b), *J. Phys. B* **8**: 1267.

Hayashi S and Kuchitshu K (1976), *Chem. Phys. Lett.* **41**: 575.

Hayes E F (1967), *J. Chem. Phys.* **46**: 4004.

Hertel I V and Stoll W (1974), *J. Phys. B* **7**: 570.

Hertel I V and Stoll W (1978), *Adv. Atom Molec. Phys.* **13**: 113.

Hertel I V, Kelley M H, and McClelland J J (1987), *Z. Phys. D* **6**: 163.

Hippler R (1990), *Phys. Lett. A* **144**: 81.

Hoffmann K R, Dababneh M S, Hsieh Y F, Kaupplia W E, Pol V, Smart J H, and Stein T S (1982), *Phys. Rev. A* **25**: 1393.

Hollywood M T, Crowe A, and Williams J F (1979), *J. Physics B* **12**: 819.

Holt A R (1969), *J. Physics B* **2**: 1209.

Holt A R and Moiseiwitsch B L (1968), *J. Physics B* **1**: 36.

Hulthen L (1944), *Kgl. Fysion. Sallsk. Lund Förth* **14**: 21.

Hulthen L (1948), *Arkiv. Mat. Ast. Fys.* **35A**: 25.

Huzinaga S (1957), *Progr. Theoret. Phys. (Kyoto)* **17**: 162.

Hwang W, Kim Y K, and Rudd M E (1996), *J. Chem. Phys.* **104**: 2956.

Inokuti M, Kim Y K, and Platzmann R L (1967), *Phys. Rev.* **164**: 55.

Jain A (1986), *Phys. Rev. A* **34**: 3707.

Jain A (1987), *J. Chem. Phys.* **86**: 1289.

Jain A (1988), *J. Physics B* **21**: 905.

Jain A and Baluja K L (1992), *Phys. Rev. A* **45**: 202.

Jain A, Baluja K L, DiMartino V, and Gianturco F A (1991), *Chem. Phys. Lett.* **183**: 34.

Jain A K and Tripathi A N (1997), *Phys. Lett. A* **231**: 224.

Jain D K and Khare S P (1976), *J. Phys. B* **9**: 1429.

Jain D K and Khare S P (1977), *Phys. Lett. A* **63**: 237.

Jansen R H J, de Heer F J, Luyken H J, van Wingerden B, and Blaauw H J (1976), *J. Phys. B* **9**: 185.

Jhanwar B L and Khare S P (1975), *J. Phys. B* **8**: 2659.

Jhanwar B L and Khare S P (1976), *J. Phys. B* **9**: L527.

Jhanwar B L, Shobha P, and Khare S P (1975), *J. Phys. B* **8**: 1228.

Jhanwar B L, Khare S P, and Kumar Jr. A (1978), *J. Phys. B* **11**: 887.

Jhanwar B L, Khare S P, and Sharma M K (1982a), *Phys. Rev. A* **25**: 1993.

Jhanwar B L, Khare S P, and Sharma M K (1982b), *Phys. Rev. A* **26**: 1392.

Jiang Y, Sun J, and Wan L (1995), *Phys. Rev. A* **52**: 398.

Jiang Y, Sun J, and Wan L (1997a), *J. Phys. B* **30**: 5025.

Jiang Y, Sun J, and Wan L (1997b), *Phys. Lett. A* **237**: 53.

Joachain C J (1987), *Quantum Collision Theory* (North Holland, Amsterdam).

Joachain C J (1990), in: AIP *XVI International Conference on the Physics of Electronic and Atomic Collisions*, Proceedings No. 203, eds. A. Dalgarno, P M Koch, M S Lubell, and T B Lucatorto (AIP Press, New York), p. 68.

Joachain C J, Vanderpoorten R, Winters K H, and Byron F W (1977), *J. Phys. B* **10**: 227.

Johnson W R and Cheng K T (1979), *Phys. Rev. A* **20**: 978.

Johnson W R and Lin C D (1979), *Phys. Rev. A* **20**: 964.

Joshipura K N (1998), *Pranama* **50**: 555.

Joshipura K N and Patel P M (1994), *Z. Phys. D* **29**: 269.

Joshipura K N and Patel P M (1996), *J. Phys. B* **29**: 3925.

Joshipura K N and Vinodkumar M (1997a), *Phys. Lett. A* **224**: 361.

Joshipura K N and Vinodkumar M (1997b), *Z. Phys. D* **41**: 133.

Joshipura K N and Vinodkumar M (1999), *Eur. Phys. J. D* **5**: 229.

Joshipura K N, Vinodkumar M, and Patel U M (2001), *J. Phys. B* **34**: 509.

Joshipura K N, Vinodkumar M, Thakar Y, and Limbachiya C (1999), *Ind. J. Phys.* **73B**: 245.

Joy H W and Parr R G (1958), *J. Chem. Phys.* **28**: 448.

Karwasz G P, Brusa R S, Gasparoli A, and Zecca A (1993), *Chem. Phys. Lett.* **211**: 529.

Karwasz G P, Brusa R S, Piazza A, and Zecca A (1999), *Phys. Rev. A* **59**: 1341.

Kauppila W E, Stein T S, Smart J R, Dababneh M S, Ho Y K, Dowing J P, and Pol V (1981), *Phys. Rev. A* **24**: 725.

Kaushik Y D, Khare S P, and Kumar A (1983), *J. Phys. B* **16**: 3609.

Kesseler J (1985), *Polarized Electrons* 2nd Edn. (Springer-Verlag, Berlin).

Kesseler J (1991), *Adv. At. Mol. Opt. Phys.* **27**: 81.

Kesseler J and Lorenz J (1970), *Phys. Rev. Lett.* **24**: 87.

Khare S P (1966a), *Phys. Rev. A* **149**: 33.

Khare S P (1966b), *Phys. Rev. A* **152**: 74.

Khare S P (1967), *Phys. Rev. A* **157**: 107.

Khare S P (1968), *Phys. Rev. A* **173**: 43.

Khare S P and Moisewitsch B L (1965), *Proc. Phys. Soc. (Lon.)* **85**: 821.

Khare S P and Moisewitsch B L (1966), *Proc. Phys. Soc. (Lon.)* **88**: 605.

Khare S P and Kumar Jr A (1978), *Pramana* **10**: 63.

Khare S P and Lata K (1984), *Phys. Rev. A* **29**: 3137.

Khare S P and Lata K (1985), *J. Phys. B* **18**: 2941.

Khare S P and Meath W J (1987), *J. Phys. B* **20**: 2101.

Khare S P and Prakash S (1985), *Phys. Rev. A* **32**: 2689.

Khare S P and Raj D (1979), *J. Phys. B* **12**: L351.

Khare S P and Raj D (1980), *J. Phys. B* **13**: 4627.

Khare S P and Raj D (1982), *Ind. J. Pure Appl. Phys.* **20**: 538.

Khare S P and Shobha P (1970), *Phys. Lett. A* **31**: 571.

Khare S P and Shobha P (1971), *J. Phys. B* **4**: 203.

Khare S P and Shobha P (1972), *J. Phys. B* **5**: 1938.

Khare S P and Shobha P (1974), *J. Phys. B* **7**: 420.

Khare S P and Tomar S (1999), in: *Trends in Atomic and Molecular Physics*, eds. K K Sud and U N Upadhayaya (Plenum, New York).

Khare S P and Vijaishri (1988), Ind. *J. Phys. B* **62**: 483.

Khare S P and Wadehra J M (1989), *Comm. At. Mol. Phys.* **23**: 55.

Khare S P and Wadehra J M (1995), *Phys. Lett. A* **198**: 212.

Khare S P and Wadehra J M (1996), *Can. J. Phys.* **74**: 376.

Khare S P, Kumar A, and Lata K (1983), *J. Phys. B* **16**: 4419.

Khare S P, Kumar A, and Vijaishri (1985), *J. Phys. B* **18**: 1827.

Khare S P, Kumar A, and Lata K (1986), *Phys. Rev. A* **33**: 2795.

Khare S P, Saksena V, and Ojha S P (1992), *J. Phys. B* **25**: 2001.

Khare S P, Saksena V, and Wadehra J M (1993), *Phys. Rev. A* **48**: 1209.

Khare S P, Sinha P, and Wadehra J M (1994a), *Phys. Lett. A* **184**: 204.

Khare S P, Sinha P, and Wadehra J M (1994b), *Hyp. Int.* **89**: 107.

Khare S P, Raj D, and Sinha P (1994c), *J. Phys. B* **27**: 2569.

Khare S P, Sinha P, and Wadehra J M (1995), *Ind. J. Phys.* **69**B: 219.

Khare S P, Sharma M K, and Tomar S (1999), *J. Phys. B* **32**: 3147.

Khare S P, Sharma M K, and Tomar S (2000), *J. Phys. B* **33**: L59.

Khayrallah G (1976), *Phys. Rev. A* **14**: 2064.

Kim Y K and Rudd M E (1994), *Phys. Rev. A* **50**: 3954.

Kim Y K, Hwang W, Weinberger N M, Ali M E, and Rudd M E (1997), *J. Chem. Phys.* **106**: 1026.

Kim Y K, Johnson W R, and Rudd M E (2000), *Phys. Rev. A* **61**: 034702.

Klienmann C J, Hahn Y, and Spruch L (1968), *Phys. Rev.* **165**: 53

Kohn W (1948), *Phys. Rev.* **74**: 1763.

Kollath K J and Lucas C B (1979), *Z. Phys. A* **292**: 215.

Kolos W and Roothan C C J (1960), *Rev. Mod. Phys.* **32**: 219.

Kolos W and Wolnicwicz L (1965), *J. Chem. Phys.* **43**: 2429.

Krishnakumar E and Srivastava S K (1990), *J. Phys. B* **23**: 1893.

Krishnakumar E and Srivastava S K (1992), *Int. J. Mass Spectrom. Ion Proc.* **113**: 1.

Krzystofowicz A M and Szmytkowski C (1995), *J. Phys. B* **28**: 1593.

Kuchitsu K and Bartell L S (1961), *J. Chem. Phys.* **35**: 1945.

Kumar P, Jain A K, Tripathi A N, and Nahar S N (1994), *Phys. Rev. A* **49**: 899.

Kumar P, Jain A K, and Tripathi A N (1995), *J. Phys. B* **28**: L387.

Kurepa M V and Vuskovic L (1975), *J. Phys. B* **8**: 2067.

Kwan C K, Hsich Y F, Kaupplia W E, Smith S J, Stein T S, and Uddin M N (1983), *Phys. Rev. A* **27**: 1328.

Kyle H L and Temkin A (1964), *Phys. Rev. A* **134**: 600.

LaBhan R W and Callaway J (1969), *Phys. Rev.* **180**: 91.

Lahmam-Bennani A (1991), *J. Phys. B* **24**: 2401.

Lane N F (1980), *Rev. Mod. Phys.* **52**: 29.

Lata (1984), Ph.D. Thesis, Meerut University.

Lindinger W and Howorka F (1985), in: *Electron Impact Ionization*, eds. T D Mark and G H Dunn (Springer, Berlin).

Liu J W and Smith V H (1973), *J. Phys. B* **6**: L275.

Liu Y and Sun J (1996), *Phys. Lett. A* **222**: 233.

Llyod C R, Teubner P J O, Weigold E, and Lewis B R (1974), *Phys. Rev. A* **10**: 175.

Long X, Liu M, Ho F, and Peng X (1990), *At. Data Nucl. Data Tables* **45**: 353.

Lucas C B (1979), *J. Phys. B* **12**: 1549.

Lucas C B and Liedtike J (1975), in: *IX International Conference on Electronic and Atomic Collision*, Book of Abstracts, eds. J S Risley and R Greballe (Seattle University Press, Seattle), p. 460.

Madison D H (1979), *J. Phys. B* **12**: 3399.

Margreiter D, Deutsch H, Schmidt M, and Mark T D (1990), *Int. J. Mass Spectrom. Ion Proc.* **100**: 157.

Margreiter D, Deutsch H, and Mark T D (1994), *Int. J. Mass Spectrom. Ion Proc.* **139**: 127.

Martyneko Y V, Firsov O B, and Chibisov M I (1963), *JETP* **17**: 154.

Maruyama T, Garwin E L, Prepost R, and Zaplace G H (1992), *Phys. Rev. B* **46**: 4261.

Massey H S W and Mohr C B O (1934), *Proc. Roy. Soc. A* **146**: 880.

Massey H S W, Burhop E H S, and Gilbody H B (1969), *Electronic and Ionic Impact Phenomena* Vol. 1 and 2 (Clarendon, Oxford).

Matese J and Oberoi R S (1971), *Phys. Rev. A* **4**: 569.

Mathur K C (1998), in: *Electron Correlations in Atoms and Solids*, eds. A N Tripathi and I Singh (Phoenix, New Delhi), p. 29.

Mayer S (1995), in: AIP *XIX International Conference on the Physics of Electronic and Atomic Collisions*, Proceedings No. 360, eds. L J Dube, J B A Mitchell, J M McConkey, and C E Brion (AIP Press, New York), p. 163.

Mayol R and Salvat F (1990), *J. Phys. B* **23**: 2117.

McCarthy I E, Noble C J, Phillips B A, and Turnball A D (1977), *Phys. Rev. A* **15**: 2173.

McClelland J J, Kelly M H, and Cellota R J (1985), *Phys. Rev. Lett.* **55**: 688.

McClelland J J, Kelly M H, and Cellota R J (1986), *Phys. Rev. Lett.* **56**: 1362.

McClelland J J, Kelly M H, and Cellota R J (1987), *Phys. Rev. Lett.* **58**: 2198.

McEachran R P and Stauffer A D (1986), *J. Phys. B* **19**: 3523.

Menedez M G, Rees J A, and Beaty E C (1980), in: *33rd Annual Gaseous Electronic Conference* (Norman, Oklahoma), Program and Abstracts, p. 27.

Morgan L A and McDowell M R C (1975), *J. Phys. B* **8**: 1073.

Morgan L A and McDowell M R C (1977), *Comm. At. Mol. Phys.* **19**: 129.

Morrison M A (1988), *Adv. Atom. Molec. Phys.* **24**: 51.

Morse P M and Allis W P (1933), *Phys. Rev.* **44**: 269.

Morse P M and Feshbach H (1953), *Methods of Theoretical Physics* (McGraw-Hill, New York).

Mott N F and Massey H S W (1965), *The Theory of Atomic Collision* 3rd Edn. (Oxford, London).

Mulliken R S and Reike C A (1941), *Rept. Progr. Phys.* **8**: 231.

Nahar S N and Wadehra J M (1991), *Phys. Rev. A* **43**: 1275.

Neerja, Tripathi A N, and Jain A K (2000), *Phys. Rev. A* **61**: 032713.

Nesbet R K (1980), *Variational Methods in Electron-Atom Scattering Theory* (Plenum, New York).

Nishimura H and Tawara H (1994), *J. Phys. B* **27**: 2063.

Nishimura H, Huo W M, Ali M A, and Kim Y K (1999), *J. Chem. Phys.* **110**: 3811.

Ochkur V I (1964), *Soviet Physics-JETP* **18**: 503.

Oda N, Nishimura F, and Tahira S (1972), *J. Phys. Soc. Jpn.* **33**: 462.

O'Malley T F and Geltman S (1965), *Phys. Rev.* **137**: 1344.

O'Malley T F, Spruch L, and Rosenberg L (1961), *J. Math. Phys.* **2**: 491.

Orient O J and Srivastava S K (1987), *J. Phys. B* **20**: 3923.

Padial N T and Norcross D W (1983), *Phys. Rev. A* **29**: 174.

Palinkas J and Schlenk B (1980), *Z. Phys. A* **297**: 29.

Pauling and Wilson (1935), *Introduction to Quantum Mechanics* (McGraw Hill, New York).

Poet R (1978), *J. Phys. B* **11**: 3081.

Probst M, Deutsch H, Becker K, and Mark T D (2001), *Int. J. Mass Spectrom. Ion Proc.* **206**: 13.

Purohit S P and Mathur K C (1993), *J. Phys. B* **26**: 2443.

Raj D (1981), Ph.D. Thesis, Meerut University.

Raj D (1991a), *J. Phys. B* **24**: L431.

Raj D (1991b), *Ind. J. Phys. B* **65**: 319.

Raj D and Kumar A (2001), *J. Phys. Lett. A* **282**: 284.

Rapp D and Englander-Golden P (1965), *J. Chem. Phys.* **43**: 1464.

Rao M V V S and Srivastava S K (1992), *J. Phys. B* **25**: 2175.

Rao M V V S, Iga I, and Srivastava S K (1995), *J. Geo. Res. Planet* **100**: 421.

Register D F, Vuskosvic L, and Trajinar S (1980), in: *33rd Annual Gaseous Electronic Conference* (Norman, Oklahoma), Program and Abstracts, p. 27.

Reike F F and Prepejchal W (1972), *Phys. Rev. A* **6**: 1507.

Reitan A (1981), *Phys. Scr.* **22**: 615.

Riley M E and Truhlar D G (1975), *J. Chem. Phys.* **63**: 2182.

Riley M E and Truhlar D G (1976), *J. Chem. Phys.* **65**: 792.

Roscoe R (1941), *Phil. Mag.* **31**: 349.

Rosenberg L, Spruch L, and O'Malley TF (1960), *Phys. Rev.* **119**: 164.

Sakae T, Sumiyoshi S, Murakami E, Matsumoto Y, Ishibashi K and Katase A (1989), *J. Phys. B* **22**: 1385.

Saksena V (1994), Ph.D. Thesis, Meerut University.

Saksena V and Kushwaha M S (1996), *Ind. J. Phys. B* **70**: 97.

Saksena V, Kushwaha M S, and Khare S P (1997a), *Int. J. Mass Spectrom. Ion Proc.* **171**: L1.

Saksena V, Kushwaha M S, and Khare S P (1997b), *Physica B* **233**: 201.

Schiff L I (1968), *Quantum Mechanics* 3rd Edn. (McGraw-Hill, Tokyo).

Schneider H, Tobehn I, Ebel F, and Hippler R (1993), *Phys. Rev. Lett.* **71**: 2707.

Schram B L, de Heer F J, van der Wiel M J, and Kistemaker J (1965), *Physica* **31**: 94.

Schram B L, Moustafa H R, Schutten J, and de Heer F J (1966), *Physica* **32**: 734.

Schutten J, de Heer F J, Moustafa H R, Boerboom A J H, and Kistemaker J (1966), *J. Chem. Phys.* **44**: 3924.

Schwartz C (1961), *Phys. Rev. A* **124**: 1468.

Scofield J H (1978), *Phys. Rev. A* **18**: 963.

Scott T and McDowell M R C (1976), *J . Phys. B* **9**: 2235.

Seaton M J (1974), *J. Phys. B* **7**: 1817.

Sethuraman S K, Rees J A, and Gibson (1974), *J. Phys. B* **7**: 1741.

Sheorey V B (1969), *J. Phys. B* **2**: 442.

Shima K, Nakagawa T, Umetani K, and Mikumo (1981), *Phys. Rev. A* **24**: 72.

Shimamura I and Takayanagi K (eds.) (1984), *Electron-Molecule Collisions* (Plenum, New York).

Shobha P (1972), Ph.D. Thesis, Meerut University.

Sienkiewicz J E and Baylis W E (1997), *Phys. Rev. A* **55**: 1108.

Silvermann S M and Lassetre E N (1964), *J. Chem. Phys.* **40**: 2922.

Simpson J A (1964), in: *Atomic Collision Processes*, ed. M R C McDowell (North Holland, Amsterdam), p. 128.

Slater J C (1930), *Phys. Rev.* **36**: 57.

Slevin J (1984), *Rep. Prog. Phys.* **47**: 461.

Slevin J A and Chwirot S (1990), *J. Phys. B* **23**: 165.

Slevin J, Porter H Q, Eminyan M, Defrance A, and Vassilev G (1980), *J. Phys. B* **13**: 3009.

Smith E R and Henry R J W (1973a), *Phys. Rev. A* **7**: 1585.

Smith E R and Henry R J W (1973b), *Phys. Rev. A* **8**: 572.

Srivastava R (1998), *Pranama* **50**: 683.

Staszewska G, Schwenke D W, Thirumalai D, and Truhlar D G (1983), *Phys. Rev. A* **28**: 2740.

Stein T S and Kauppila W E (1986), in: *Electronic and Atomic Collisions*, eds. D C Lorentz, W E Meyerhof, and J R Peterson (Academic, New York), p. 105.

Steph N C and Golden D E (1980), *Phys. Rev. A* **21**: 759.

Sternheimer R M (1954), *Phys. Rev.* **96**: 951.

Sternheimer R M (1957), *Phys. Rev.* **107**: 1565.

Stier H (1932), *Z. Phys.* **76**: 439.

Strand T G and Bonham R A (1964), *J. Chem. Phys.* **40**: 1686.

Stratton J A, Morse P M, Chu L J, Little J D C, and Carbato F J (1956), *Spheroidal Wave Functions* (John Wiley, New York).

Sueoka O (1987), in: *Atomic Physics with Positrons*, NATO ASI series B Vol. 169, eds. J W Humberston and E A G Armour (Plenum, New York), p. 41.

Sueoka O and Mori S (1986), *J. Phys. B* **19**: 4035.

Sueoka O and Mori S (1989), *J. Phys. B* **22**: 963.

Sun J, Jiang Y, and Wan L (1994), *Phys. Lett. A* **195**: 81.

Sutcliffe V C, Haddad G N, Steph N C, and Golden D E (1978), *Phys. Rev. A* **17**: 100.

Szmytkowski C (1989), *Z. Phys. D* **13**: 69.

Szmytkowski C (1991), *J. Phys. B* **24**: 3895.

Szmytkowski C (1993), *J. Phys. B* **26**: 535.

Szmytkowski C and Maciago K (1986), *Chem. Phys. Lett.* **129**: 321.

Szmytkowski C and Sienkiewicz J E (1994), *J. Phys. B*: **27**: 2277.

Szmytkowski C and Krzystofowicz A M (1995), *J. Phys. B* **28**: 4291.

Szmytkowski C, Zecca A, Karwas G, Oss S, Maciago K, Marinkovic B, Brusa R S, and Grisenti R (1987), *J. Phys. B* **20**: 5817.

Szmbtkowski C, Mozejko P, and Kasperski G (1997), *J. Phys. B* **30**: 4364.

Tan K H, Fryar J, Farago P S and McConkey J W (1977), *J. Phys. B* **10**: 1073.

Tanaka H, Boesten L, Sato H, Kimura M, Dillon M A, and Spense D (1990), *J. Phys. B* **23**: 577.

Tate J T and Smith P T (1932), *Phys. Rev.* **39**: 270.

Temkin A (1962), *Phys. Rev.* **126**: 130.

Temkin A and Lemkin J C (1961), *Phys. Rev.* **121**: 788.

Temkin A, Shertzer J, and Bhatia A K (1998a), *Phys. Rev. A* **57**: 1091.

Temkin A, Wang Y D, Sullivan E C, and Bhatia A K (1998b), *Can. J. Phys.* **76**: 23.

Trajmar S, Register D F, and Chutjian A (1983), *Phys. Rep.* **97**: 219.

Tsai J S, Lebow L, and Paul D A L (1976), *Can. J. Phys.* **54**: 1741.

Twomey T R, Griffith T C, and Heyland G R (1977), in: Abstr. *X ICPEAC* (Paris), p. 808.

Ulsh R C, Wellenstein H F, and Bonham R A (1974), *J. Chem. Phys.* **60**: 103.

Vanderpoorten R (1975), *J. Phys. B* **8**: 926.

van Wingerden B, Weigold E, de Heer F J and Nygaard K J (1977), *J. Phys. B* **10**: 1345.

van Wingerden B, Wagenaara R W, and de Heer F J (1980), *J. Phys. B* **13**: 3481.

Veigele W J (1973), *At. Data Nucl. Data Tables* **5**: 51.

Verma S and Srivastava R (1998), *Pranama* **50**: 355.

Vuskovic L and Kurepa M L (1976), *J. Phys. B* **9**: 837.

Wadehra J M and Khare S P (1993), *Phys. Letts. A* **172**: 433.

Wagenaar R W and de Heer F J (1980), *J. Phys. B* **13**: 3481.

Walker D W (1971), *Adv. Phys.* **20**: 257.

Wallace S J (1973), *Ann. Phys.* **78**: 190.

Walters H R J (1985), *Phys. Rep.* **116**: 1.

Walters H R J, Kernoghan A A, and McAlinden M T (1995), in: AIP *XIX International Conference on the Physics of Electronic and Atomic Collisions*, Proceedings No. 360, eds. L J Dube, J B A Mitchell, J M McConkey, and C E Brion (AIP Press, New York), p. 397.

Watson G N (1958) *Theory of Bessel Functions* (Cambridge University Press, Cambridge).

Weinbaum S (1933), *J. Chem. Phys.* **1**: 317.

Weissbluth M (1978), *Atoms and Molecules* (Academic, New York).

Whelan C T, Walters H R J, Lahmam-Bennani A, and Ehrhardt H (eds.) (1993), *(e,2e), and Related Processes* (Kluwer ,Dordrecht).

Williams J F (1975), *J. Phys. B* **8**: 2191.

Williams J F and Crowe A (1975), *J. Phys. B* **8** :2233.

Williams J F and Willis B A (1975), *J. Phys. B* **8**: 1670.

Williams J F and Trajmar S (1978), *J. Phys. B* **11**: 2021.

Williams K G (1969), in: Abstr. *VI International Conference on Physics of Electronic and Atomic Collisions* (MIT, Boston), p. 735.

Yates A C (1974), *Chem. Phys. Lett.* **25**: 480.

Younger S M and Mark T D (1985), in: *Electron Impact Ionization*, eds. T D Mark and G H Dunn (Springer-Verlag, Vienna).

Yuan J, Zhang Z, and Wang H (1990a), *Phys. Rev. A* **41**: 4732.

Yuan J, Zhang Z, and Wang H (1990b), *Phys. Rev. A* **42**: 5363.

Yuan J, Zhang Z, and Wang H (1991), *Phys. Lett. A* **160**: 81.

Yuan J, Zhang Z, and Wang H (1992), *Phys. Lett. A* **168**: 291.

Zecca A, Karwasz G P, Brusa R S, and Szmytkowski C (1991), *J. Phys. B* **24**: 2747.

Zecca A, Karwasz G P, and Brusa R S (1992), *Phys. Rev. A* **45**: 2777.

Zecca A, Noguerra J C, Karwasz G P, and Brusa R S (1995), *J. Phys. B* **28**: 477.

Zecca A, Melissa R, Brusa R S, and Karwasz G P (1999), *Phys. Lett. A* **257**: 75.

Zeiss G D, Meath W J, MacDonald J C F, and Dawson D J (1975), *Radiat. Res.* **63**: 64.

Zeiss G D, Meath W J, MacDonald J C F, and Dawson D J (1977), *Can. J. Phys.* **55**: 2080.

# Index

# Series Publications

Below is a chronological listing of all the published volumes in the *Physics of Atoms and Molecules* series.

ELECTRON AND PHOTON INTERACTIONS WITH ATOMS
Edited by H. Kleinpopper and M. R. C. McDowell

ATOM–MOLECULE COLLISION THEORY: A Guide for the Experimentalist
Edited by Richard B. Bernstein

COHERENCE AND CORRELATION IN ATOMIC COLLISIONS
Edited by H. Kleinpoppen and J. F. Williams

VARIATIONAL METHODS IN ELECTRON–ATOM SCATTERING THEORY
R. K. Nesbet

DENSITY MATRIX THEORY AND APPLICATIONS
Karl Blum

INNER-SHELL AND X-RAYS PHYSICS OF ATOMS AND SOLIDS
Edited by Derek J. Fabian, Hans Kleinpoppen, and Lewis M. Watson

INTRODUCTION TO THE THEORY OF LASER–ATOM INTERACTIONS
Marvin H. Mittleman

ATOMS IN ASTROPHYSICS
Edited by P. G. Burke, W. B. Eissner, D. G. Hummer, and I. C. Percival

ELECTRON–ATOM AND ELECTRON–MOLECULE COLLISIONS
Edited by Juergen Hinze

ELECTRON–MOLECULE COLLISIONS
Edited by Isao Shimamura and Kazuo Takayanagi

ISOTOPE SHIFTS IN ATOMIC SPECTRA
W. H. King

AUTOIONIZATION: Recent Developments and Applications
Edited by Aaron Temkin

ATOMIC INNER-SHELL PHYSICS
Edited by Barnd Crasemann

COLLISIONS OF ELECTRONS WITH ATOMS AND MOLECULES
G. P. Drukarev

THEORY OF MULTIPHOTON PROCESSES
Farhad H. M. Faisal

PROGRESS IN ATOMIC SPECTROSCOPY, Parts A, B, C, and D
Edited by W. Hanle, H. Kleinpoppen, and H. J. Beyer

RECENT STUDIES IN ATOMIC AND MOLECULAR PROCESSES
Edited by Arthur W. Kingston

QUANTUM MECHANICS VERSUS LOCAL REALISM: The Einstein-Podolsky-Rosen Paradox
Edited by Franco Selleri

ZERO-RANGE POTENTIALS AND THEIR APPLICATIONS IN ATOMIC PHYSICS
Yu. N. Demkov and V. N. Ostrovskii

COHERENCE IN ATOMIC COLLISION PHYSICS
Edited by H. J. Beyer, K. Blum, and J. B. West

ELECTRON–MOLECULE SCATTERING AND PHOTOIONIZATION
Edited by P. G. Burke and J. B. West

ATOMIC SPECTRA AND COLLISIONS IN EXTERNAL FIELDS
Edited by K. T. Taylor, M. H. Nayfeh, and C. W. Clark

ATOMIC PHOTOEFFECT
M. Ya. Amusia

MOLECULAR PROCESSES IN SPACE
Edited by Tsutomu Watanabe, Isao Shimamura, Mikio Shimizu, and Yukikazu Itikawa

THE HANLE EFFECT AND LEVEL CROSSING SPECTROSCOPY
Edited by Giovanni Moruzzi and Franco Strumia

ATOMS AND LIGHT: INTERACTIONS
John N. Dodd

POLARIZATION BREMSSTRAHLUNG
Edited by V. N. Tsytovich and I. M. Ojringel

INTRODUCTION TO THE THEORY OF LASER–ATOM INTERACTIONS (Second Edition)
Marvin H. Mittleman

ELECTRON COLLISIONS WITH MOLECULES, CLUSTERS, AND SURFACES
Edited by H. Ehrhardt and L. A. Morgan

THEORY OF ELECTRON–ATOM COLLISIONS, Part 1: Potential Scattering
Philip G. Burke and Charles J. Joachain

POLARIZED ELECTRON/POLARIZED PHOTON PHYSICS
Edited by H. Kleinpoppen and W. R. Newell

INTRODUCTION TO THE THEORY OF X-RAY AND ELECTRONIC SPECTRA OF FREE
ATOMS
Romas Karazija

VUV AND SOFT X-RAY PHOTOIONIZATION
Edited by Uwe Becker and David A. Shirley

DENSITY MATRIX THEORY AND APPLICATIONS (Second Edition)
Karl Blum

SELECTED TOPICS ON ELECTRON PHYSICS
Edited by D. Murray Campbell and Hans Kleinpoppen

PHOTON AND ELECTRON COLLISIONS WITH ATOMS AND MOLECULES
Edited by Philip G. Burke and Charles J. Joachain

COINCIDENCE STUDIES OF ELECTRON AND PHOTON IMPACT IONIZATION
Edited by Colm T. Whelan and H. R. J. Walters

PRACTICAL SPECTROSCOPY OF HIGH-FREQUENCY DISCHARGES
Sergei A. Kazantsev, Vyacheslav K. Khutorshchikov, Günter H. Guthöhrlein, and Laurentius
Windholz

IMPACT SPECTROPOLARIMETRIC SENSING
S. A. Kazantsev, A. G. Petrashen, and N. M. Firstova

NEW DIRECTIONS IN ATOMIC PHYSICS
Edited by Colm T. Whelan, R. M. Dreizler, J. H. Macek, and H. R. J. Walters

ELECTRON MOMENTUM SPECTROSCOPY
Erich Weigold and Ian McCarthy

POLARIZATION AND CORRELATION PHENOMENA IN ATOMIC COLLISIONS: A Practical
Theory Course
Vsevolod V. Balashov, Alexei N. Grum-Grzhimailo, and Nikolai M. Kabachnik

RELATIVISTIC HEAVY-PARTICLE COLLISION THEORY
Derrick S. F. Crothers

INTRODUCTION TO THE THEORY OF COLLISIONS OF ELECTRONS WITH ATOMS AND
MOLECULES
S. P. Khare

The manufacturer's authorised representative in the EU is Springer
Nature Customer Service Centre GmbH, Europaplatz 3, 69115 Heidelberg,
Germany. If you have any concerns regarding our products, please
contact ProductSafety@springernature.com

Printed and bound by CPI Group (UK) Ltd, Croydon, CR0 4YY

23/04/2026

02095625-0008